International Encyclopaedia of ENERGY RESOURCES

International Encyclopaedia of ENERGY RESOURCES

— *Volume 2* —

— *Edited By* —

Dr. M.P. Singh

Head of the Department of Forest Sciences

Birsa Agricultural University

Ranchi - 834 006 (Jharkhand)

2005

DAYA PUBLISHING HOUSE

Delhi - 110 035

ISBN 81-7035-357-2 (SET)
ISBN 81-7035-358-0 (Vol. 1)
ISBN 81-7035-359-9 (Vol. 2)
ISBN 81-7035-360-2 (Vol. 3)
ISBN 81-7035-361-0 (Vol. 4)

Published by : **Daya Publishing House**
1123/74, Deva Ram Park
Tri Nagar, Delhi - 110 035
Phone: 27383999
Fax: (011) 23244987
e-mail: dayabooks@vsnl.com
website: www.dayabooks.com

Showroom : 4762-63/23, Ansari Road, Darya Ganj
New Delhi - 110 002
Phone: 23245578, 23244987

Laser Typesetting : **Classic Computer Services**
Delhi - 110 035

Printed at : **Chawla Offset Printers**
Delhi - 110 052

PRINTED IN INDIA

Preface

This book is for people who still think that there are energy problems ahead, and who are not satisfied with an energy future dominated by coal and nuclear power.

Fortunately, the time has come when one can examine other alternatives seriously–in particular, energy conservation and renewable energy–without being relegated at once to the outer fringes. Southern California Edison, Pacific Gas and Electric, and numerous other progressive utilities, having embraced these alternatives, are a bit too formidable to be drummed out of the corps. The Office of Technology Assessment, the Gas Research Institute, and the Harvard Business School likewise are not thought of as outposts of a lingering counterculture.

Still, not enough people in the energy supply industries, in government, or in private agencies concerned with energy policy are aware of the full range of possibilities in energy conservation and in the numerous renewable energy technologies. For them, this book is an attempt to lay out, in an ordered fashion, the basis in fact, logic, and judgment, for an energy -efficient, renewable energy future.

One cluster of views about energy may be characterised as "supply oriented." The response to the oil shortage was quite expectedly framed in terms of increasing energy supply, especially in forms other than oil. The basic premises motivating supply-side arguments included the following assumptions:

Energy demand will grow with output. A healthy and growing economy requires a growing supply of energy. In early formulations the rate of energy supply growth was about equal to the growth rate of the economy.

While this volume does provide a great deal of information about energy sources, its more appropriate use is as a stimulus for discussion. Towards this end the book often raise more questions than they answer. The complexity of the ethical dilemmas posed by energy sources is such that we are still trying to identify whai questions should be asked. The volume recognises that the first step in making the types of ethical decisions which are increasingly demanded of us is searching for the right questions. The book vary greatly in both length and orientation. This will make it useable by a wide range of groups–from students in the classroom, to professionals such as scientists, medical doctors, business people, lawyers, and ethicists, as well as environmental and other socially active groups. It is hoped that this volume will contribute in a better way to a society.

Dr. M.P. Singh

Contents

— Volume 2 —

38

Great Lakes Region: Burundi, Kenya, Rwanda, Tanzania, and Uganda

The following provides a brief overview of the energy sectors of the Great Lakes region–Burundi, Kenya, Rwanda, Tanzania, and Uganda.

General Background

Africa's Great Lakes region is located in eastern Africa and includes countries surrounding Lake Kivu, Lake Tanganyika, and Lake Victoria. Combined, Burundi, Kenya, Rwanda, Tanzania, and Uganda have a population of 108 million people, and an estimated gross domestic product (GDP) in 2002 of about $24.4 billion.

Kenya, Tanzania, and Uganda recently re-established the East African Community (EAC), which had been dissolved in 1977 due to political and economic differences. Under the EAC treaty signed in 1999 and officially implemented in January 2001, the three countries will negotiate the framework for a customs union over the next four years and will then move to establish a common market. Officials also envisage a monetary and political union in a process which they say could take up to 20 years. Both Rwanda and Burundi have expressed interest in joining the EAC.

Besides EAC, Burundi, Kenya, Rwanda, and Uganda are members of the Community of East and Southern African States (COMESA), formed in 1994 and composed of 20 countries in Eastern and Southern Africa. Tanzania withdrew from COMESA in 2000, citing a high trade imbalance with

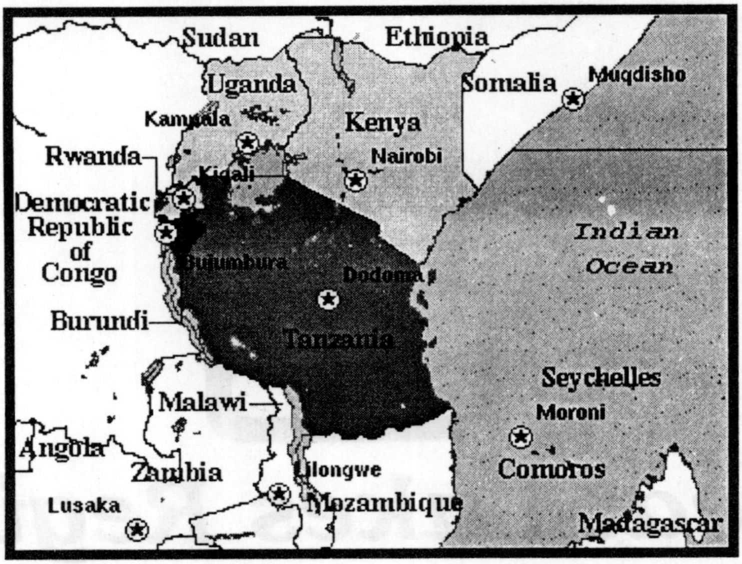

COMESA-member countries, particularly Kenya. In October 2000, COMESA launched its free trade program, creating schedules for internal tariff reductions. Rwanda declined to participate in the COMESA free trade arrangement, citing the potential loss of government revenue. In May 2002, Burundi announced a timetable for its entry into the COMESA free trade zone. By 2003 it plans to reduce tariffs by 80 per cent, and to implement a 100 per cent tariff reduction the following year. By 2004, COMESA aims to have a common external tariff structure for all third party trade, among other regional and cooperative goals. Tanzania is a member of the *Southern African Development Community (SADC)*, an economic regional organization similar to COMESA (see Table 1 for the size and population of each country in the Great Lakes region).

Table 38.1: Great Lakes Country Overview

Country	*Area (sq. mi)*	*Population, millions (2001E)*	*Head of State*
Burundi	27,830	7.0	President Pierre Buyoya
Kenya	224,961	31.9	President Daniel Arap Moi
Rwanda	26,338	8.2	President Paul Kagame
Tanzania	364,899	36.8	President Benjamin William Mkapa
Uganda	91,135	24.8	President Yoweri Museveni
Great Lakes Region Total	735,163	102.2	

The economies of the Great Lakes region states have different structures, and are at various stages of development. Uganda's economy grew at an estimated 5.1 per cent in 2001, while Kenya's real GDP grew at 0.9 per cent for the same year. Among the countries of the region, the rate of inflation ranged from 0.8 per cent in 2001 in Kenya, to 9.2 per cent in Burundi. All of the states in the Great Lakes region are dependent on foreign aid, with Rwanda, Tanzania and Uganda classified as Heavily Indebted Poor Countries (HIPCs) by the World Bank.

Table 38.2: Great Lakes Region Economic Overview

Country	*Nominal GDP ($billion) 2001E*	*Real GDP Growth Rate 2001E*	*Inflation Rate 2001E*	*U.S. Merchandise Trade ($million) 2001*		*Currency*	
				Exports to	*Imports from*	*Name*	*Exchange Rate 10/21/02 US$1 =*
Burundi	0.7	3.3%	9.2%	5.30	2.80	Franc	1,022.79
Kenya	11.0	0.9%	0.8%	577.70	128.30	Shilling	78.97
Rwanda	1.9	2.7%	6.6%	17.30	7.20	Franc	476.19
Tanzania	8.9	5.1%	5.1%	44.70	32.20	Shilling	970.18
Uganda	6.1	5.1%	6.1%	31.70	17.70	Shilling	1,800.86
Great Lakes Region Total	**27.7**	–	–	**676.7**	**188.2**		
				U.S. Trade Balance 488.4			

Energy Overview

In 2000, the Great Lakes region nations collectively consumed 0.26 quadrillion British thermal units (Btu) of commercial energy (less than 0.1 per cent of the world's total) and generated about 3.45 million metric tons of carbon emissions (also less than 0.1 per cent of the world's total). Kenya accounted for almost 60 per cent of the region's commercial energy consumption in 2000. Due to the relatively small urbanized population in the region, access to commercial energy sources is limited. The majority of the Great Lakes region population still relies on biofuel (wood, animal waste, etc.) as their primary fuel source. According to the World Bank, the number of regional inhabitants with access to electricity in 1996 ranged from 2.1 per cent in Burundi to 8 per cent in Kenya. Commercial energy resources in the region include coal, natural gas (which is not yet being exploited), hydroelectricity, geothermal, and possibly some oil. In 2000, 79.6 per cent of electricity in the region was generated by hydroelectric plants and 16.4 per cent by geothermal. Oil prices in the Great Lakes Region have a significant impact on local economies, since the area is heavily dependent on oil imports.

Oil

Great Lakes regional oil consumption averaged about 84,000 barrels per day (bbl/d) in 2001, almost all of which was imported. Kerosene is used extensively in rural areas for lighting and, in urban areas, for cooking and lighting. The countries of the region are attempting to discourage the use of wood fuel through joint provision of and promotion of the use of liquefied petroleum gas (LPG).

Kenya has no known oil reserves of its own, and must import all of its 54,000 bbl/d consumption (see Table 3). Most Kenyan oil companies import their oil from Abu Dhabi's National Oil Corporation (Adnoc). Recently there has been talk of importing oil from Sudan, though the idea has been highly controversial because of concerns that the Sudanese government could be using oil revenues to finance its civil war against southern rebels. In March 2002, Nigeria cancelled its crude oil supply contracts with Kenya after cutting its oil production to meet lower OPEC quotas. The Kenyan government has spent about $169 million exploring for oil and natural gas over the past 15 years. Over 30 wells have been drilled so far, but with no success. In August 2002, the Kenyan government executed agreements that grant exclusive exploration rights of blocks L6, L8, and L9 to Pancontinental Oil & Gas and Afrex

Limited. Altogether, the three blocks comprise a 21,400-square-kilometer (12,840-mile) area on or near the Kenyan coastline. The joint venture is operated by Afrex, which holds a 60 per cent share, while Pancontinental holds the remaining 40 per cent.

Table 38.3: Great Lakes Region Petroleum Overview

Country	*Estimated Proven Crude Reserves–1/02 (Million Barrels)*	*Crude Oil Production (Barrels per Day) 2002E*	*Crude Oil Refining Capacity (Barrels per Day) 1/02*	*Oil Consumption (Barrels per Day) 2002E*
Burundi	0	0	0	2,000
Kenya	0	0	90,000	56,000
Rwanda	0	0	0	5,000
Tanzania	0	0	14,900*	16,000
Uganda	0	0	0	7,000
Great Lakes Region Total	**0**	**0**	**104,900**	**86,000**

Note: Tanzania's refinery is no longer operational, and is now being used as an oil storage depot.

Tanzania also relies exclusively on imports for its oil needs. Since 1980, foreign oil prospectors have invested more than $293 million in search of oil in Tanzania without success. The government in Dar es Salaam continues to encourage oil exploration in the country, seeing oil as a key to economic development. State-owned Tanzania Petroleum Development Corporation (TPDC) has signed several joint venture agreements with foreign oil exploration companies in recent years.

In August 2000, TPDC and a UK joint venture firm, Western Geophysical, completed the first comprehensive survey of oil deposits on the Indian Ocean floor off Tanzania's coastline. Western Geophysical brought in a specialized prospecting vessel to conduct the survey, which was designed to shed light on the possibilities of striking oil offshore. Experts have long suspected that there are considerable hydrocarbon reserves off Tanzania's coast. Data from the $11 million project was sold to interested oil companies as part of a new licensing round, launched in September 2000. The licensing round, which closed in April 2001, will open up over 24,000 square miles of deepwater acreage to exploration by dividing the Mafia Offshore Deepsea Basin, stretching from Mnazi Bay in the south to Dar es Salaam in the north, into six licensing areas. A second licensing round covering deepwater acreage to the north of Dar es Salaam was concluded in August 2002, calling for bids on 12 blocks. Bids were received from only two companies, Shell and Global Resources, both interested in blocks 9-12. In October 2002, exploration rights were granted to Shell for all four blocks. Shell plans to conduct 2D seismic surveys of the areas, then use the results to conduct 3D seismic surveys.

Uganda is especially vulnerable to rising oil prices, since it imports most of the oil it consumes (about 7,000 bbl/d) from Kenya's Mombasa refinery, which in turn imports crude oil from abroad. The oil is transported from Nairobi to Uganda by large tanker trucks. An alternative supply route through Tanzania has been developed running from Dar es Salaam to Mwanza by railroad and on to Jinja by lake ferry. The imported oil is stored in one of 20 refined product storage depots, with a total capacity of 365,000 barrels (two-thirds owned by oil companies and one-third by the government). This inefficient transportation, along with high taxes on fuel imports, account for Uganda's high oil prices. The tax on oil also accounts for nearly 25 per cent of the country's total revenue. Uganda's oil industry was deregulated in 1994.

The rising price of imported oil products has accelerated Uganda's efforts to develop its domestic oil resources. Although Uganda has no proven oil or gas resources, some exploration is taking place, and there are signs that the western Rift Valley may contain hydrocarbon deposits. In September 2002, Heritage Oil Corporation announced that exploratory drilling operations had begun in Block 3, located in the Semiliki Valley in western Uganda. This is the first petroleum well ever drilled in the country, and is intended to confirm the results of Heritage Oil's seismic studies from 1999. That survey revealed a potential of nearly one billion barrels of oil in the Semliki Valley as well as the Pakwach basin area, northern and southern Lake Albert basin, and Lake George basin area. While financial difficulties forced Hardman Petroleum to abandon its license for the Lake Albert region in January 2000, other international oil companies, such as the China National Petroleum Corporation, have shown a growing interest in oil exploration in Uganda.

Pipelines

In May 1999, the Kenyan and Ugandan governments announced plans for an oil pipeline from Eldoret in western Kenya to Kampala, Uganda. The plans were finalized in a formal agreement signed by the two nations in October 2000. Construction of the pipeline extension was set to begin in April 2002, although that date has been postponed indefinitely due to lack of investor interest. The extension would be constructed over a four year period, under the auspices of the EAC. The $80 million, 16,500-bbl/d pipeline will supply Uganda, Rwanda, Burundi, northwestern Tanzania and eastern Democratic Republic of Congo, and is expected to cut the price of oil in Uganda by almost half. Uganda currently imports most of its oil through Kenya's Mombasa refinery. While waiting for construction to begin, Uganda has started to import some cheaper refined petroleum products from overseas, rather than rely solely on the inefficient Mombasa refinery.

In August 2002, Tanzania and Zambia announced that they are considering privatizing their jointly owned oil pipeline. A feasibility study is under way to determine whether the 1,710-kilometer (1,060-mile) pipeline should remain under government control. At present, Tanzania holds a 33 per cent share in Tanzania-Zambia oil pipeline (TAZAMA), while Zambia holds the remaining 67 per cent.

Refining

There are two oil refineries in the Great Lakes region–one in Kenya and one in Tanzania. The Kenya Petroleum Refinery-Mombasa unit has crude processing capacity of 90,000 bbl/d, and Tanzania's Kigamboni Dar es Salaam-Tanzanian & Italian Petroleum Refining facility has 14,900 bbl/d. Kenya's refinery has been operating well below capacity and has struggled financially since the 1994 liberalization of the of the oil sector, which allowed local oil marketers to import finished petroleum products directly into the country. Kenya's energy minister has asked to review the base oil rule, which requires oil firms in Kenya to process 70 per cent of their crude oil at the Mombasa refinery. In August 2002, the energy minister also pledged to make a decision on whether the refinery will begin to produce unleaded fuel, which it currently is unable to do. Although investors have spent approximately $45 million to upgrade and automate the refinery since 1997, it has continued to operate at 60 per cent below capacity. In June 2000, the Kenyan government, which owns a 50 per cent stake in the refinery, announced plans to divest its shares.

Tanzania's refinery recently was closed after the World Bank urged the move, on the grounds that the refinery was no longer competitive following the liberalization of the country's petroleum industry. The refinery, which had been closed temporarily in November 1999 due to a payment dispute with TPDC, continues to act as an oil storage depot.

Natural Gas

Rwanda and Tanzania currently are the only two nations in the Great Lakes region with proven natural gas reserves (see Table 4). Although natural gas is not produced or consumed in the region at present, several projects have been identified for expanded use of this resource. In Tanzania, a project to exploit natural gas in the country's largest known field–on Songo Songo Island, located in the Indian Ocean southeast of Dar-es-Salaam–will soon be entering its final phase of development thanks to a proposed $200-million World Bank loan, the terms of which were finalized in October 2001. Construction of the pipeline is expected to begin before the end of 2002, and the project is expected to be completed in 2004. In August 2000, TransCanada Pipeline sold its 49 per cent stake in Songo Songo to a subsidiary of the U.S.-based AES Corporation. In October 2002, the firm Larsen and Toubro was awarded the contract to collect natural gas for processing. The project involves a 140-mile pipeline from Songo Songo to a gas-fired plant in Dar-es-Salaam to provide fuel for a major thermal power plant and other industrial users. Natural gas will be collected from two onshore wells and from three offshore wells, then processed at a gas plant on Songo Songo Island. Once the project is completed, the five liquid fuel turbines at the 112-megawatt (MW) Ubungo power plant will be converted to gas and the power generated by the plant will be fed directly into the national grid. Songo-Songo's reserves are estimated at 1 trillion cubic feet (Tcf).

Table 38.4: Great Lakes Region Natural Gas Overview

Country	*Estimated Proven Gas Reserves (Billion Cubic Feet) 1/02*	*Gas Production (Billion Cubic Feet) 2000E*	*Gas Consumption (Billion Cubic Feet) 2000E*
Burundi	0	0	0
Kenya	0	0	0
Rwanda	2,000	0	0
Tanzania	800	0	0
Uganda	0	0	0
Great Lakes Region Total	**2,800**	**0**	**0**

Coal

Tanzania is the only country in the Great Lakes region with significant coal resources. Total recoverable coal reserves in the Lake Victoria region amount to 220 million short tons (see Table 5). The primary use of coal in the region is in the generation of electricity. In May 1999, China's Hunan International Economic and Technical Cooperation Company bought 62 per cent of Tanzania's Kiwira coal mine. The Tanzanian government kept a 38 per cent share. Kiwira, situated in southern Tanzania near the border with Malawi, could become East Africa's largest coal producer. The new company, Tanzania-China Kiwira Coal and Power, plans to triple annual output to 300,000 short tons both for domestic use and for export, mainly to China. Large coal deposits also exist in the Ruhuhu river area, and are estimated to be the best quality reserves Tanzania possesses.

In November 2000, the Kenyan Ministry of Energy announced that it was budgeting $500,000 for coal prospecting in order to diversify its fuel sources for electricity generation. Hydroelectric power currently accounts for over 70 per cent of Kenya's electricity generation. The Ministry of Energy is searching for coal in the Mui and Mutitu basins in Mwingi and Kitui districts, although no deposits have been found to date.

Table 38.5. Great Lakes Region Coal Overview

Country	*Estimated Recoverable Coal Reserves (Million Short Tons)*	*Coal Production (Million Short Tons) 2000E*	*Coal Consumption (Million Short Tons) 2000E*
Burundi	0	0	0
Kenya	0	0	0.08
Rwanda	0	0	0
Tanzania	220	0.01	0.01
Uganda	0	0	0
Great Lakes Region Total	**220**	**0.01**	**0.09**

Electricity

In 2000, installed electric generating capacity for the Great Lakes region totalled about 1,900 megawatts (MW). Electricity generation for the region in 2000 was 9.25 billion kilowatthours (Bkwh.), the majority of which– 7.36 Bkwh–was hydroelectricity (see Table 6). Total Great Lakes region electricity consumption was 8.7 Bkwh. Kenya, Tanzania and Uganda are developing plans to share power supplies, including a regional energy interconnectivity plan that will enable any EAC country to connect with another nation's electricity supply. Among the nations of the region, Uganda has the biggest hydropower potential (from the Nile River) and would play a major part in any power-sharing project.

Table 38.6: Great Lakes Region Electricity Overview

Country	*Installed Capacity (megawatts) 1/00*	*Net Generation (Billion Kilowatt Hours) 2000E*			*Consumption (Billion Kilowatt Hours) 2000E*	*Trade (Billion Killowatt Hours) 2000E*	
		Total	*Thermal*	*Hydroelectric*		*Imports*	*Exports*
Burundi	49	0.15	0.001	0.15	0.17	0.03	0
Kenya	933	4.62	1.0	3.25	4.43	0.15	0
Rwanda	30.6	0.11	0.003	0.11	0.17	0.07	0.001
Tanzania	620	2.77	0.5	2.27	2.62	0.05	0
Uganda	280	1.6	0.015	1.58	1.31	0.001	0.18
Great Lakes Region Total	**1,912.6**	**9.25**	**1.519**	**7.36**	**8.7**	**0.301**	**0.11**

In March 2002, Kenya's Kipevu II power plant was officially inaugurated. The $86 milion plant has the capacity to generate 74 MW of hydroelectricity for the nation, which suffered from severe energy shortages between 1999 and 2001. Owned by Tsavo Power Company, the plant will increase Kenya's installed electric capacity by about 25 per cent by providing thermal energy. Kenya's recent energy shortages stem from its over-reliance on hydroelectricity and the prolonged droughts the nation has faced. The Kipevu II power plant will be subject to Kenya's new environmental regulations, as well as environmental requirements set by the World Bank. Funding by the World Bank had previously been withheld due to Kenya's failure to meet with certain World Bank criteria.

Other electric expansion projects include Olkaria II and the Sondu Miriu project. Olkaria II will provide Kenya with 64 MW of geothermal electricity when it is completed. At present, the first 32-MW turbine is scheduled for completion in November 2002. Funding for the project is provided by the World Bank, the European Investment Bank (EIB), Kreditanstalt for Wiederaufbau, and Kenya Electricity Generating Company Ltd. The Sondu Miriu project will consist of building a 60-MW hydroelectric power station located in Kenya's Nyanza province along Lake Victoria. It is being constructed by Konoike Construction JV, Vieddeke Heavy Construction Company of Norway, and Murray & Roberts Contractors International of South Africa. Construction began in 1999 and is expected to be completed in 2005. Funded largely by the Japan Bank for International Cooperation, the project has been on hold since June 2001 when funds were suspended due to environmental concerns. Japan may release the remaining money to complete the project if Kenya meets World Bank and IMF requirements.

In June 2000, Nairobi agreed to the accelerated construction of two, 55-MW, diesel-fired power plants at a cost of $135 million. Located in the towns of Lanet and Eldoret, the two plants are still under construction, despite assurances that they would come online by 2001. In August 2000, contracts were signed for construction of two thermal power plants with a combined capacity of 105-MW in Embakasi and Ruaraka. The government also announced plans to signficantly expand the storage capacity of the Masinga Dam, ensuring hydroelectric generation for over a year.

Tanzania's sole producer and supplier of electricity, Tanzania Electric Supply Company (Tanesco) continues to face a severe budgetary shortfall, due largely to a multi-million dollar bill for past electricity consumption owed to the company by the Tanzanian government. In July 2002, the company cut off the electricity to a number of government and residential buildings in an attempt to increase revenues by forcing consumers to pay their past due bills. Tanzania recently hired external managers to run the state-owned power company. The managers from Net Group Solutions in South Africa have met with strong opposition from Tanesco workers since taking over in May 2002. Concerns center around whether the small firm will be able to effectively manage Tanzania's electricity grid. This is the first step to the eventual privatization of the nation's largest energy firm.

Tanesco has been working to develop a number of power projects. Construction was completed on the Kihansi hydroelectric dam in April 2000. Kihansi, located 341 miles southwest of Dar Es Salaam, has a capacity of 180 MW but has been operating at nearly 50 per cent capacity due to on-going efforts to save the Kihansi spray toad, which is threatened by the dam. Other projects include the Malaysian 100-MW IPTL power project (currently on hold), the 37-MW Songo Songo gas project, and U.K.'s Agrreko 100-MW power project. In late January 1998, the Tanzanian government signed an agreement with Tornado Resources for the development of a $32.5-million natural gas-fired, electric power project in southern Tanzania. The project uses gas from the 500-Bcf Mnazi Bay gas field in the town of Mtwara, transported through a 16-mile gas pipeline, to a 20-MW power plant.

In January 2001, Tanzania, Kenya and Zambia agreed to implement cross-border electricity trade, a two-year project estimated to cost over $153.5 million. The project will entail the construction of 372 miles of power lines from Zambia to Mbeya, and from Arusha to Nairobi. Given its connection to the SADC power grid, Zambia will be able to sell low-cost and reliable electricity to Kenya and Tanzania.

Uganda will be exporting more electricity to Kenya to relieve that nation's electricity shortages. Once the Bujagali plant is completed, it will be able to export an additional 80 MW of the 200 MW generated by the new plant. Uganda also played a role in alleviating the power crisis in Kenya during the drought by exporting electricity to the East African country from the Kiyara Power Station, a $230 million extension project in Jinja completed in June 2000. In its final phase, Kiyara will have an installed capacity of 240 MW. President Museveni predicted in May 2000 that Uganda would significantly raise its electricity exports to Tanzania and Rwanda.

Uganda is seeking $1.69 billion in foreign investment over the next ten years to upgrade its energy sector. In April 1999, British consultants carried out a national energy plan for Uganda which later received government approval. The plan listed five electricity generation units, including the Bujagali hydroelectric facility (250 MW-2,000 MW), Karuma dam (100 MW-200 MW), promoted by Norpak, and the rehabilitation of the Nalubaale dam (formerly the Owens Falls dam) (180 MW) and its extension (five units of 40 MW each). Under the plan, Uganda's government is to retain ownership of existing power stations, but will be able to cede them to private operators under leasing agreements. In August 2002, the South African energy firm Eskom Enterprises made the sole bid for management of the Uganda Electricity Generation Concession (UEGC). The firm offered to charge $41.7 million and is seeking a 12 per cent rate of return on any investments. Eskom plans to spend $35.8 million in generation and maintenance costs.

The controversial Bujagali dam project was put on hold in July 2002, when the World Bank decided to suspend its support pending further study of the project's environmental consequences. The $550 million hydroelectric dam would be East Africa's largest foreign direct investment project. Opponents say that the dam would be damaging to the environment and would promote excessively high electricity prices in the country.

A major loan from Nigeria intended to upgrade Burundi's power sector was put on hold in July 2002, in part due to concerns over the ongoing civil war. The $243 million loan would have been borrowed by Nigeria from the African Development Bank and the Arab Bank, which would in turn loan the money to Burundi. Burundi would then have used the loan to clear its debts with the African Development Bank and the Arab bank, as well as finance various power projects, including the first phase of the Mpanda power station. That project would involve the construction of a dam, a 10.4-MW power station, and a transmission line to connect the station with the national grid. The loan would have financed smaller projects, such as installing electricity in 14 villages, doubling the capacity of the Buhinga hydroelectric station, and renovating the the medium voltage network in Bujumbura.

In July 2002, the Israeli Electric Company announced that it was conducting a preliminary study on using Lake Kivu's methane gas reserves to supply a power station in Rwanda. If built, the power station would have a capacity of 25 MW, or 2.5 times the amount of electricity currently produced by Rwanda's national utility company, Electrogaz. One consultancy has stated that it believes Kivu's methane gas reserves could generate as much as 200 MW. A similar project has been under development since 2001, in which a local company, Gisenyi Electric and Gas Company hopes to construct a plant that would have an initial capacity of 2.5 MW and eventually increase to 10 MW.

Rwanda's cabinet ratified its entry into the African Energy Commission (AFREC) in May 2002, citing the country's need for access to low electricity rates to enhance its development.

In November 2000, Rwanda and Burundi announced plans for the joint construction of the Ruzizi III hydroelectric dam on the Ruzizi River along the Rwanda-Burundi border. In July 2000, the Burundi National Assembly voted to liberalize its electricity sector, ending the monopoly of the state electric utility.

Renewable Energy

Kenya has two steam stations, the Olkaria renewable power station (45 MW) and the Kipevu Thermal Station (45.5 MW). Nairobi is promoting additional geothermal power, and hopes to increase its production threefold by 2003 when two new geothermal power plants, costing $155 million, are completed at Olkaria. Construction of the 64-MW Olkaria II plant began in January 2000 and by October 2000 was injecting 13 MW into the national grid. The Olkaria III geothermal plant has been constructed and is expected to have a generating capacity of 64 MW by July 2003. Kenya currently produces about 1.0 bkwh of geothermal power, nearly 30 per cent of net electricity generation. The Kenyan Energy Ministry estimates that Kenya has a potential for over 2,000 MW of installed capacity of geothermal electricity, second only to New Zealand.

In November 2000, China pledged to fund a $180,000-solar-energy project designed to provide energy to rural Kenyan educational institutions over the next two years. Similarly, U.S.-based NGOs donated $600,000 worth of solar equipment to Uganda and other countries of the region. Ugandan banks also launched a program in July 2000 designed to extend credit to local entrepreneurs needing solar equipment.

39

Greece

Greece is an important potential transit site for energy exports from the Caspian/Caucasus regions, with limited energy reserves of its own.

General Background

Greece is among the smallest of the economies in the European Union (EU), but has enjoyed fairly strong growth over the past few years with relatively low inflation. In 2001 and 2002, for instance, Greece's real gross domestic product (GDP) grew by an estimated 4.1 per cent and 4.0 per cent, respectively. For 2003, Greece's real GDP growth is expected to slow slightly, to 3.7 per cent, with inflation at 3.1 per cent. Greece's unemployment rate has been trending downwards in recent years, from 11.9 per cent in 1999 to 9.9 per cent in 2002. For 2003, unemployment is forecast at 9.6 per cent. Greece hopes to achieve long-term real GDP growth of 5.0 per cent -5.5 per cent over the next 15 years, but outside analysts believe that this is overly optimistic, with the *Economist Intelligence Unit (EIU)*, for instance, forecasting a 3.4 per cent annual growth rate from 2003 to 2007, followed by 3.2 per cent yearly increases from 2008 to 2012.

Greece's economic growth is being driven in part by infrastructure construction for the 2004 Olympic Games, which are to be held in Athens in August 2004. Greece joined the "Eurozone" (its 12th member) in January 2001, and the general government budget experienced a surplus in 2001 for the first time in over three decades. In spite of Olympic-related investment, Greece has not attracted as much foreign direct investment (FDI) as it would like, in part due to security concerns related to the left-wing terror group called "November 17," which has carried out numerous attacks on foreign-owned businesses around Athens (note: the Greek government launched a major crackdown on the group in July 2002). In addition, Greece has not been seen as attractive to foreign investors because of its small domestic market, underdeveloped communications and transport infrastructure, relatively high tax burden (some tax breaks were announced in early 2003), and distance from other EU markets. As a result of all this, Greece's current ratio of FDI to GDP is lower than any other EU country.

Meanwhile, since the mid-1990s, Greece has embarked on a series of macroeconomic and structural reforms, including measures aimed at reducing the size of the bureaucracy, expanding privatization, and attracting foreign investment. The International Monetary Fund (IMF) and OECD both have called for continued efforts in these areas. In 2002, Greece reportedly sold 2.5 billion Euro worth of assets. In February 2003, however, the Greek privatization effort suffered a setback, with the collapse of deals to sell off controlling stakes in state oil refiner Hellenic Petroleum and state airline Olympic Airways. Still, the government says it expects the privatization program to pick up in coming months, with other companies slated for sale including: the Public Gas Corporation (DEPA); Athens Water and Sewerage; Piraeus Port Authority; Postal Savings Bank; and Agricultural Bank. According to the *EIU*, one obstacle to Greece's privatization program has been a prolonged and nasty slump on the Athens Stock Exchange (ASE). Another has been the Greek government's reluctance to cede control (the "golden share") in key public utilities and other areas. Greece's next general election must be held by April 2004, but could be called earlier by current Prime Minister Costas Simitis.

Relations between Greece and Turkey have improved recently, allowing for discussion of economic and energy cooperation. In March 2002, Greece and Turkey began discussions on resolving a decades-long disagreement over Aegean Sea boundaries. The two countries also are increasing cooperation in the economic and energy areas. Greece is a major investor in the former Yugoslavia and its energy infrastructure is being integrated with that of the Balkan states. Improved relations with neighboring states could help Greece significantly in achieving its ambitious goal of becoming the major Balkan energy (oil, natural gas, electricity) hub by 2010. The situation in the Former Yugoslav Republic of Macedonia, although relatively stable currently, remains a cause for concern since a spate of serious unrest and violence by rebels in 2001.

Oil

Greece has limited oil reserves of just 9 million barrels. The country produces 7,162 barrels per day (bbl/d) and is highly import reliant for its 408,000 bbl/d oil consumption (2002 estimate). Oil is Greece's most important fuel source, accounting for 61 per cent of total energy consumption in 2001, a percentage that has remained fairly stable since the mid-1980s. Oil is imported primarily from Iran, Saudi Arabia, Russia, Libya and Egypt. The Middle East is expected to remain the major source of Greek oil supplies in coming years, although Russia could become more important as new pipelines are constructed.

Greece's oil industry is dominated by state-owned Hellenic Petroleum (HP), which was formed in 1998 from the former state oil company, Public Petroleum Corporation (DEP). HP conducts oil exploration, imports crude and products, operates three large refineries (one in the Former Yugoslav Republic of Macedonia) accounting for 56 per cent of oil product output in the country, distributes and markets oil products, and owns a 30 per cent share in the Public Gas Corporation. HP has been partially privatized in stages, and is now 41.6 per cent privately held. A merger with domestic rival, the private company Petrola, is expected to be completed by October 2003, leaving the new company with 80 per cent of the Greek refining market.

In May 2001, the Greek government announced its intention to sell another 23.17 per cent of HP, and three firms–Austria's OMV, Russia's Yukos and Lukoil–expressed interest in bidding. A joint venture between Lukoil and Greece's Latsis Group–which owns Greece's third-largest refiner Petrola–offered to pay 454 million Euro for the stake, with Russia's Yukos and Austria's OMV withdrawing from the bidding in early 2002. In early 2003, the Greek Development Ministry said that the Lukoil-Latsis bid was "not judged to be acceptable for the national interest," and would now seek "alternatives."

Greece's oil production comes from the Prinos area in the Aegean Sea, off the coast of Kavala. The Prinos fields, which began production in 1996, are operated by the U.S., Greek, and Canadian North Aegean Petroleum Company (NAPC) consortium. In February 2001, a new oilfield was found offshore the Aegean island of Thasos (also near Kavala) by Kavala Oil.

Greece's first oil exploration licensing round was held in 1996 and awarded six concessions. In May 2002, Greece announced that it would hold its second oil exploration licensing by early 2003. The round is to include both offshore and onshore areas in northwestern and southwestern Greece, plus unexplored blocks in the Ionian Sea. Exploration in the Aegean Sea is complicated by lack of agreement between Greece and Turkey delineating continental shelf boundaries.

HP is developing a $100 million, 143-mile pipeline to carry crude oil from the northern port city of Thessaloniki to HP's newly-acquired Okta refinery near Skopje, in the Former Yugoslav Republic of Macedonia. Construction of the pipeline by HP's subsidiary El Pet Balkaniki began in November 1999. This pipeline will have the capacity to carry about 50,200 bbl/d. The pipeline will be managed in partnership with the Former Yugoslav Republic of Macedonia, and will carry crude that currently is shipped by rail from Thessaloniki to Okta.

In January 1997, Greece, Bulgaria and Russia agreed on a plan to build an oil pipeline linking the Bulgarian Black Sea port of Burgas with Alexandropoulis on the Mediterranean coast of Greece. The proposed 178-mile, underground, Trans Balkan pipeline would allow Russia to export oil through the Black Sea while bypassing Turkey's Bosporus and Dardanelles Straits. However, the $600-$700 million project has been stalled by a wide range of technical and economic disputes. Russia has affirmed that the pipeline, with proposed capacity ranging from 600,000 bbl/d to 800,000 bbl/d, will work at least at 50 per cent of its capacity, and Russian oil major Yukos has expressed its interest in the project, which may ease concerns over filling the pipeline. Greece also has discussed with Kazakhstan the possibility of shipping oil through the pipeline. In May 2002, Bulgaria said that it hoped an agreement on the pipeline could be signed by the end of 2002, with the main outstanding issue being the percent stake that each country will take in the project. In June 2003, Greece said that it was hoping for a positive decision on the matter.

Downstream

HP owns about half of Greek refining capacity, which totaled 405,500 bbl/d as of January 1, 2002. HP's Aspropyrgos facility refines about 140,000 bbl/d, while the Thessaloniki refinery has a 66,500-bbl/d capacity. Two private refineries, owned by Motor Oil Corinth Refineries and Petrola Hellas, are export-oriented, selling only limited volumes to the national market. Motor Oil, which had been majority-owned by Saudi Aramco, was partially sold off to Petroventure of Jersey in July 2001, a joint venture of Aramco and Vardinoyannis, which separately each own 16.4 per cent of Motor Oil. HP also is the largest player in the Greek retail oil market at a 26 per cent market share. U.S.-based Texaco and British-Dutch Royal Dutch/Shell decided in 2000 to trade Texaco's Greek retail assets for some of Shell's United Kingdom retail assets.

In December 2002, HP announced plans to purchase 200 gasoline stations in the Balkans region during 2003. The stations would be operated under the EKO Hellenic Petroleum brand name in Albania, Bulgaria, Macedonia, Romania, and Serbia.

In October 2002, the European Commission instructed Greece to increase the level of its strategic oil reserves to at least 90 days worth of consumption. The Commission claimed that Greece had fallen below this level over the past two years, and warned that the country could face fines if it did not comply. Greece argued that it currently had 109 days worth of strategic oil product stocks on hand.

Natural Gas

With reserves of only 18 billion cubic feet (Bcf), Greece produces negligible amounts of natural gas. Consumption, however, has increased significantly over the past few years, increasing from only 1 billion cubic feet (Bcf) in 1996 to 76 Bcf in 2001. Consumption is expected to continue increasing, possibly tripling over the next ten years. More than two-thirds of Greece's natural gas imports currently come from Russia, with the rest coming from Algeria. Greece is looking to diversify its gas import sources to such countries as Iran and Azerbaijan. Azeri gas would be transported via Turkey through the South Caucasus Gas Pipeline and a planned Turkey-Greece link.

The Greek natural gas industry is controlled by the state-owned Greek Public Gas Company (DEPA), which was created in 1988 in an attempt to diversify the primary energy supply by increasing the role of natural gas. DEPA is 35 per cent owned by HP, with the Greek government owning the rest. In September 2002, 35 per cent of DEPA's government share was offered for sale, with nine buyers (Total, Eni, Electricite de France, Gaz de France, Edison, Enel, Sonatrach, Gazprom, Ruhrgas) originally expressing interest (Total, Gaz de France, Electricite de France and Edison reportedly dropped out of the bidding in April 2003; Spain's Gas Natural offered a 260 million Euro bid in June 2003). Eventually, the plan is for DEPA to be only minority-owned by the state, and for the new majority owner to help build a natural gas pipeline network from the Caspian, via Turkey, through Greece and onwards to the rest of Europe. In May 2003, Greece's Finance Minister, Nikos Christodoulakis, said that the Greek government was nearing the final stages of selecting a strategic investor for DEPA.

DEPA began importing natural gas from Russia via Bulgaria in July 1997 through a Bulgarian pipeline (Greece's only operational gas pipeline), and the company has contracts to supply natural gas to electric utilities and industrial companies. In March 2001, DEPA and Gazexport (part of Gazprom of Russia) agreed on an importation deal for 2002 which includes a 5 per cent price increase. Under current Greek law, companies other than DEPA can only import gas to generate electric power destined for export.

Recent improvements in Greek-Turkish relations are facilitating discussions of energy cooperation. For instance, Greece and Turkey agreed in July 2000 to work together to develop connections between their natural gas networks. This commitment was reaffirmed at the "EU and Black Sea Economic Cooperation Conference" in September 2001 by the Turkish officials at the meeting. Senior Greek and Turkish officials have signed an agreement at EU headquarters to study how best to develop natural gas connections. The two countries have agreed to work with the EU-sponsored Interstate Oil Gas Transport to Europe (INOGATE) project, which provides technical assistance to modernize oil and gas transport in central Europe and Asia in order to work toward European pipeline linkage to Caucasus and Central Asian oil and gas resources.

Along these lines, Greece and Turkey signed a binding agreement in February 2003 to build a $300 million natural gas pipeline extension connecting Karacabey, Turkey (near Bursa) with Komotini in northern Greece. This deal follows up on a March 28, 2002 memorandum of understanding (MOU) which called for extending the Iran-Turkey natural gas pipeline into Greece. Reportedly, the 175-mile-long pipeline extension (125 miles in Turkey, 50 miles in Greece) could be completed by 2005. The pipeline initially will transport around 17.7 Bcf of natural gas per year, but this could be expanded significantly. Eventually, natural gas from Iran–and possibly the Caspian region (e.g., Azerbaijan)–could be transported to southeastern Europe and possibly beyond via Turkey and Greece.

In other news, in October 2002, a twin 866-mile natural gas pipeline running from Russia under the Black Sea to Turkey was completed, with natural gas flows starting in February 2003, about one year behind the original schedule. The $3.2 billion "Blue Stream" pipeline runs from Izobilnoye in southern Russia, to Dzhugba on the Black Sea, then under the Black Sea for about 247 miles to the Turkish port of Samsun, and on to Ankara. Over the course of the 25-year agreement signed in December 1997, Turkey is to import 14.1 Tcf of natural gas from Russia via Blue Stream, with the pipeline eventually extended to other Mediterranean countries, including Greece.

Greece received its first liquefied natural gas (LNG) shipment in November 1999, beginning a 21-year contractual agreement between Algeria and Greece under which DEPA will purchase gas from Algeria's Sonatrach. Greece has one LNG terminal at Revithoussa, near Athens, with a capacity of 23 Bcf per year. A feasibility study is underway to determine whether to construct an underwater gas pipeline connecting Italy and Greece; if this pipeline does not come to fruition, it is possible that another LNG terminal will be constructed in Greece or that the terminal at Revithoussa will be expanded.

Coal

Lignite ("brown coal"), a brownish-black coal of low quality used almost exclusively for steam-electric power generation, is Greece's only significant fossil fuel source. Greece's lignite reserves total 3,168 million short tons (Mmst). The largest deposits are at Ptolemais and Amintaio, in northern

Greece. The country has no hard coal reserves, and imports hard coal from South Africa, Russia, Venezuela, and Colombia. Domestic production has been partly opened to private companies, but the Public Power corporation is still the largest producer, as well as consumer.

Electricity

In 2001, Greece generated 49.8 billion kilowatthours (Bkwh) of electricity, around 90 per cent of which was thermal, 10 per cent hydropower, and 1 per cent solar (percentages do not equal 100 per cent due to rounding). Most of the thermal is lignite-fired, with some oil-fired plants. New plants will be gas-fired for the most part. Electricity demand has been growing rapidly–nearly 50 per cent over the last decade–meaning (according to the Energy Regulatory Authority, RAE) that some 6,000 megawatts (MW) of additional capacity will be needed to guarantee adequate supply through 2015. Greek authorities are concerned that electrical generation capacity will be insufficient for the 2004 Olympic Games in Athens.

Greece's Public Power Corporation (PPC) is a state-owned monopoly that controls electric production, transmission, and distribution in the country. The company owns and operates 33 power plants on the mainland, and numerous smaller plants on the country's many isolated islands. In February 2001, with the partial liberalization of Greece's power market (35 per cent was opened to competition) in accordance with the EU's Electricity Directive, PPC lost its legal monopoly on electricity generation but remains the sole distributor. In November 2000, 20 per cent of PPC was floated on the Athens stock exchange. In October 2002, the European Commission decided not to oppose Greece taking measures to compensate PPC for "stranded costs" due to liberalization of the country's power market.

Other EU member countries had to open up their electricity markets by February 1999, but Greece was granted a two-year waiver in recognition of its unique situation: it borders no other member state, and much of its territory is comprised of islands that cannot be linked into the national grid. Given the lead time on construction of new power plants, it will be a while before competing generators are functioning, so PPC still has an effective production monopoly. The OECD has urged Greece to break up PPC, and plans are for the Greek power sector to be completely liberalized by 2007 (in July 2003, workers at PPC staged a 48-hour walkout to protest liberalization of the Greek power sector). According to *Power Engineering International*, foreign companies interested in Greece's power sector include Electricite de France, Belgium's Tractebel, Italy's Enel and Edison, UK's National Power, and Germany's RWE.

Greece's power network currently is connected with the networks of Albania, the Former Yugoslav Republic of Macedonia, and Bulgaria, allowing Greece to export electricity to Kosovo in Yugoslavia, through Albania and the Former Yugoslav Republic of Macedonia (although transmission problems in those countries have sometimes prevented much of this electricity from reaching its intended recipients). In June 2001, energy ministers from Albania, Bosnia and Herzegovina, Bulgaria, Greece, Macedonia, and Romania signed a memorandum for the creation of a competitive energy market in the Balkans. Greece would like to upgrade its link with Bulgaria and to Europe via Croatia and Bosnia. Greece is involved in a number of projects to link its electric grid with neighboring countries. In July 2002, Greece and Italy completed work on a 500-megawatt (MW) cable (in both directions) under the Ionian Sea to link their national power grids. The 102-mile cable links Otranto, Italy, and Aetos, Greece. The project is a joint venture between ENEL (75 per cent) and PPC (25 per cent). In May 2003, Bulgaria's National Electric Company signed deals with Greece and Macedonia to export 150 MW of power, and reportedly is negotiating sale of 600 MW to Greece.

Improved Greek-Turkish relations also are affecting the Greek electricity sector. In January 2000, a Greek-Turkish-U.S. (Copelouzos-Gama-ExxonMobil) consortium announced plans to construct a gas-fired power plant in Greece. The plant will have a capacity between 400 MW and 600 MW and will be used to export electricity to Turkey in addition to helping supply increasing Greek domestic demand. Electricity will be exported via the new 400-kilovolt (kV) transmission line to be constructed between Filippoi (Greece) and Hamidabad (Turkey). Greece and Turkey also hope to reach agreement by 2006 on linking the two countries' power grids.

Renewable electricity generation projects are on the rise in Greece, and the government has established the Centre for Renewable Energy Sources (CRES), under the Development Ministry, to promote renewable energy. CRES estimates that 15 per cent of the country's electricity needs can be produced by wind farms, with installed wind-power capacity possibly expanding from 270 MW at present to 2,000 MW by 2010. The EU requires that member states produce 12 per cent of their electricity from renewable sources by 2010, and this appears easily achievable for Greece–assuming that regulatory hurdles and technical problems (*i.e.*, the need to extend Greece's power grid to the islands where wind power is generated) can be overcome. Already, wind farms exist on a number of Greek islands (Crete, Evia, Andros, Samos, etc.) and 20 per cent of households use solar water heaters. Islands in the western part of Greece are connected to the mainland system by submarine cables.

The use of solar power in Greece reduces the need for conventionally generated energy by about 1.4 billion kilowatthours per year. A 50-MW parabolic trough-type solar power plant is under construction in Crete. DEH is planning a 100-kilowatt photovoltaic (PV) park for the island of Gavdos, in addition to already-existing PV capability on the island. U.S.-based Energy Photovoltaics (EPV) announced in July 2001 that it would build a $22 million manufacturing facility for PV panels in Kilkis, with annual production capacity of 5 MW. The company is also implementing a 400-kilowatt photovoltaic power station on the roof of its manufacturing unit building.

In February 2002, a report by Merrill Lynch estimated that wind power capacity additions in Greece would be 26 MW in 2002, 19 MW in 2003, 54 MW in 2004, and 56 MW in both 2005 and 2006. In June 2001, Gemesa of Spain signed an agreement with Hellenic Energy and Development Company to invest 420 million Euro to develop wind power plants with a total capacity of 500 MW by 2008. Windforce of the UK plans to develop $800 million of projects in the EU, including three wind farms in Greece (at Makronisos, Kilkis, and Lakonia) with a total capacity of 650 MW.

Country Overview

Independence: 1829 (from the Ottoman Empire)

Population (2003E): 10.7 million

Location/Size: Southern Europe, bordering the Aegean, Ionian and Mediterranean Seas/131,940 sq. km. (51,146 sq. mi) ; roughly the size of Alabama

Major Cities: Athens (capital), Thessaloniki, Piraeus, Patras

Languages: Greek (official), English, French

Ethnic Groups: Greek (98 per cent); other (2 per cent)

Religion: Greek Orthodox (98 per cent), Muslim (1.3 per cent), other (0.7 per cent)

Defense (8/98): Army (116,000), Navy (19,500), Air Force (33,000), Conscripts (112,700)

Economic Overview

Currency: Euro

Market Exchange Rate (7/24/03): US$1 = 1.1485 Euro

Nominal Gross Domestic Product (GDP, 2000E): $113 billion

Real GDP Growth Rate (2002E): 4.0 per cent *(2003F):* 3.7 per cent

Unemployment Rate (2002E): 9.9 per cent *(2003F):* 9.6 per cent

Inflation Rate (2002E): 3.6 per cent *(2003F):* 3.1 per cent

Major Trading Partners: Germany, Italy, other OECD Europe

Major Export Products: Manufactures, food and beverages, petroleum products

Major Import Products: Manufactured consumer goods, capital goods, crude oil, food products

Merchandise Exports (2002E): $10.4 billion

Merchandise Imports (2002E): $33.1 billion

Net Merchandise Trade Deficit (2002E): $22.7 billion

Current Account Deficit as a per cent of GDP (2002E): 6.8 per cent *(2002F):* 6.1 per cent

Energy Overview

Proven Oil Reserves (1/1/03E): 9 million barrels

Oil Production (2002E): 7,162 barrels per day (bbl/d), of which 3,162 bbl/d is crude oil

Oil Consumption (2002E): 408,000 bbl/d

Net Oil Imports (2002E): 400,838 bbl/d

Crude Oil Refining Capacity (1/1/03E): 406,500 bbl/d

Major Crude Oil Import Sources: Persian Gulf OPEC

Natural Gas Reserves (1/1/03E): 18 billion cubic feet (Bcf)

Natural Gas Production (2001E): 0.1 Bcf

Natural Gas Consumption (2001E): 71 Bcf

Coal Reserves (2000E): 3,168 million short tons (all lignite)

Coal Production (2001E): 74.5 million short tons (Mmst)

Coal Consumption (2001E): 75.9 Mmst

Electric Generation Capacity (2001E): 10.3 gigawatts

Net Electricity Generation (2001E): 49.8 billion kilowatthours (Bkwh)

Environmental Overview

Total Energy Consumption (2001E): 1.39 quadrillion Btu* (0.3 per cent of world total energy consumption)

Energy-Related Carbon Emissions (2001E): 28.1 million metric tons of carbon (0.4 per cent of world total carbon emissions)

Per Capita Energy Consumption (2001E): 131.4 million Btu (vs U.S. value of 341.8 million Btu)

Per Capita Carbon Emissions (2001E): 2.65 metric tons of carbon (vs U.S. value of 5.5 metric tons of carbon)

Energy Intensity (2001E): 9,623 Btu/ $1995 (vs U.S. value of 10,736 Btu/ $1995)**

Carbon Intensity (2001E): 0.19 metric tons of carbon/thousand $1995 (vs U.S. value of 0.17 metric tons/thousand $1995)**

Fuel Share of Energy Consumption (2001E): Oil (61.4 per cent), Coal (29.6 per cent), Natural Gas (5.3 per cent)

Fuel Share of Carbon Emissions (2001E): Oil (58.0 per cent), Coal (38.2 per cent), Natural Gas (3.8 per cent)

Status in Climate Change Negotiations: Annex I country under the United Nations Framework Convention on Climate Change (ratified August 4th, 1994). Under the negotiated Kyoto Protocol (signed on April 29th, 1998, ratified along with EU in May 2002), Greece has agreed to limit greenhouse gas increase to 25 per cent above 1990 levels by the 2008-2012 commitment period. Within the EU, each country has a different commitment.

Major Environmental Issues: Air pollution and water pollution.

Major International Environmental Agreements: A party to Conventions on Air Pollution, Air Pollution-Nitrogen Oxides, Air Pollution-Sulphur 94, Antarctic-Environmental Protocol, Antarctic Treaty, Biodiversity, Climate Change, Desertification, Endangered Species, Environmental Modification, Hazardous Wastes, Law of the Sea, Marine Dumping, Nuclear Test Ban, Ozone Layer Protection, Ship Pollution, Tropical Timber 83, Tropical Timber 94 and Wetlands. Has signed, but not ratified, Air Pollution-Persistent Organic Pollutants, Air Pollution-Volatile Organic Compounds.

Oil and Gas Industries

Organization: Hellenic Petroleum–the state petroleum company; DEPA–the state-controlled gas company; Public Power Corporation–the state-owned utility

Major Refineries (capacity–bbl/d, 1/1/03E): HP Aspropyrgos (140,000), Motor Oil Aghii Theodori (100,000), Petrolas Hellas Elefsis (100,000), HP Thessaloniki (66,500)

Major Ports: Piraeus,Thessaloniki, Patras

40

Horn of Africa

The following provides a brief overview of the energy sectors of the Horn of Africa region–Djibouti, Eritrea, Ethiopia and Somalia.

DJIBOUTI

The French Territory of the Afars and the Issas became Djibouti in 1977. In November 1991, the mainly Afar-supported Front for the Restoration of Unity and Democracy (FRUD) began fighting the Issa-dominated government. Sporadic attacks continued in 1997 and 2000. On May 12, 2001, President Ismail Omar Guelleh presided over the signing of the final peace accord officially ending the decade-long civil war. France maintains one of its largest military bases outside France in Djibouti with some 2,800 troops as well as warships, aircraft and armored vehicles. In addition to French troops, Djibouti has become a launching point for the US war on terrorism in the Middle East. United States troops have been stationed and passing through the country since late-2002. In exchange for use of territory, the United States has offered monetary assistance to help Djibouti improve its counterterrorism operations.

Djibouti's main economic asset is its strategic location. The city of Djibouti, capital and home to nearly two-thirds of the country's population, is a major transshipment port and bunkering facility. Good transportation infrastructure with the country and links to neighboring African states earns Djibouti much-needed transit taxes and harbor fees. Trade through Djibouti increased significantly during the Ethiopian-Eritrean war when Djibouti became the only significant port for landlocked Ethiopia.

Djibouti is working to significantly expand the capacity of its ports by building a new oil jetty to accommodate oil products, LPG, edible oils, and bitumen on vessels up to 120,000 deadweight tons (DWT) and have the capacity to handle 100,000 cubic meters of petroleum products. This first phase of port expansion will begin in September 2003, with operations beginning a year later. The second

phase of the project adds a $300 million, 2,000 meter long, mega container port. The new additions to Djibouti's ports, supported by Emirates National Oil and Dubai Ports International, will enable the port of Djibouti to meet growing cargo requirements over the next 20 years.

Djibouti's real gross domestic product (GDP) is expected to grow 2.8 per cent in 2003, following estimated growth of 2.4 per cent in 2002, and 1.9 per cent in 2001. Strong growth has been fueled by the transportation and communication sectors as well as the trade and tourism sectors, which together make up more than 40 per cent of GDP. An unemployment rate of 40 per cent to 50 per cent continues to be a major problem for Djibouti's economy. In December 2002, the International Monetary Fund (IMF) approved the third disbursement of funds ($6 million) from a Poverty Reduction and Growth Facility (PRGF) signed with Djibouti in 1999.

Oil

Although there is currently no upstream (exploration or production) oil activity in Djibouti, the government has tried to generate interest in offshore oil exploration without success. The downstream oil sector, however, is an important aspect of Djibouti's economy, given the role the capital city plays as a significant regional bunkering and refueling facility. Three companies–ExxonMobil, Shell and Total–handle refueling at Djibouti's port. The companies, along with ChevronTexaco, also distribute and market petroleum products in the country. Storage capacity at the port facility is 1.26 million barrels (200,000 cubic meters). The Dubai Ports Authority (DPA) was awarded a 20-year contract in June 2000 to manage the port. DPA hopes to increase Djibouti's handling capacity from 125,000 metric tons to 300,000 metric tons per year and to make it the leading transshipment point on the African continent.

Table 40.1: Economic and Demographic Indicators

Country	Gross Domestic Product (GDP) 2001E (Billions of U.S. $–PPP)	Real GDP Growth Rate, 2002 Estimate	Real GDP Growth Rate, 2003 Projection	Per Capita GDP, 2001E (PPP)	Population 2002E (Millions)
Djibouti	$0.6	2.4%	2.8%	$1,400	0.65
Eritrea	$3.2	8.8%	7.1%	$740	4.50
Ethiopia	$46.0	4.3%	4.0%	$700	67.2
Somalia	$4.1	NA	NA	$550	7.8
Regional Total/Average	**$53.9**	**4.2%**	**3.9%**	**$670**	**80.1**

PPP = Purchasing Power Parity; NA = Not Available.

Table 40.2: Energy Consumption and Carbon Dioxide Emissions, 2001

Country	Total Energy Consumption (Quadrillion Btu)	Petroleum	Natural Gas	Coal	Nuclear	Hydro-Electric	Other Electricity	Net Electricity Imports	Carbon Dioxide Emissions (Million metric tons of carbon)
Djibouti	0.025	100.0%	0.0%	0.0%	0.0%	0.0%	0.0%	0.0%	0.504
Eritrea	0.012	100.0%	0.0%	0.0%	0.0%	0.0%	0.0%	0.0%	0.242
Ethiopia	0.066	73.1%	0.0%	0.0%	0.0%	26.3%	0.0%	0.0%	0.945
Somalia	0.008	100.0%	0.0%	0.0%	0.0%	0.0%	0.0%	0.0%	0.163
Regional Total/Average	**0.111**	**84%**	**0.0%**	**0.0%**	**0.0%**	**15.6%**	**0.0%**	**0.0%**	**1.854**

Note: Percentages may not add up to 100%due to rounding.

Table 40.3: Energy Supply Indicators

Country	Crude Oil Reserves, 1/1/03 (Million Barrels)	Natural Gas Reserves, 1/1/03 (Billion Cubic Feet)	Coal Reserves (Million Short Tons)	Petroleum Production, 2001 (Thousand Barrels Per Day)	Natural Gas Production, 2001 (Billion Short Tons)	Coal Production, 2001 (Million Short Tons)	Electric Generating Capacity, 2001 (Gigawatts)	Crude Oil Refining Capacity, 1/1/03 (Thousand Barrels Per Day)
Djibouti	0	0	0	0	0	0	0.09	0
Eritrea	0	0	0	0	0	0	0.06	14.6
Ethiopia	0.428	880	0	0	0	0	0.53	0
Somalia	0	200	0	0	0	0	0.07	10.0
Regional Total	**0.428**	**1,080**	**0**	**0**	**0**	**0**	**0.76**	**24.6**

Electricity

Djibouti currently has installed electricity generating capacity of 85 megawatts (MW), all of which is thermal (oil-fired). In January 2001, U.S.-based Geothermal Development Associates (GDA) announced that it had completed a feasibility study on the development of a 30-MW geothermal power plant in Djibouti. The study, which commenced in August 2000, established the commercial viability of the proposed generating facility. The $115 million plant, to be located in the Lake Assal region west of the capital, will be constructed on the build own operate (BOO) financing scheme. The Global Environmental Facility (GEF), a joint initiative of the World Bank and the United Nations (UN), has approved a $280,000 financing package to pay for contract negotiations required for the project. To date, however, these funds have not been released. At the same time, however, Electricite de Djibouti, the national electric company, has begin to remove aging diesel-fired generating units. To continue to provide power to rural residents, the government, with the help of a grant from a number of Arab financial institutions, is installing solar and wind capacity. The primary goal of the project is to replace old diesel powered rural water pumps with new ones powered by renewable resources, but excess energy will be used for electrification.

ERITREA

In 1952, a UN resolution federating the former Italian colony of Eritrea with Ethiopia went into effect. In 1962, Emperor Haile Selassie unilaterally dissolved the Eritrean parliament and annexed the country. The Eritrean fight for independence continued even after Haile Selassie was ousted in a coup in 1974. A 30-year struggle for independence ended in 1991, with Eritrean rebels defeating the governmental forces led by Mengistu Haile Miriam. On April 23-25, 1993, Eritreans voted overwhelmingly for independence from Ethiopia in a UN-monitored free and fair referendum. The Eritrean authorities declared Eritrea an independent state on April 27, 1993. The National Assembly was elected in national, freely contested election and then chose former liberation movement leader Isaias Afwerki as President.

When Eritrea and Ethiopia separated amicably in 1993, several border issues remained unresolved. Fighting broke out between the two countries in May 1998 and continued until June 2000 when both sides accepted an Organization of African Unity (OAU) peace proposal. The formal treaty ending the war was signed on December 12, 2000 and an independent entity, the Eritrea Ethiopia Boundary Commission (EEBC) was established to determine the new border. The EEBC released its report on April 13, 2002, but Ethiopia is contesting some of the commission's decisions.

Growth of the Eritrean economy was hampered by the war. Eritrea's real GDP growth in the two years prior to the conflict averaged 7.4 per cent. Real GDP growth fell to 4.0 per cent in 1998, 0.8 per cent in 1999 and -8.2 per cent in 2000. With the cessation of hostilities, real GDP growth of 9.7 per cent was observed in 2001, and growth of 8.8 per cent in 2002. Continued growth of 7.1 per cent is expected in 2003. Inflation is an ongoing problem, it was 24 per cent in 2002, but it is expected to decline to less than 15 per cent in 2003. Multilateral and bilateral donors have pledged nearly $132 million towards the demobilization of 200,000 Eritrean soldiers and the social and economic recovery of the country. Donor development funding is seen as crucial in the improvement of the country's economy.

Oil

Hydrocarbon exploration, primarily offshore in the Red Sea, began in the 1960's when Eritrea was still federated with Ethiopia. In 1995, Eritrea signed a production sharing contract (PSC) with U.S.-based Anadarko Petroleum (Anadarko) for the offshore Zula Block. Anadarko signed a second

PSC for the offshore Edd Block, located south of the Zula Block, in September 1997. Anadarko announced, in December 1997, that it had reached an agreement with ENI/Agip (Agip) to swap interests in exploration acreage. Anadarko received a 25 per cent interest in a Tunisian block operated by Agip, and Agip received a 30 per cent share in the 6.7-million acre Zula Block and 30 per cent interest in the Edd Block. Burlington Resources, a U.S.-based independent, later joined the consortium by acquiring a 20 per cent interest in both acreages. Anadarko's first two exploration wells, both drilled on the Zula Block, were unsuccessful. In January 1999, a third dry well, Edd-1 on the Edd Block, was drilled. Citing the disappointing exploration results, Anadarko and its partners ceased exploration activities and relinquished their rights to the offshore blocks.

A more recent attempt also did not meet with success. In May 2001, U.S.-firm CMS Energy signed an exploration agreement for a 14,000-square kilometer block in northeastern Eritrea but the company has released its claim as it put its entire oil exploration and production unit up for sale in May 2002.

Bab el-Mandeb (Mandab)

The Bab el-Mandeb is a narrow waterway situated between Eritrea, Djibouti and Yemen that connects the Red Sea with the Gulf of Aden and the Arabian Sea. In 2000, it was estimated that around 3.2-3.3 million barrels per day (bbl/d) of oil flowed through the Bab el-Mandeb. Disruptions or closure of the Bab el-Mandeb could keep tankers from the Persian Gulf from reaching the Suez Canal/Sumed Pipeline complex, diverting them around the southern tip of Africa (the Cape of Good Hope). This would add greatly to transit time and cost, and effectively tie up spare tanker capacity. The Bab el-

Mandeb could be bypassed (for northbound oil traffic) by utilizing the East-West oil pipeline, which traverses Saudi Arabia and has a capacity of about 4.8 million bbl/d. However, southbound oil traffic would still be blocked. In addition, closure of the Bab el-Mandeb would effectively block non-oil shipping from using the Suez Canal, except for limited trade within the Red Sea region.

In December 1995 and again in August 1996, Eritrean and Yemeni forces clashed over control of the Hanish Islands, located just north of the Bab el-Mandeb. In October 1996, the two countries signed an agreement over the islands. The two sides agreed to put their case before an international court of arbitration (Permanent Court of Arbitration-PCA). The court will then issue two rulings; one on who has sovereignty over the disputed area, and one on the demarcation of the two sides' maritime boundary. In October 1998, the PCA ruled that the Hanish Islands are subject to the territorial sovereignty of Yemen. In December of 1999, the PCA issued its ruling on the maritime boundary.

Downstream and Refining

Eritrea has crude refining capacity of 18,000 bbl/d, but the refinery located in the Red Sea port of Assab has been shut down since 1997 due to the high operating and maintenance costs. Ethiopia and Eritrea, joint operators of the facility, decided to close the Assab refinery in August 1997 and import refined petroleum products to meet domestic needs.

Eritrea's petroleum consumption was estimated at 6,000 bbl/d in 2001. Marketing and distribution of petroleum products is performed by ExxonMobil, Shell and Total. In June 2000, Shell purchased the downstream operations of Agip in several African countries including Eritrea and Ethiopia. The Eritrean assets included service stations, two petroleum product depots and an LPG (liquefied petroleum gas) filling station.

In August 2000, Sudan's National Petroleum Company announced plans to lay pipelines to supply Eritrea and Ethiopia with petroleum products from its Khartoum refinery. Eritrea has not yet benefited from Sudanese oil and relations between the two countries soured in April 2003 when Sudan accused Eritrea of supporting Sudanese rebels in the eastern part of Sudan. Although Eritrea denies the charges, future trade relationships are unlikely under the current climate.

Electricity

Eritrea has approximately 60 MW of diesel-fired generating capacity. The Eritrean Electricity Authority (EEA) handles generation, transmission and distribution of electricity. In 1997, South Korean firms Daewoo and Hanjung signed an agreement to build a heavy oil-fired plant in at Hirgigo, just outside of Massawa. The plant, nearly completed, was damaged in a bombing raid by Ethiopia in 2000. In 2001, Eritrea signed loan agreements with the United Arab Emirates and Saudi Arabia for the facility's repair. The 88 MW facility came online in March 2003, but many industry experts fear that the new capacity could overload Eritrea's dated grid system. Both the European Development Fund and the World Bank have considered projects to update the transmission lines, but firm contracts have not been negotiated.

Electricity is only available in Eritrea's larger cities and towns, leaving about 80 per cent of the Eritrean population without access to electricity. Some smaller villages have community diesel generators which can provide small amounts of electricity to households. Photovoltaic (PV) electricity generation is being used in special applications throughout the country. Twenty-six rural health centers are each supplied with 2-kilowatt (kW) solar photovoltaic power systems for refrigeration, lighting, operating theaters, fans, and laboratory equipment. Additionally, the majority of the 140 rural clinics are equipped with solar powered vaccine refrigerators. Approximately 3 per cent (about

60 villages) of Eritrea's villages have been supplied with PV systems (0.8 to 1.2kW) to power water pumps to supply drinking water. Each system serves a minimum of 300 households. Over 70 rural schools (out of 700) have been provided with PV systems for lighting and power.

ETHIOPIA

Ethiopia is the oldest independent country in Africa. Unique among African countries, Ethiopia maintained its freedom from colonial rule, except during the Italian occupation of 1936-41. In 1974, a military junta, the Derg, deposed Emperor Haile Selassie, who had ruled since 1930, and established a socialist state. The Derg was toppled by a coalition of rebel forces, the Ethiopian People's Revolutionary Democratic Front (EPRDF), in 1991. A constitution was adopted in 1994 and Ethiopia's first multiparty elections were held in 1995. A two-year border war with Eritrea ended when a peace treaty was signed in December 2000. Disagreement on the location of the border between the two countries is ongoing, despite the final ruling of a commission charged with identifying the border.

Ethiopia's economy is primarily agrarian, with the agricultural sector accounting for 45 per cent of GDP and 80 per cent of the workforce. Coffee, Ethiopia's primary export crop, accounted for 58 per cent of total exports in 1999, and has averaged two-thirds of all export earnings over the last 20 years. Other important agricultural exports include qat (khat), a mild stimulant from the leaves of the Catha Edulis shrub, pulses, oilseeds, live animals and hides. Ethiopia's real GDP growth was 5.4 per cent in 2000 and increased to 7.7 per cent in 2001. Growth in 2002 slowed to 4.3 per cent as a severe drought decimated agricultural production. Growth forecasts for 2003 suggest that the economy will expand by 4.0 per cent

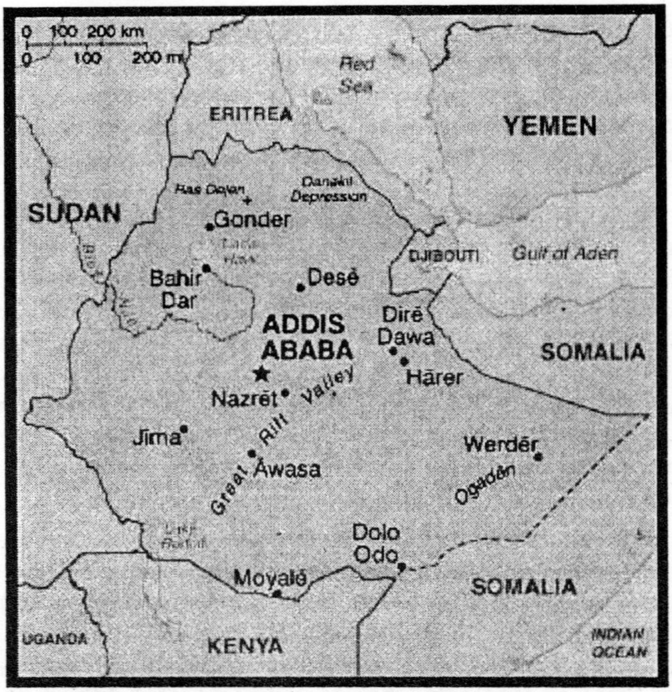

Although continued donor support is seen as the crucial element in Ethiopia's economic reform, both the IMF and World Bank suspended new lending to Ethiopia during the border war with Eritrea. The suspension was lifted after the signing of the peace accord in December 2000. The World Bank approved a $400 million loan to finance emergency recovery, military demobilization and reintegration projects. In July 2001, the IMF approved a $112 million PGRF to support Ethiopia's economic program. In November 2001, the IMF and World Bank announced that Ethiopia was eligible for a $1.9 billion debt relief package under the Heavily Indebted Poor Countries (HIPC) Initiative, becoming the 24th country to qualify for debt relief under the HIPC's enhanced framework. The savings in debt service resulting from the HIPC are substantial, amounting to about $96 million per year on average until 2021. The resources made available by debt relief provided under the HIPC will be allocated to key anti-poverty programs. Poverty-targeted expenditures are projected to increase steadily, from 10.9 per cent of GDP in 2001, to 14.7 per cent in 2002, and 15.5 per cent in 2003.

Oil and Natural Gas

Ethiopia's current proven hydrocarbon reserves are minimal, but the potential to increase reserves to commercial viability is seen as promising. The country's geology is similar to that of its oil-producing neighbors to the east (on the Arabian peninsula) and the west (Sudan). In April 2001, the Ministry of Mines and Energy reported that hydrocarbon seeps had been discovered in several regions. The government plans to conduct feasibility studies to establish the extent and viability of the deposits.

Hydrocarbon exploration in Ethiopia's Ogaden Basin began over 80 years ago (Standard Oil in 1920). The Ethiopian government formed the Calub Gas Share Company (CGSC) to develop the fields. In 1994, the World Bank approved a $74 million loan to develop the Ogaden Basin fields. The Ethiopian Privatization Agency (EPA) put the CGSC up for privatization in 1998, but the EPA, citing weak bids, withdrew the tender. In December 1999, Houston-based Sicor announced that it had signed a $1.4 billion joint-venture deal to develop the Calub natural gas project. Under the terms of the agreement, Gasoil Ethiopia Project (GEP), the joint-venture firm, will acquire 95 per cent of the CGSC under the Ethiopian government's privatization law. Currently, 5 per cent of the CGSC is held by local private investors. The Ethiopian government will hold a 20 per cent interest in GEP with Sicor holding the remaining share. GEP plans to construct a 375-mile, 24-inch pipeline to transmit natural gas to the town of Awash, which is approximately 75 miles east of the capital Addis Ababa. At Awash, plans call for construction of a cryogenic liquids plant and two gas-to-liquids process systems with capacity to process 200 million cubic feet per day (Mmcf/d) of natural gas. The end products will be synthetic fuels and petrochemical feedstocks plus steam that will generate electricity and help produce 20,000 bbl/d of potable water. A planned refinery will produce products including diesel, gasoline, kerosene and jet fuels. The gas-to-liquids system will also produce some 500 tons of ammonia per day as feedstock for a urea plant to be constructed. Although construction on the pipeline was scheduled to begin in the summer of 2002, gas development in Ogaden has not yet begun.

In June 2003, the Ethiopian government signed an oil exploration deal with Petronas for 5,800 square mile tract in Gambela, in the far western part of the country. The region is closely related the Sudan oil fields. Petronas has committed to investing in regional infrastructure, employing local staff, improving health services, and developing the skills of the ministry of Mines. Petronas is also interested in natural gas exploration in Ogaden, but no official plans have yet been made.

Downstream

Ethiopia's petroleum consumption was estimated at 23,000 bbl/d in 2001. With the closure of the Assab refinery in 1997, Ethiopia is totally reliant on imports to meet its petroleum requirements. Some petroleum imports are received at the port of Djibouti, and shipped via rail and tanker truck to Ethiopia. With the recent development of oil in Sudan, however, Ethiopia has begun importing oil which, under COMESA, is not subject to tariffs. Oil imports from Sudan began in January 2003 transported by tanker trucks along a new road between the two countries. Since the trade began, however, oil shipments, which are expected to meet 85 per cent of Ethiopia's gasoline requirements, have halted twice. Nevertheless, the two countries have agreed to upgrade the road along this important transit corridor to increase commerce.

Marketing and distribution of petroleum products is performed by ExxonMobil, Shell and Total. In June 2000, Shell purchased the downstream operations of Agip in several African countries including Ethiopia. The Ethiopian assets included over 100 service stations, two depots and four LPG filling plants.

Electricity

Ethiopia has approximately 529 MW of installed generating capacity. The vast majority of Ethiopia's existing capacity (85 per cent) is hydroelectric. The Ethiopian Electric Power Corporation (EEPCO), the state-owned firm responsible for electricity generation, plans to construct several new generating facilities to provide electricity to Ethiopia. Currently, less than half of Ethiopia's towns have access to electricity though EEPCO electrified more than eighty towns between 2001 and 2003. Since most of Ethiopia's electricity is generated from hydroelectric dams, the country's power system is vulnerable to extended droughts. Ethiopia recently endured more than six months of power cuts due to low water levels in dams around the country. Initially blackouts were scheduled once a week, but as the drought wore on, customers lost power for 15 hours two days a week, a situation that strained the resources of many businesses in urban centers.

EEPCO is rapidly expanding their generating capacity. The 73-MW Tis Abay 2 facility, located on the Blue Nile (Abay) came online in 2001. U.S.-based Harza Engineering (now MWH Global) is overseeing the construction of an additional 34-MW unit at the Finchaa hydroelectric facility in western Ethiopia. EEPCO also plans to open the new 180-MW Gilgel Gibe hydroelectric facility in October 2003. Gilgel Gibe, located on the Omo River in southwestern Ethiopia, will increase the country's power capacity to 700-MW. EEPCO has begun development of Ethiopia's largest generating facility at Tekeze. The 300-MW hydroelectric facility will be located in northern Ethiopia and will cost about $350 million.

Construction of Ethiopia's first Independent Power Project (IPP) was set to commence in early 2002. The Gojeb IPP will consist of a 150-MW hydroelectric facility in western Ethiopia. The project is being developed by Mohammed International Development Research Organization & Companies (Midroc). When completed, Midroc will sell the output from Gojeb to EEPCO. Agreements on additional IPP projects were signed in June 2001. The largest facility will be the 162-MW Genale hydroelectric facility located on the border between the Oromia Region and the Southern Peoples Nationalities Regional State. The plants will be built under the Build-Operate-Transfer (BOT) system. ENERCO will operate the facilities for 30 years, which would be renewable for another 30 years.

In April 2001, Ethiopia signed agreements to export electricity to neighboring Djibouti. Negotiations are ongoing and exports are expected to begin in 2004, following the interconnection of the countries' electric grids.

SOMALIA

The Somali Republic gained independence on July 1, 1960. Somalia was formed by the union of British Somaliland and Italian Somaliland. A socialist state was established following a coup led by Major General Muhammad Siad Barre. Rebel forces ousted the Barre regime in 1991, but turmoil, factional fighting, and anarchy ensued. The Somali National Movement (SNM) gained control of the north, while in the capital of Mogadishu and most of southern Somalia the United Somali Congress achieved control.

In 1992, responding to the political chaos and humanitarian disaster in Somalia, the United States and other nations launched peacekeeping operations to create an environment in which assistance could be delivered to the Somali people. By March 1993, the potential for mass starvation in Somalia had been overcome, but the security situation remained fragile. On October 3, 1993 U.S. troops received significant causalities (19 dead over 80 others wounded) in a battle with Somali gunmen. When the United States (in 1994) and the UN withdrew (in 1995) their forces from Somalia, after suffering significant casualties, order still had not been restored.

A Transitional National Government (TNG) was created in October 2000 with the three-year mandate of creating a permanent national Somali government. Although they declared their independence, the TNG does not recognize Somaliland and Puntland as independent republics but has been unable to reunite the country. Peace talks began in late-2002 and are ongoing. Somaliland has refused to participate in these talks saying that while it would welcome peace in former Italian Somalia, Somaliland is an independent country awaiting international recognition.

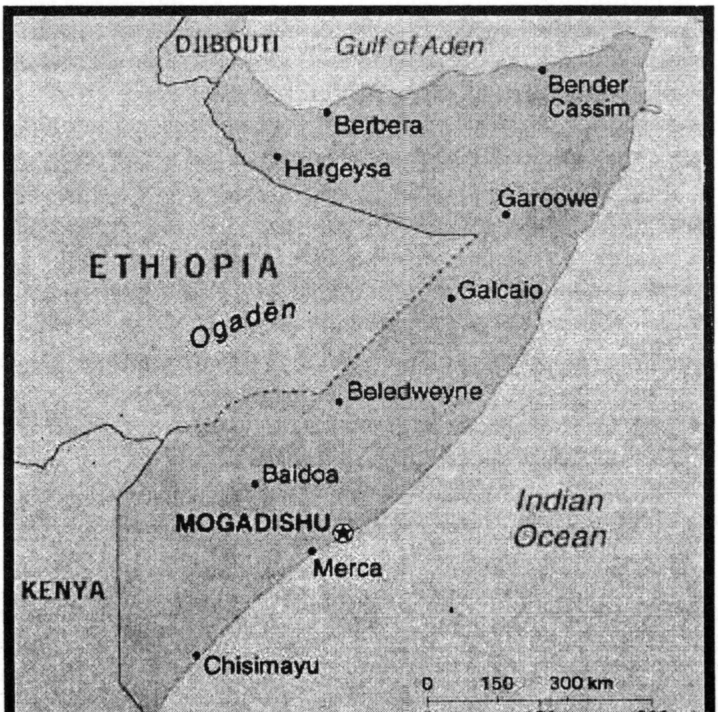

Somalia's economy, one of the world's least developed, has been further hampered by the country's ongoing internal strife. Reliable economic data is scarce, and the TNG cannot manage the national economy while it struggles to gain control over the country. Livestock production (cattle, goats & sheep) is the mainstay and largest foreign exchange earner of the Somali economy. An outbreak of Rift Valley Fever (RVF) in southern Saudi Arabia and Yemen (the first reported outside Africa) in 2000 led six Gulf States–Saudi Arabia, Bahrain, Oman, Qatar, Yemen and the United Arab Emirates–to ban livestock imports from the Horn of Africa. Middle Eastern countries are considering lifting the ban on livestock in early 2004. Another significant portion of the Somali economy, foreign remittances, have fallen significantly following the US government's closure of the Al-Barakat transfer company, which has been accused of transferring funds on behalf of Osama bin Laden and the Al-Qaida terrorist network. Remittances from abroad are estimated to be $200-$500 million annually.

While the TNG is in the process of reestablishing Somalia's Central Bank. Somalia is unable to receive IMF, and other multilateral aid due to the lack of institutions/financial infrastructure in place.

Oil and Natural Gas

Somalia has no proven oil reserves, and only 200 billion cubic feet of proven natural gas reserves. Somalia currently has no hydrocarbon production. Oil seeps were first identified by Italian and British geologists during the colonial era. Exploration activities were focused in northern Somalia, and several foreign firms, including Agip, Amoco, Chevron, Conoco and Phillips, held concessions in the area. The firms all declared force majeure following the collapse of the central government.

Exploration activity remains hindered by the internal security situation, and the multiple sovereignty issues. In February 2001 Total signed an exploration agreement with the TNG. The twelve-month agreement grants Total the rights to explore in the Indian Ocean off southern Somalia. Hassan Farah, TNG's Minister for Water and Mineral Resources, stated that the government would provide security during the exploration activities. Several factional leaders have denounced the agreement, and stated that the TNG did not have the authority to sanction the agreement, nor the power to guarantee the safety and security of the exploration operations.

In May 2001, Somaliland signed an agreement with U.K.-registered Rovagold and two Chinese firms, CPEC and CPC, for the right to explore for oil. Dubai-based Zarara Energy also signed an exploration agreement with Somaliland. The Somaliland government has said it will honor, until they expire, the existing contracts foreign companies signed with the Barre regime that are in their territory. None of the firms have resumed operations in Somaliland.

Somalia's petroleum consumption was an estimated 4,000 bbl/d in 2001. The organization officially responsible for all petroleum product distribution and retailing is the cooperative Iskash. The state-owned Iraqsoma Refinery Corporation operated a 10,000-bbl/d refinery outside of Mogadishu, but it has been inoperative since 1991. Total is involved in the downstream sector in Somaliland. It rehabilitated and manages the operations of the oil terminal in Berbera, Somaliland's primary port. Total also supplies fuel to airports located in Berbera and Somaliland's capital of Hargeisa.

Electricity

Somalia currently has installed electricity generating capacity of 70 megawatts (MW), all of which is diesel-fired. Ente Nazionale Energia Elettrica (ENEE) is the entity responsible for generation, transmission and distribution of electricity in Somalia. Electrical infrastructure has been damaged and destroyed, and the ongoing strife has hindered the development of new electric resources. A planned hydroelectric facility on the Juba River has been delayed due to the continued fighting.

Studies have indicated that the Horn of Africa, especially Somalia, is a prime location for harnessing wind for electricity generation. Plans for wind generation have been proposed, but were derailed following the ouster of the Barre regime.

In October 2001, WorldWater Corp., a U.S.-based water management and solar engineering company, signed agreements with the TNG to become the master consultant and contractor for all water and energy programs in Somalia. Under the three-year agreement WorldWater would develop, manage and oversee contracting for the country's water resources and incorporate renewable energy projects such as solar power into Somalia's infrastructure. This includes locating and managing groundwater sources in municipal and rural areas, delivering water for drinking and for irrigation using the WorldWater's solar pumping systems and generating independent electricity with its solar power systems.

41

India

The Republic of India (India), the world's sixth largest energy consumer, plans major energy infrastructure investments to keep up with increasing demand–particularly for electric power. India also is the world's third-largest producer of coal, and relies on coal for more than half of its total energy needs.

General Background

India's economic growth is currently recovering from a mild slowdown in 2002, which was mainly attributable to weak demand for manufactured exports and the effects of a drought on agricultural output. Real growth in the country's gross domestic product (GDP) was 4.8 per cent for 2002, and is projected to rise to 5.7 per cent in 2003. The economic effects of recent political tensions in the region have been quite modest. (The Indian fiscal year for economic statistics begins on April 1.)

After many years of pursuing economic policies based on import substitution and state ownership of key industries, India's government embarked on a series of economic reforms in the mid-1990s. Included in the reforms have been a relaxation of restrictions on foreign ownership in some sectors, and the privatization of some industrial enterprises. After much delay, the Indian government decided in January 2003 to move forward with the sale of majority stakes in two downstream oil companies, which has been seen as a major development.

India has a longstanding territorial dispute with Pakistan over the ownership of Kashmir, which has led to a tense relationship between the two countries since the partition of British India in 1947. After a large-scale mobilization of military forces along their border during most of 2002, tensions eased somewhat late in the year, and both sides have pulled back some of their forces from the border. India's rivalry with Pakistan has direct relevance to the country's energy sector, as it impedes plans for regional natural gas and/or oil pipelines (*i.e.*, from Central Asia).

India has implemented a series of policy changes since the mid-1990s to encourage foreign investment. Tariffs on imported capital goods have been lowered, and in some cases eliminated (such

as equipment for large scale power generation projects). Restrictions on foreign ownership have been relaxed. Previously, foreign ownership usually had been limited to a minority ownership stake. Now, in many sectors, majority foreign ownership is permitted. In some areas, however, reform has been slow–particularly the energy sector. Electricity consumption is still supported by heavy subsidies, and the formal end of the Administered Pricing Mechanism (APM) for petroleum products in April 2002 did not completely end government controls on petroleum product prices. Annual foreign direct investment (FDI) in India has hovered in the range of $3-4 billion over the last several years, though, compared to roughly $40 billion per year of FDI in China.

Oil

Oil accounts for about 30 per cent of India's total energy consumption. The majority of India's roughly 5.4 billion barrels in oil reserves are located in the Bombay High, Upper Assam, Cambay, Krisha-Godavari, and Cauvery basins. The offshore Bombay High field is by far India's largest producing field, with production of 203,000 barrels per day (bbl/d) in 2001. India's average oil production level (total liquids) for 2002 was 759,000 bbl/d. India had net oil imports of over 1.2 million bbl/d in 2002.

Future oil consumption in India is expected to grow rapidly, to 3.2 million bbl/d by 2010, from 2.0 million bbl/d in 2002. India is attempting to limit its dependence on oil imports somewhat by expanding domestic exploration and production. To this end, the Indian government is pursuing the New

Exploration Licensing Policy (NELP), first announced in 1997, which permits foreign involvement in exploration, an activity long restricted to Indian state-owned firms. While the initial response to the 1999 tender was disappointing, with no bids received from the major multinational oil companies (causing an extension of the deadline for submission of bids), India proceeded with the award of 25 oil exploration blocks in early January 2000. The largest winner in the bidding round was India's domestic Reliance Industries, in partnership with independent Niko Resources of Canada, which received 12 blocks. British independent Cairn Energy, Russia's Gazprom, the U.S. firm Mosbacher Energy, and Geopetrol of France were all awarded single blocks in partnership with Indian firms. India's state-owned Oil and Natural Gas Corporation (ONGC) was awarded eight blocks, three of which it will hold in partnership with other public-sector Indian firms. A second round of bidding, with a total of 25 blocks offered, concluded in March 2001. Sixteen of the blocks have been awarded to ONGC, and four blocks to Hardy Oil of the United Kingdom, in partnership with India's Reliance Petroleum. The others were either awarded to smaller independent firms or failed to receive bids. As with the first round, no bids were received from major international oil companies. Bids for the third round were received in August 2002, with a total of 27 blocks offered. Awards under this third round were made in February 2003, with domestic Indian firms receiving most of the blocks. Reliance Industries received nine offshore blocks, one adjacent to the Krishna-Godavari Basin. ONGC was awarded 13 blocks, five offshore and eight onshore. The Gujarat State Petroleum Corporation received one.

Low drilling recovery rates are a major part of the oil supply problem for India. Recovery rates average only around 30 per cent in currently producing Indian fields, well below the world average. It is hoped that allowing foreign investment will bring in technology that is not available to Indian state firms, thereby increasing overall recovery rates. ONGC currently is undertaking a project to increase recovery rates in the Bombay High offshore field and several others as well, aiming to boost the overall recovery rate for its production assets from 28 per cent to 40 per cent. Recent experience does not support an optimistic view about India's prospects for s sharp increase in oil production as no major new finds have been made in recent years. Analysts consider it likely that most of India's easily recoverable oil has been discovered. The main cause for hope is offshore exploration, and in particular deep water exploration. One onshore area which also has shown promise is western Rajasthan, and a small find was reported in early 2001 by Cairn Energy offshore from Gujarat which is believed to hold about 200 million barrels of recoverable reserves. Most recent drilling, however, has found natural gas rather than crude oil.

BG purchased a 30 per cent stake in the Panna, Mukta, and Tapti offshore oil and gas fields in February 2002, which had previously been held by Enron. A dispute between BG and ONGC, which owns a 40 per cent interest in the fields, over which firm would operate them was resolved in February 2003 with a "joint operatorship agreement." Reliance Industries holds the other 30 per cent stake.

Downstream/Refining

For most of the 1990s, India imported a large quantity of refined products, lacking the refining capacity to keep up with growing demand. In 1999, refinery construction allowed India to close the gap. At the end of 2002, India had a total of 2.1 million bbl/d in refining capacity, an increase of 970,000 bbl/d since 1998. In late summer 1999, Reliance Petroleum's huge Jamnagar refinery came onstream. It has since reached its full capacity of 540,000 bbl/d. Jamnagar does not have its own retail distribution network, but sells its product through three of the state-owned firms. It plans to build a retail network of its own in coming years. Refinery construction has been encouraged by regulatory changes by the Indian government, including a five-year tax holiday for refineries completed by 2003.

Another major downstream infrastructure development is the construction of pipelines being undertaken by Petronet India, a company created by an agreement in 1998 between India's state-owned refineries, which will add 500,000 bbl/d to India's current 325,000 bbl/d capacity for pipeline transportation of refined products. Pipelines between refineries and major urban centers are replacing rail as the main mode of transportation.

While retail gasoline sales are still controlled by state firms, several multinationals have entered the Indian lubricants market, which was deregulated five years ago. Over one-third of the market is currently held by such firms as Shell, ExxonMobil, and Caltex. While these operations are relatively small, they are seen as allowing the majors to study the Indian market, establish brand recognition, and prepare for the eventual deregulation of the Indian retail petroleum products sector. Still, a requirement that foreign firms invest at least $400 million before entering the downstream market has served to limit their entry into petroleum products retailing.

Industry Restructuring and Price Deregulation

The Indian government officially ended the Administered Pricing Mechanism (APM) for petroleum product prices in April 2002. Prior to this deregulation, the Indian government had tried to offset the effects of price changes in crude oil by maintaining an Oil Pool Account, which was to build financial reserves when crude oil prices fell and release them back as increased subsidies when crude oil prices rose. In practice, though, the April 2002 reforms have not completely removed government influence on petroleum product prices. Subsidies have been maintained on some products, such as kerosene, which is commonly used as a cooking fuel by low-income households in India. State-owned downstream companies also still must submit proposed price changes to the Ministry of Petroleum and Natural Gas for approval. There have been some indications that this approval process has been used in late 2002 and early 2003 to curb increases in petroleum product prices as crude oil prices have risen.

After an extended delay due to internal policy debates, the Indian government announced a decision in December 2002 to sell off majority stakes in two of the largest state-owned downstream oil companies, Hindustan Petroleum (HPCL) and Bharat Petroleum (BPCL). HPCL is slated to be sold off to a "strategic" buyer–another major oil company. BPCL's shares will be offered through a public stock offering. The Indian federal government currently owns 51 per cent of HPCL and 66 per cent of BPCL. The divestments are planned for the second half of 2003.

Natural Gas

Indian consumption of natural gas has risen faster than any other fuel in recent years. From only 0.6 trillion cubic feet (Tcf) per year in 1995, natural gas use was nearly 0.8 Tcf in 2000 and is projected to reach 1.2 Tcf in 2005 and 1.6 Tcf in 2010. A major development in December 2002 was the announcement by Reliance Industries of its discovery of a large amount of natural gas in the Krishna-Godavari Basin offshore from Andhra Pradesh along India's southeast coast. New reserves from this find are estimated at about 5 Tcf. Cairn Energy also reported finds in late 2002 offshore from Andhra Pradesh as well as Gujarat, which contain reserves estimated at nearly 2 Tcf.

Even with these new reserves, India's domestic natural gas supply is not likely to keep pace with demand, and the country will have to import much of its natural gas, either via pipeline or as LNG. The main market impacts from the new finds will be on India's east coast, which currently lacks extensive natural gas infrastructure. While EIA's current forecast in the International Energy Ouloook 2002 predicts a 6.1 per cent annual growth rate in natural gas consumption, this reflects a substantial downward revision from previous forecasts, which had projected consumption of 2.7 Tcf per year by 2010. Problems with financing LNG import projects have dimmed some of the previous prospects for explosive growth in natural gas consumption in India, and helped to revive interest in pipeline import options. Financial problems in the power sector, the main consumer of natural gas, also have had a negative effect.

Most of India's current natural gas production takes place in the Bombay High basin and the state of Gujarat. Current projects include enhancing natural gas production at the Tapti fields and recovering previously flared natural gas at the Bombay High oilfield.

India is investing heavily in the infrastructure required to support increased use of natural gas. Gas Authority of India Limited (GAIL), a government-owned entity, is in the process of doubling the throughput capacity on its main Hazira-Bijaipur-Jagdishpur (HBJ) Pipeline. Work on the capacity expansion began in 2002, and will eventually raise the capacity of the line from about 1.1 billion cubic feet per day (Bcf/d) to 2.1 Bcf/d. GAIL also plans a new distribution network in West Bengal and a pipeline which would connect Calcutta with Chennai. Shell has signed a memorandum of understanding with the state government of Uttar Pradesh in northern India for the development of a natural gas distribution infrastructure.

India's Foreign Investment Promotion Board (FIPB) had approved 12 prospective LNG import terminal projects in the mid-to-late-1990s, but it was never considered likely that all would be built in the near future, as their combined capacity would have exceeded even the most optimistic demand projections. The Indian govenment froze approvals of new LNG terminals in 2001, and the payment problems at the Enron-backed Dabhol Power Plant in Maharashtra have led many to question the financial viability of some of the LNG import projects. Since the main consumers of the imported gas would be power producers, the poor financial condition of most of the state power boards which purchase power and run the transmission grids is likely to be a major constraint on the development of LNG imports.

The largest state sector projects are to be conducted by Petronet, a joint venture between ONGC, IOC, the Gas Authority of India Ltd. (GAIL), the National Thermal Power Corporation (NTPC), and Gaz de France. Under the current plan, each of the state firms would own a 10 per cent stake, the Gujarat state government will own a 5 per cent stake, and the rest will be offered to private investors, possibly including an equity stake for Qatar's RasGas, the main supplier of LNG for the project. Petronet plans two import terminals, one at Dahej and the other at Cochin. The import terminal at Dahej currently is under construction, and is expected to be completed in late 2003. RasGas is to begin supplying LNG to Petronet when the Dahej terminal is completed. The Dahej terminal has had advantages over some of the other proposed projects, being tied in with the main state-owned natural gas company, GAIL, and the existing HBJ pipeline network. It is unclear whether sufficient demand will exist to support a planned 2007 completion date for the Cochin import terminal.

Shell also has begun construction of its LNG import terminal at Hazira in Gujarat, and has contracted for LNG supplies from Oman. The facility is scheduled to begin operation in 2005. Like the Petronet Dahej terminal, it is to be linked into existing natural gas pipelines.

The Dabhol LNG terminal was nearly finished at the time construction was halted in June 2001, and it will likely be completed by another firm once a buyer is found for the now-bankrupt Enron's 65 per cent share of the project. Two other American firms involved in the project, General Electric and Bechtel, which each now own 10 per cent, are reported to be in the final stages of negotiations to acquire the stake formerly held by Enron.

In the wake of the problems with Dabhol, firms backing several other LNG projects pulled out in the second half of 2001. Dhaksin Bharat Energy, a consortium including CMS Energy and Unocal, also announced the cancellation of its planned LNG project at Ennore. TotalFinaElf has suspended further action on its planned LNG import terminal at Trombay. These LNG projects were cancelled largely in response to the Indian government's decision not to extend sovereign payment guarantees to power projects which were to have been among their largest customers. Another proposed project at Kakinada on India's east coast may be jeopardized by cheaper natural gas supplies which will become available once Reliance Industries offshore new offshore finds are developed.

Aside from LNG imports, imports of natural gas by pipeline may play a role eventually in satisfying India's gas needs. One possibility would supply India with natural gas from Iran's huge South Pars field via a pipeline, either subsea or through Pakistan. Iran has discussed the proposal with India and Pakistan. Australia's Broken Hill Proprietary (BHP) is the main foreign backer of the idea. An offshore route bypassing Pakistan is under study by Snamprogetti of Italy. Pakistan had said in early 2001 that it would allow supplies to cross its territory, and Iran would bear the contractual responsibility for assuring gas supplies to India, but the project does not appear likely to be implemented in the near future due to the obvious security concerns for India.

Another possible import route would link the natural gas reserves of Bangladesh into the Indian gas grid. Current proven reserves of natural gas in Bangladesh are at least 14 Tcf, but the foreign firms involved in natural gas exploration in Bangladesh, which includes Unocal, believe that reserves are higher. Shell, which backs exports to India, has estimated Bangladeshi natural gas reserves at 38 Tcf, and a study by the U.S. Geological Survey put the country's probable reserves at 32 Tcf. Bangladesh has been reluctant to approve exports to India, however, until all questions about reserves and its domestic supply have been resolved. Shell reportedly has been in negotiations with Unocal about possible imports of Bangladeshi gas for its distribution projects in Uttar Pradesh, and Unocal made a formal proposal to the Bangladeshi government for gas exports in October 2001. As of February 2003, the Bangladeshi government is still considering the idea, and has not reached a final decison. The new natural gas reserves discovered off Andhra Pradesh in 2002 could compete with imports from Bangladesh, thus increasing the pressure to reach a decision in the near future.

India's government has been considering reforms in its natural gas pricing mechanism, which is currently set by the government. Deregulation has bee delayed several times, and the recent increase in domestic reserves may lead to a tilt toward less regulation of prices, making LNG importers compete with domestic natural gas suppliers.

Coal

Coal is the dominant commercial fuel in India, satisfying more than half of India's energy demand. Power generation accounts for about 70 per cent of India's coal consumption, followed by heavy industry. Coal consumption is projected in the International Energy Annual 2002 to increase to 450 million short tons (Mmst) in 2010, up from 369 million short tons in 2000. This is a substantial increase in the rate of growth projected in previous forecasts. India is the world's third largest coal producer (after China and the United States), so most of the country's coal demand is satisfied by domestic supplies. Indian coal generally has a high ash content and low calorific value, so most coking coal must be imported. Major Indian coal fields are found in Bihar, West Bengal, and Madhya Pradesh.

The Indian government controls almost all coal production, which has been plagued by low productivity, distribution problems, and loss of markets to higher quality, less expensive imports.

Nearly all of India's 390 mines are under Coal India Ltd. (CIL), which accounts for about 90 per cent of the country's coal production. Current policy allows private mines only if they are "captive" operations which feed a power plant or factory. The current government has called off plans for further coal-sector liberalization in the face of strong opposition from labor unions.

Electricity

India is trying to expand electric power generation capacity, as current generation is seriously below peak demand. Although about 80 per cent of the population has access to electricity, power outages are common, and the unreliability of electricity supplies is severe enough to constitute a constraint on the country's overall economic development. The government had targeted capacity increases of 100,000 megawatts (MW) over the next ten years. As of January 2001, total installed Indian power generating capacity was 112,000 MW.

The drive to increase the country's generating capacity, along with the general trend toward economic liberalization in India in the 1990s, led to much interest among foreign investors in setting up Independent Power Producers (IPPs) in India. While dozens of projects were approved, most of the largest projects have been stalled by delays in regulatory approvals and in some cases failure to secure adequate financing. India's state electricity boards (SEB's), which run the power distribution infrastructure and own most current generating capacity, are in very poor financial shape, with many of them technically insolvent. One reason is the sale of power at subsidized rates, which does not cover costs (particularly in the agricultural sector). Other problems include the high level of transmission and distribution losses and widespread power theft. Since the SEBs would be the main purchasers of power from IPP projects, resolving their financial problems is critical to attracting the capital necessary to ensure the country an adequate supply of electric power.

While India currently does not have a unified national power grid, the country plans to link the SEB grids eventually, and has set up a state company, Powergrid, to oversee the unification. India also plans to establish national and state level regulatory bodies to set tariffs and promote competition.

In July 1998, the Indian government announced an easing of rules related to foreign investment in the power sector. Proposals for investments up to 15 billion rupees (about $350 million) involving up to 100 per cent foreign equity now will be approved automatically. Automatic approval will be given for investments in generation or distribution from hydroelectric, coal, lignite, oil, or gas power plants, but not for nuclear plants and associated distribution networks. The earlier policy had allowed for only up to 74 per cent foreign equity. Still, the financial problems of the SEBs have prevented substantial foreign investment from flowing into India's electric power sector.

India's government approved a large number of "mega-projects," defined as plants with capacity of more than 1,000 MW for thermal plants and more than 500 MW for hydroelectric plants, from the mid-to-late-1990s, but project approvals have often not led to construction. The 740-MW initial phase of the Dabhol LNG-fired power plant began operation in May 1999, and Phase II, which would add 1,440-MW of capacity, is about 90 per cent complete. Payment problems with the Maharashtra State Electricity Board (MSEB), however, prompted Enron-backed Dabhol Power Corporation (DPC) to serve notice of breach of contract on MSEB in May 2001. Construction on Phase II was halted in June 2001. General Electric and Bechtel, each of which own existing 10 per cent stakes in DPC, have reportedly been negotiating in recent months for the purchase of the 65 per cent stake held by the now-bankrupt Enron, as well as payment guarantees from the Indian federal government.

Due to the financial problems of the SEBs, a large number of foreign firms cancelled or delayed power generation projects in India between 1999 and 2001:

1. The largest of the power projects which had obtained government approval, the $5 billion, 3,960-MW coal-fired Hirma Power Plant, was canceled by Mirant Corporation in December 2001.
2. A 1040-MW coal-fired plant at Vishakapatnam was planned by Hinduja Power and National Power (UK). In June 2001, however, the Industrial Development Bank of India (IDBI) announced that it was withdrawing its loan to the project, throwing its future into doubt.
3. Electricite de France has quit the coal-fired 1072-MW Bhadrawati project in Maharashtra state.
4. The 1,886-MW LNG-fired unit at Ennore, with an associated LNG import terminal, was canceled by CMS Energy in June 2001. CMS Energy also announced in October 2001 that it was pulling out of several smaller projects.
5. India's National Thermal Power Company was planning a 2,000-MW LNG-fired plant at Pipavav, but the project has been shelved after BG withdrew from the LNG import terminal project in June 2001.
6. Powergrid was planning a 1,320-MW coal-fired plant planned for Cuddalore, which was delayed indefinitely in early 2001.
7. Cogentrix cancelled the 1,000-MW Mangalore coal-fired project in December 1999.
8. South Korea's Daewoo Power and ABB Lummus cancelled plans for a 1,400-MW plant in Madhya Pradesh in August 2000.

The DPC controversy is seen by many analysts as a test case for India's power sector, as it has demonstrated the lack of creditworthiness of the SEBs. The Dabhol plant, valued at $2.9 billion, is the largest single foreign investment in India. No major foreign-owned projects have been launched in the past year.

The Indian government is attempting to implement reforms which would increase the financial strength of the SEBs in an attempt to attract capital back to electricity generation projects. New legislation on power sector reform was approved by the Indian cabinet in February 2002, and is to be introduced in the parliament.

Environment

India, the world's second most populous nation, has seen its population explode from 300 million in 1947 to approximately one billion today. This rapidly growing population has placed a strain not only on India's infrastructure, but also on its environment. According to the World Health Organization, New Delhi is one of the top ten most polluted cities in the world. Two primary sources of air pollution in India are vehicular emissions and untreated industrial smoke.

Coal is a major commercial energy source in India. Increased coal consumption over the past four decades has led to a nine-fold increase in energy-related carbon emissions . Environmental effects due to the relatively high use of coal in the energy mix are exacerbated by the low energy efficiency of coal-based electricity generating plants. Inefficient plants are one of the contributing factors to a steadily increasing energy consumption per unit of output (ie. energy intensity). With the high costs associated with replacing existing coal-based plants, it is realistic to assume that these plants will continue running for the next couple of decades.

India's per capita energy use and carbon emissions, while lower than the world average, result in a substantial percentage of world energy use and carbon emissions, due to the country's large population and heavy reliance on coal. Increased use of renewable energy is one means of reducing carbon emissions. Two major sources of renewable energy in India are wind power and hydroelectric plants. India's five year plan for 2002-2007 calls for 10 per cent of new electric generating capacity to come from renewable sources.

India faces great challenges in energy and environment as it enters the 21st Century . A rapidly growing population will continue to increase demands for electricity generation and will place greater pressures on the environment to absorb increasing vehicular emissions.

Country Overview

Independence: August 15, 1947 (from the United Kingdom)

Population (07/2002E): 1.0 billion (2nd most populous country)

Location/Size: Southern Asia/3.3 million square kilometers 1.3 million square miles), one-third the size of the United States

Major Cities: New Delhi (capital), Mumbai (Bombay), Calcutta, Chennai (Madras), Hyderabad, Bangalore, Ahmedabad

Languages: Hindi, 17 other official languages, English

Ethnic Groups: Indo-Aryan (72 per cent), Dravidian (25 per cent), Mongoloid, other (3 per cent)

Religions: Hindu (80 per cent), Muslim (14 per cent), Christian (2.4 per cent), Sikh (2 per cent), Buddhist (0.7 per cent), Jains (0.5 per cent), other (0.4 per cent)

Defense (8/98): Army (980,000), Air Force (110,000), Navy (55,000), Jammu/Kashmir Border Security Force (185,000)

Economic Overview

Currency: Rupee

Exchange Rate (2/24/02): US$1 = 47.8 rupees

Gross Domestic Product (GDP, FY2002E): $514.4 billion

Real GDP Growth Rate (FY2002E): 4.8 per cent *(FY2003E):* 5.7 per cent

Inflation Rate (FY2002E): 3.6 per cent *(FY2003E):* 4.6 per cent

Current Account Balance (FY2002E): -$0.9 billion

Major Trading Partners: United States, Japan, United Kingdom, Germany, Russia

Merchandise Trade Balance (FY2002E): -$7.3 billion

Merchandise Exports (FY2002E): $48.8 billion

Merchandise Imports (FY2002E): $56.1 billion

Major Export Products: Gems and jewelry, engineering goods, clothing, cotton textiles, leather and leather manufactures, iron ore, chemicals, software

Major Import Products: Petroleum and petroleum products, machinery, iron and steel, edible oils, chemicals, fertilizers

Monetary Reserves (FY 2002, non-gold): $46.8 billion

External Debt (FY2002E): $101.0 billion

NOTE: FY (Fiscal Year) (FY 2002 April 1, 2002 to March 31, 2003)

Energy Overview

Proven Oil Reserves (1/1/03E): 5.4 billion barrels

Oil Production (2002E): 759,000 barrels per day (bbl/d), of which 665,000 bbl/d was crude oil

Oil Consumption (2002E): 2.0 million bbl/d

Net Oil Imports (2002E): 1.2 million bbl/d

Crude Oil Refining Capacity (1/1/03E): 2.1 million bbl/d

Natural Gas Reserves (1/1/03E): 26.9 trillion cubic feet (Tcf)

Natural Gas Production (2001E): 803 Bcf

Natural Gas Consumption (2001E): 803 Bcf

Recoverable Coal Reserves (12/31/96E): 82.4 billion short tons

Coal Production (2001E): 360 million short tons (Mmst)

Coal Consumption (2001E): 339 Mmst

Net Coal Imports (2001E): 21 Mmst

Electric Generation Capacity (1/1/01E): 112 gigawatts (GW), including 83 GW thermal, 25 GW hydro, 2 GW nuclear

Electricity Generation (2001E): 533 billion kilowatthours (79 per cent conventional thermal; 18 per cent hydro; 2 per cent nuclear)

Environmental Overview

Total Energy Consumption (2001E): 12.8 quadrillion Btu* (3.2 per cent of world total energy consumption)

Energy-Related Carbon Emissions (2001E): 251.3 million metric tons of carbon (3.8 per cent of world total carbon emissions)

Per Capita Energy Consumption (2001E): 12.6 million Btu (vs U.S. value of 341.8 million Btu)

Per Capita Carbon Emissions (2001E): 0.25 metric tons of carbon (vs U.S. value of 5.5 metric tons of carbon)

Energy Intensity (2001E): 25,307 Btu/$1995 (vs U.S value of 10,736 Btu/ $1995)**

Carbon Intensity (2001E): 0.50 metric tons of carbon/thousand $1995 (vs U.S. value of 0.17 metric tons/thousand $1995)

Sectoral Share of Energy Consumption (1998E): Industrial (41.0 per cent), Transportation (9.5 per cent), Residential (47.3 per cent), Commercial (2.2 per cent)

Sectoral Share of Carbon Emissions (1998E): Industrial (67.3 per cent), Transportation (15.6 per cent), Residential (13.7 per cent), Commercial (3.3 per cent)

Fuel Share of Energy Consumption (2001E): Coal (50.9 per cent), Oil (34.4 per cent), Natural Gas (6.5 per cent)

Fuel Share of Carbon Emissions (2001E): Coal (64.5 per cent), Oil (30.3 per cent), Natural Gas (5.2 per cent)

Renewable Energy Consumption (1998E): 9,015 trillion Btu* (2 per cent increase from 1997)

Number of People per Motor Vehicle (1998): 142.9 (vs U.S. value of 1.3)

Status in Climate Change Negotiations: Non-Annex I country under the United Nations Framework Convention on Climate Change (ratified November 1st, 1993). Not a signatory to the Kyoto Protocol.

Major Environmental Issues: Deforestation; soil erosion; overgrazing; desertification; air pollution from industrial effluents and vehicle emissions; water pollution from raw sewage and runoff of agricultural pesticides; tap water is not potable throughout the country; huge and rapidly growing population is overstraining natural resources.

Major International Environmental Agreements: A party to the Antarctic-Environmental Protocol, Antarctic Treaty, Biodiversity, Climate Change, Desertification, Endangered Species, Environmental Modification, Hazardous Wastes, Law of the Sea, Nuclear Test Ban, Ozone Layer Protection, Ship Pollution, Tropical Timber 83, Tropical Timber 94, Wetlands and Whaling.

The total energy consumption statistic includes petroleum, dry natural gas, coal, net hydro, nuclear, geothermal, solar, wind, wood and waste electric power. The renewable energy consumption statistic is based on International Energy Agency (IEA) data and includes hydropower, solar, wind, tide, geothermal, solid biomass and animal products, biomass gas and liquids, industrial and municipal wastes. Sectoral shares of energy consumption and carbon emissions are also based on IEA data.

ENERGY INDUSTRY

Organization: Petroleum–Oil and Natural Gas Corporation (ONGC), Oil India Ltd. (OIL), Indian Oil Corporation (IOC); *Natural Gas*–Gas Authority of India Limited (GAIL); *Coal*–Coal India Limited (CIL); *Electric Power*–National Thermal Power Corporation (NTPC), National Hydroelectric Power Corporation, State Electricity Boards

Major Oil Fields (2001 production): Bombay Offshore: Bombay High (203,355 bbl/d), B-38/Heera & S. Heera (48,000 bbl/d), Neelam (29,895 bbl/d); *Eastern*: Lakwa-Lakhmani (14,660 bbl/d); *Western*: Gandhar (40,000 bbl/d); *Southern*: Tiruvarur (51,165 bbl/d)

Major Oil Terminals: Bombay, Cochin, Haldia, Kandla, Madras, Vizag

Major Oil Refineries (1/1/03 capacity): Reliance-Jamnagar, 540,000 bbl/d, Koyali-Gujarat, 185,100 bbl/d; Mangalore, 180,000 bbl/d, Mathura-Uttar Pradesh, 156,000 bbl/d; Mahul-Bombay (Bharat Petroleum), 120,000 bbl/d; Madras, 130,660 bbl/d, Mahul-Bombay (Hindustan Petroleum), 111,700 bbl/d

Major Pipelines: Oil–Salaya-New Delhi, Barauni-Digboi, Kandla-Bhatindu (products); *Natural Gas*–Hazira-Bijapur-Jagdishpur (HBJ)

42

Indonesia

Indonesia is important to world energy markets because of its OPEC membership and substantial, but declining, oil production. Indonesia also is the world's largest liquefied natural gas (LNG) exporter.

General Background

Indonesia's economic growth increased modestly in 2002 despite the continuing global economic slowdown. Indonesia's real gross domestic product (GDP) grew at a rate of 3.7 per cent in 2002, up from 3.1 per cent in 2001. Real GDP growth is forecast at 3.8 per cent for 2003, a sign that the nation's economy will continue its recovery from the fiscal crisis of 1997.

With about 75 per cent of Indonesian businesses in technical bankruptcy following the country's economic collapse in 1998, the government was forced to turn to the International Monetary Fund (IMF) for an emergency debt-relief package totaling $43 billion. The IMF recommended that Indonesia implement an economic reform program in order to help save its economy. IMF recommendations included creating greater transparency in the issuing of government loans and subsidies, and stricter enforcement of laws and regulations in the area of government procurement. The government has announced several reform initiatives since receiving the IMF bailout package, including the planned privatization of several sectors of the economy, but progress has been slow. A follow-on loan package, from several lenders including the United States and the IMF, was approved in October 2000. Installments of IMF funds were delayed in 2001 due to delays in economic reforms, but in September 2001, the IMF resumed transfers and released a $395 million loan tranche under the follow-on loan package. In March 2003, the IMF reported that Indonesia had made good progress in the institution of reforms, and disbursed $469 million in further aid. The review cited Indonesia's continued economic growth, decreasing inflation rates, and strengthened banking sector as examples of the progress made, while mentioning that more reforms will be necessary in 2003 (the final year of the assistance program) to increase the country's attractiveness to foreign investors.

President Megawati Sukarnoputri has been in power since July 2001, assuming the presidency after her predecessor, President Abdurrahman Wahid, was removed from office by the national legislature. The regional challenges facing the Indonesian government remain the same: a separatist movement in Aceh; an oil and gas rich province in north Sumatra which abuts the strategically important Strait of Malacca; and a separatist movement in Irian Jaya, a gas-rich province at the eastern end of the country.

One of the key areas in which energy and politics intersect in Indonesia is the distribution of oil and gas revenues between the central government in Jakarta and regional governments in areas which produce oil and gas. Since Indonesia's transition to democracy, the country's regional governments have been pressing for a greater share of oil and gas revenues. In particular, the separatist movement in Aceh has caused security problems for oil and gas companies in that region, although recent government talks ended in a peace deal in December 2002. However, some violence has continued despite the agreement, and international monitors were withdrawn in April 2003 following threats and intimidation attacks against their offices. An unrelated bombing of nightclubs in Bali, which claimed about 200 lives, prompted the Indonesian government to impose increased security measures on some energy and mining facilities.

Oil

Indonesia currently holds proven oil reserves of 5 billion barrels, which represents a 14 per cent decline in proven reserves since 1994. Much of Indonesia's proven oil reserve base is located onshore. Central Sumatra is the country's largest oil producing province and the location of the large Duri and Minas oil fields. Other significant oil field development and production is located in accessible areas such as offshore northwestern Java, East Kalimantan, and the Natuna Sea. Indonesian crude oil varies widely in quality, with most streams having gravities in the 22° to 37 ° API range. Indonesia's two main export crudes are Sumatra Light, or Minas, with a 35 ° API, and the heavier, 22° API Duri crude. A study released in August 2002 by Indonesia's Directorate General of Oil and Gas shows that oil reserves in the Cepu block alone, located in Central/East Java, are close to 500 million barrels, about half of which is considered recoverable.

During 2002, Indonesian crude oil production averaged 1.1 million barrels per day (bbl/d), having decreased from just over 1.2 million bbl/d in 2001. The recent declines in production are due mainly to the natural decline of aging oil fields, which recent oil discoveries have been too small to offset. Besides crude oil, Indonesia also produces approximately 230,000 bbl/d of natural gas liquids and lease condensate, which are not part of its OPEC quota, bringing the country's total oil production to around 1.3 million bbl/d. Indonesia is the only Southeast Asian member of OPEC, and its current OPEC crude oil production quota is 1.27 million bbl/d.

The majority of Indonesia's producing oil fields are located in the central and western sections of the country. Therefore, the focus of new exploration has been on frontier regions, particularly in eastern Indonesia. Sizable, but as of yet unproven, reserves may lie in the numerous, geologically complex, pre-tertiary basins located in eastern Indonesia. These regions are much more remote and the terrain more difficult to explore than areas of western and central Indonesia.

One significant oil find made by Indonesian state oil company Pertamina occurred in March 2003. The three wells drilled, located in Prabumulih, in South Sumatra, have hit reserves that are estimated to hold about 230 million barrels of oil. An earlier significant oil and natural gas discovery was made by Pertamina in December 2002. Oil and gas deposits located in the village of Gunung Kemala, in South Sumatra, are estimated to include 230 million barrels of oil, and 590 billion cubic feet (bcf) of natural gas. Pertamina hopes to make the deposits productive during 2004.

In December 2002, Italian-based firm ENI and Unocal signed a production-sharing agreement with Indonesia to explore for oil and gas offshore South Sulawesi. Each company holds a 50 per cent stake, and will spend a total of $35 million over the next six years to explore the Mauru Bakau block, which is approximately 700 square miles in size and located in the Makassar Strait.

China National Offshore Oil Corporation (CNOOC) became the largest offshore oil producer in Indonesia in January 2002, after purchasing nearly all of Repsol-YPF's assets in the country for $585 million. The assets acquired by CNOOC are as follows: a 65.34 per cent stake in the South-East Sumatra PSC, a 36.72 per cent stake in the North-West Java PSC, located offshore, a 50 per cent stake in the Poleng PSC, a 25 per cent stake in the West Madura PSC, and a 16.7 per cent stake in the Blora PSC. Pertamina is one of CNOOC's partners in all of those PSCs. Repsol-YPF retained ownership of a 25 per cent stake in a joint venture with Pertamina at the Jambi-Merang natural gas field.

Companies producing from existing fields are investing in programs to increase recovery rates and to prolong the life of the fields. Caltex, which has the largest operation of any multinational oil company in Indonesia, is undertaking a steam injection project at the Duri field on Sumatra.

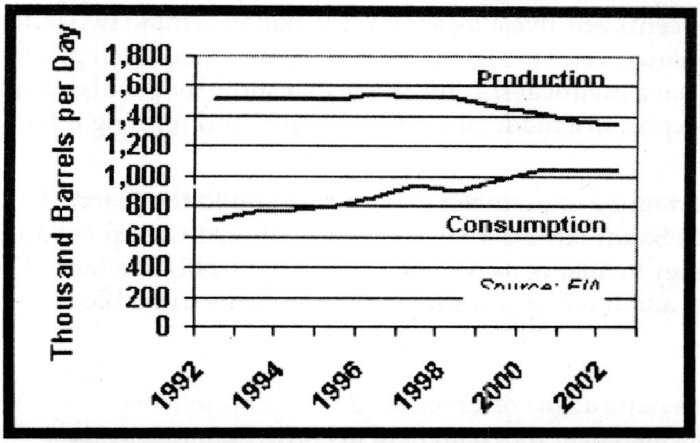

INDONESIA'S TOTAL OIL PRODUCTION, 1992-2002

Three major new oil projects are expected to begin production before 2004. Unocal's West Seno field, under development offshore from East Kalimatan, is expected to be producing 40,000 bbl/d by May 2003 and eventually peaking at 60,000 bbl/d. An additional 100,000 bbl/d of production capacity is to be added by Conoco's Belanek project in the West Natuna Block B by 2004. ExxonMobil's Banyu Urip field, in Java, is expected to come onstream in 2003 and reach its peak production capacity of 100,000 bbl/d in 2004. Even with these new fields, though, Indonesia's oil production is not likely to rise markedly, due to the continuing decline of matûre fields. The country could possibly become a net oil importer by the end of the current decade.

Oil Sector Reforms

The liberalization of Indonesia's downstream oil and gas sector has been under discussion for several years. In October 2001, the Indonesian legislature passed legislation which will remove Pertamina's monopoly on upstream oil development (which requires it to be included in all PSCs) by the end of 2003. Its monopoly on the distribution of petroleum products is to be terminated by 2004.

Indonesia's Ministry of Mines and Energy will take over the function, currently carried out by Pertamina, of awarding and supervising production sharing contracts with foreign oil companies. Foreign firms also are to be freed from some of the regulatory approval requirements which they argue hinder their efficiency. Subsidies for domestic petroleum products consumption have been reduced in a bid to reduce the government's budget deficit and restrain increases in domestic consumption. One concern foreign oil companies have with the new law is the granting of a limited authority to regional governments to tax oil companies' profits.

Pertamina is planning to adopt a franchise system for its liquified petroleum gas (LPG) by 2006. The new system will require Pertamina to sever ties with its current retailers, who act as sub-agents of Pertamina now. The state-run company is concerned about proper distribution of LPG to end-users, an area that has been problematic for Pertamina in the past.

Refining

Indonesia has eight refineries, with a combined capacity of 992,745 bbl/d. The largest refineries are the 348,000-bbl/d Cilacap in Central Java, the 240,920-bbl/d Balikpapan in Kalimantan, and the 125,000-bbl/d Balongan, in Java.

PT Kilang Minyak Intan Nusantara, a joint venture of Al-Banader International Group of Saudi Arabia (40 percent), China National Electrical Equipment Corporation (40 per cent) and PT Intanjaya Agromegah Abadi (20 per cent), are investing a total of $6 billion to build two Indonesian oil refineries, one in Pare-Pare, South Sulawesi and the other in Batam Island, Riau. Both projects are expected to be operational in 2005, and have identical specifications that show they will each process 300,000 bbl/d. The refineries will be export-oriented, taking Saudi crude and refining it for sale primarily to the Chinese market.

In February 2003, Pertamina announced that it would build the planned Tuban refinery alone, after its partnership with Saudi Arabia's Hi-Tech International Group collapsed. The Saudi firm failed to raise enough money to finance its portion of the 100,000 bbl/d refinery. Construction costs are estimated at $250 million, and it will process oil mainly from the Cepu Block.

Natural Gas

Indonesia has proven natural gas reserves of 92.5 trillion cubic feet (Tcf). Most of the country's natural gas reserves are located near the Arun field in North Sumatra, around the Badak field in East

Kalimantan, in smaller fields offshore Java, the Kangean Block offshore East Java, a number of blocks in Irian Jaya, and the Natuna D-Alpha field, the largest in Southeast Asia. Despite its significant natural gas reserves and its position as the world's largest exporter of liquefied natural gas (LNG), Indonesia still relies on oil to supply about half of its energy needs. About 70 per cent of Indonesia's LNG exports go to Japan, 20 per cent to South Korea, and the remainder to Taiwan. As Indonesia's oil production has leveled off in recent years, the country has tried to shift towards using its natural gas resources for power generation. However, the domestic natural gas distribution infrastructure still is not extensive.

A particularly significant Indonesian gas field, Natuna, is located in the South China Sea, 683 miles north of Jakarta and 140 miles northeast of Natuna Island. Discovered in 1970 by Italy's Agip, the field contains an estimated 46 Tcf of recoverable reserves. Agip relinquished its concession, however, and the field has been developed only recently. In January 1999, SembGas of Singapore signed a contract to purchase 325 million cubic feet per day (Mmcfd) of natural gas from the West Natuna Gas Consortium, a joint venture consisting of Pertamina, Conoco, and Gulf Indonesia Resources. A contract for the pipeline was awarded to McDermott International, which began supplying natural gas to Singapore in mid-2001. In November 1999, Conoco reported a new gas discovery at West Natuna which raised reserves by about 1 Tcf.

Four major gas pipelines went on stream during 2002, all owned by PT Perusahaan Gas Negara (PGN). The first is a 50 million cubic feet per day (mcf/d) distribution line going from Jambi to the Lampung region. The second and third are 500 mmcf/d pipelines going from Pagardewa to Cilegon, and from Cilegon to Cimanggis, respectively. The fourth pipeline is a distribution line in the western area of West Java that has a capacity of 550 mmcf/d. PGN is slated to be privatized, and has six more pipeline projects to build, with capacities ranging from 50 mmcf/d to 1000 mmcf/d.

In September 2002, PT Perusahaan Gas Negara (PGN), Indonesia's natural gas pipeline operating and distribution company, announced its intention to sell a 40 per cent stake in PT Transgasindo (a subsidiary also called Transco I) to a consortium called Trans Asia Pipeline Company for $180 million. The consortium's members include Malaysia's Petronas and ConocoPhillips. Transgindo currently owns and operates a 335 mile natural gas pipeline that connects South Sumatra to the Duri steamflood project in Central Sumatra. It also has plans to build another natural gas pipeline from South Sumatra to Singapore, which would deliver about $9 billion worth of natural gas over a 22-year period.

Also in September 2002, China awarded an $8.5 billion contract to Indonesia to supply China's planned terminal in Fujian Province with LNG for 25 years, beginning in 2007. The LNG will be supplied from the Tannguh natural gas field, which is operated by BP. Construction of the terminal will begin in 2004, and completed in 2007. Included in the deal was an agreement for BP to sell 12.5 per cent of its stake in the Tangguh field to CNOOC. The purchase took place in February 2003, and cost CNOOC $275 million, and included the Berau PSC, the Muturi PSC, and the Wiriagar PSC.

A major related project in the planning stages is BP's Tangguh LNG project in Irian Jaya, based on over 14 Tcf of natural gas reserves found onshore and offshore the Wiriagar and Berau blocks. The project would involve two trains with a combined capacity of 7 million tons per year. BP's current plans call for the project to be completed by 2007. Three bids have been offered by different consortia for the design and construction of the project's onshore facilities: one by the Bechtel group, another led by Chiyoda and Technip, and a third by Kellogg, Brown & Root and JGC Corporation. The bids were valued between $1.4 and 1.5 billion each. The project is considered risky due to an active separatist movement in Irian Jaya. BP holds a 48 per cent interest in the Berau block with partners Mitsubishi at

22.87 per cent, Nippon Oil Exploration, 17.14 per cent and KG Berau Petroleum, 12 per cent. For the Wiriagar block, Arco holds an 80 per cent interest and KG Wiriagar Petroleum Ltd. holds the remaining 20 per cent. BP announced in November 2001 that it had signed an agreement to supply LNG to the Philippines beginning in 2006, and in December 2001 announced that it had made a formal proposal to China for supplies of gas to its planned Guangdong LNG regasification terminal. El Paso Energy has been reported to have had discussions on possible supplies of LNG from Tangguh to future import terminals in the United States or Mexico.

In another possible use for Indonesia's gas resources, Shell is examining the possibility of building a gas-to-liquids (GTL) plant in Indonesia. The plant, if the project goes forward, would produce 70,000 bbl/d of diesel and other middle distillates using the Fischer-Tropsch GTL process.

Coal

Indonesia has 5.92 billion short tons of recoverable coal reserves, of which 85 per cent is lignite and 15 per cent is anthracite. Sumatra contains roughly two-thirds of Indonesia's total coal reserves, with the balance located in Kalimantan, West Java, and Sulawesi. In 2001, Indonesia exported 60.6 million short tons (Mmst), or about 60 per cent of its coal production. The majority of these exports are destined for Japan, South Korea, Taiwan.

Indonesia plans to double coal production over the next five years, mostly for exports to other countries in East Asia and India. The new capacity will come primarily from private mines. The Clough Group of Australia was awarded a $215 million contract for improvements at the Indonesian firm GBP's Kutai mine in East Kalimatan. Another foreign firm with major interests in Indonesian coal mining is Australia's Broken Hill Proprietary (BHP).

In keeping with Indonesia's plan to increase coal production, Banpu Pcl announced in March 2003 that it plans to invest $40 million by the end of 2003 to increase its own level of coal production.

Electricity

Indonesia has installed electrical generating capacity estimated at 21.4 gigawatts, with 87 per cent coming from thermal (oil, gas, and coal) sources, 10.5 per cent from hydropower, and 2.5 per cent from geothermal sources. Prior to the Asian financial crisis, Indonesia had plans for a rapid expansion of power generation, based mainly on opening up Indonesia's power market to Independent Power Producers (IPPs). The crisis led to severe financial strains on state-utility Perusahaan Listrik Negara (PLN), which made it difficult to pay for all of the power for which it had signed contracts with IPPs. PLN has over $5 billion in debt, which has grown markedly in terms of local currency due to the decline in the value of the rupiah. The Indonesian government has been unwilling to take over the commercial debts of PLN.

In February 2003, it was reported that PLN was seeking a $180 million loan from the World Bank to help develop electricity transmission networks in the provinces of Sumatra and Kalimantan. The loan would also help develop an information technology system. The World Bank will discuss the loan in July.

In January 2003, the World Bank announced that it was planning to build three micro-hydropower plants in the Indonesian province of Papua (also known as Irian Jaya). A feasibility study on all of the area's water sources has already been conducted by the Bank, and the results are being studied. By building these facilities, the World Bank hopes to improve services to the local population as well as to encourage development activities in the province.

PLN has announced that it is ready to install up to 175,000 new electricity lines during 2003 to meet Indonesia's ever-increasing electricity demands. The lines would be installed by PLN's distribution unit for the provinces of West Java and Banten, and would be located in those provinces.

Indonesia has plans to build a new 600-MW power plant northeast of Jakarta, in an effort to relieve power shortages in the country's main power grid, Java-Bali. Little new foreign investment has been made in recent years, prompting government action.

In 2002, Indonesia's government undertook measures to liberalize the nation's electricity market in order to make it more interesting for foreign investment. Competition for power generation will be open on the islands of Batam, Java, and Bali by 2007. In 2008, retail competition in the electricity market will begin under the terms of the nation's new electricity law, approved in September 2002. The law requires an end to PLN's monopoly on electricity distribution within five years, after which time private companies (both foreign and domestic) will be permitted to sell electricity directly to consumers. However, all companies will need to use PLN's existing transmission network.

Environment

Indonesia's major environmental challenges involve supporting its large population. Air and water pollution have reached critical levels, especially on the most populated island of Java. Indonesia's carbon emissions remain low, but there is concern that an increase in the use of indigenous coal will increase Indonesia's carbon emissions in the coming years. Indonesia is well endowed with renewable energy potential, especially geothermal energy. Indonesia's renewable resouces are not yet fully exploited.

In March 2003, the Asian Development Bank approved a $600,000 grant to help combat Jakarta's air pollution problem. The technical assistance grant will be used primarily to promote a clean vehicle fuel program, known as the "Blue Skies" project. Indonesia is also phasing out the use of leaded gasoline, with a complete ban set to come into force in 2005.

Country Overview

Independence: Proclaimed independence on August 17, 1945. On December 27, 1949, Indonesia became independent from the Netherlands.

Population (2002E): 231.3 million

Location/Size: Southeastern Asia/735,310 sq. mi., slightly less than three times the size of Texas

Major Cities: Jakarta (capital), Surabaya, Bandung, Medan, Semarang, Palembang, Ujung Pandang

Languages: Bahasa Indonesia (official), English, Dutch, local dialects including Javanese

Ethnic Groups: Javanese (45 per cent), Sundanese (14 per cent), Madurese (7.5 per cent), coastal Malays (7.5 per cent), other (26 per cent)

Religions: Muslim (88 per cent), Protestant (5 per cent), Roman Catholic (3 per cent), Hindu (2 per cent), Buddhist 1 per cent), other (1 per cent)

Economic Overview

Currency: Rupiah

Exchange Rate (3/24/03): US$1 = 8,950 rupiah

Gross Domestic Product (2002E): $181.9 billion *(2003E):* $213.1 billion

Real GDP Growth Rate (2002E): 3.7 per cent *(2003E):* 3.8 per cent

Inflation Rate (Consumer Price Index) (2002E): 11.9 per cent *(2003E):* 7.7 per cent

Merchandise Exports (2002E): $57.0 billion

Merchandise Imports (2002E): $31.2 billion

Merchandise Trade Balance (2002E): $25.8 billion

Major Export Products: Manufactured goods, petroleum, natural gas and related products, foodstuffs, raw materials

Major Import Products: Capital equipment, raw and intermediate materials, consumer goods, petroleum products

Major Trading Partners: Japan, United States, Singapore, Hong Kong, Britain, Australia

Energy Overview

Proven Oil Reserves (1/1/03E): 5.0 billion barrels

Oil Production (2002E): 1.3 million barrels per day (bbl/d), of which 1.1 million bbl/d was crude oil

OPEC Production Quota (since 2/1/03): 1.27 million bbl/d

Oil Consumption (2002E): 1.0 million bbl/d

Net Oil Exports (2002E): 305,000 bbl/d

Major Oil Customers: Japan, United States, South Korea, China, Australia, Taiwan, Singapore, Thailand

Crude Oil Refining Capacity (1/1/03E): 992,745 bbl/d

Natural Gas Reserves (1/1/03E): 92.5 trillion cubic feet (tcf)

Natural Gas Production (2001E): 2.44 trillion cubic feet (tcf)

Natural Gas Consumption (2001E): 1.28 tcf

Net Gas Exports (2001E): 1.16 Tcf

Major LNG Customers (2002): Japan, South Korea, Taiwan

Coal Reserves (12/31/01): 5.92 billion short tons of recoverable reserves of which 85 per cent is lignite and 15 per cent is anthracite

Coal Production (2001E): 99.6 million short tons (Mmst)

Coal Consumption (2001E): 39.0 Mmst

Net Coal Exports (2001E): 60.6 Mmst

Major Coal Customers (2001): Japan, Taiwan, South Korea, the Philippines

Electric Generation Capacity (1/1/01E): 21.4 gigawatts

Electricity Production (2001E): 95.8 billion kilowatt hours

Electricity Consumption (2001E): 89.1 billion kilowatt hours

Environmental Overview

Total Energy Consumption (2001E): 4.63 quadrillion Btu* (1.1 per cent of world total energy consumption)

Energy-Related Carbon Emissions (2001E): 87.13 million metric tons of carbon (1.3 per cent of world total carbon emissions)

Per Capita Energy Consumption (2001E): 21.5 million Btu (vs U.S. value of 341.8 million Btu)

Per Capita Carbon Emissions (2001E): 0.41 metric tons of carbon (vs U.S. value of 5.5 metric tons of carbon)

Energy Intensity (2001E): 21,441 Btu/ $1995 (vs. U.S. value of 10,736 Btu/ $1995)

Carbon Intensity (2001E): 0.40 metric tons of carbon/thousand $1995 (vs. U.S. value of 0.17 metric tons/thousand $1995)

Fuel Share of Energy Consumption (2001E): Oil (46.9 per cent), Natural Gas (30 per cent), Coal (19.7 per cent)

Fuel Share of Carbon Emissions (2001E): Oil (48 per cent), Natural Gas (26.1 per cent), Coal (25.9 per cent)

Status in Climate Change Negotiations: Non-Annex I country under the United Nations Framework Convention on Climate Change (ratified August 23rd, 1994). Signatory to the Kyoto Protocol (signed July 13th, 1998–not yet ratified).

Major Environmental Issues: Deforestation; water pollution from industrial wastes, sewage; air pollution in urban areas.

Major International Environmental Agreements: A party to Conventions on Biodiversity, Climate Change, Endangered Species, Hazardous Wastes, Law of the Sea, Nuclear Test Ban, Ozone Layer Protection, Ship Pollution, Tropical Timber 83, Tropical Timber 94 and Wetlands. Has signed, but not ratified, Desertification and Marine Life Conservation.

The total energy consumption statistic includes petroleum, dry natural gas, coal, net hydro, nuclear, geothermal, solar, wind, wood and waste electric power. The renewable energy consumption statistic is based on International Energy Agency (IEA) data and includes hydropower, solar, wind, tide, geothermal, solid biomass and animal products, biomass gas and liquids, industrial and municipal wastes. Sectoral shares of energy consumption and carbon emissions are also based on IEA data.

Oil and Gas Industries

Organizations: Perusahaan Pertambangan Minyak dan Gas Bumi Negara (Pertamina)–oil exploration, production, transportation, and marketing; Perum Gas Negara (PGN) -gas distributor and transmission company

Major Producing Oil Fields: Duri, Minas, Belida, Ardjuna, Arun, KG/KRA, Widuri, Nilam, Attaka

Oil Refineries (1/1/03): Cilacap, Central Java (380,000); Pertamina-Balikpapan, Kalimantan (240,920); Musi, South Sumatra (109,155); EXOR-1, Balongan, Java (125,000); Dumai, Central Sumatra (114,000); Sungai Pakning, Central Sumatra (47,500); Pangakalan Brandan, North Sumatra (4,750); Cepu, Central Java (3,420)

Product Pipelines: Trans-Java (serving the Surabaya market)

Oil Tanker Terminals: Java: Cilegon, Cilacap, Surabaya, Ardjuna B (offshore) *Sumatra:* Pangkalan Brandan, Belawan, Dumai, Musi, Perlak, Palembang, Tanjung Uban (offshore) *Kalimantan:* Balikpapan Sulawesi: Ujung Pandang *Irian Jaya:* Sorong, Jaya Seram: Bula Natuna Sea: Ikan Pari

Major Gas Fields: Sumatra: Arun, Alur Siwah, Kuala Langsa, Musi, South Lho Sukon, Wampu *East Kalimantan:* Attaka, Badak, Bekapai, Handil, Mutiara, Nilam, Semberah, Tunu *Natuna Sea:* Natuna *Java:* Pagerungan, Terang/Sirasun *Irian Jaya:* Tangguh

Major Gas Pipelines: Sumatra: Pangkalan Brandan-Dumai

LNG Plants: Arun, Bontang

43

Iran

Iran is OPEC's second largest oil producer and holds 9 per cent of the world's oil reserves. It also has the world's second largest natural gas reserves.

General Background

Iran's economy, which relies heavily on oil export revenues (around 80 per cent of total export earnings, 40 per cent -50 per cent of the government budget, and 10 per cent -20 per cent of GDP), was hit hard by the plunge in oil prices during 1998 and early 1999, but with the rebound in oil prices since then, has recovered somewhat. For 2002, Iran's real GDP grew by around 4 per cent ; for 2003 it is expected to grow at a slightly higher, 4.3 per cent rate. Relatively high oil export revenues the past year or two have allowed Iran to set up an oil stabilization fund. For 2003, Iran's budget anticipates a price of around $18.50 per barrel, well below current levels.

Despite relatively high oil export revenues, Iran continues to face budgetary pressures, a rapidly growing, young population with limited job prospects and high levels of unemployment; heavy dependence on oil revenues; significant external debt (including a high proportion of short-term debt); high levels of poverty; expensive state subsidies (billions of dollars per year) on many basic goods; a large, inefficient public sector and state monopolies (bonyads, which control at least a quarter of the economy and constitutionally are answerable only to supreme leader Ayatollah Ali Khamenei); international isolation and sanctions.

Iran is attempting to diversify by investing some of its oil revenues in other areas, including petrochemicals. Iran also is hoping to attract billions of dollars worth of foreign investment to the country by creating a more favorable investment climate (*i.e.*, reduced restrictions and duties on imports, creation of free-trade zones). In May 2002, the country's Expediency Council approved the "Law on the Attraction and Protection of Foreign Investment," which aims at encouraging foreign investment by streamlining procedures, guaranteeing profit repatriation, and more. This Law, which was sent to

the government for implementation in January 2003, represents the first foreign investment act passed by Iran's legislature since the 1978/79 revolution. The legislation was delayed for several years due to disagreements between reformers and conservatives. In June 2001, the Council of Guardians had rejected the bill as passed by the Majlis the previous month. In November 2001, the Majlis had passed a second, heavily amended, version of the bill. Although this version was far weaker than the first bill, the Council of Guardians again rejected it

Sanctions

In March 2003, President Bush extended sanctions originally imposed in 1995 by President Clinton for another year, citing Iran's "support for international terrorism, efforts to undermine the Middle East peace process, and acquisition of weapons of mass destruction." The 1995 executive orders prohibit U.S. companies and their foreign subsidiaries from conducting business with Iran, while banning any "contract for the financing of the development of petroleum resources located in Iran." In addition, the U.S. Iran-Libya Sanctions Act (ILSA) of 1996 (renewed for 5 more years in July 2001) imposes mandatory and discretionary sanctions on non-U.S. companies investing more than $20 million annually in the Iranian oil and natural gas sectors. In May 2002, the United States announced that it would review a contract by Canada's Sheer Energy to develop an Iranian oilfield to determine whether or not it violates ILSA.

Oil

As of late 2002, Iran held 90 billion barrels of proven oil reserves, or roughly 9 per cent of the world's total. The vast majority of Iran's crude oil reserves are located in giant onshore fields in the southwestern Khuzestan region near the Iraqi border and the Persian Gulf. Most of Iran's current oil production is accounted for by the following fields: Ahwaz-Bangestan (150,000-170,000 bbl/d current production, with plans to increase to 400,000 bbl/d), Marun, Gachsaran, Agha Jari, and Bibi Hakimeh. Most of Iran's crude oil is low in sulfur, with gravities in the 30°-39° API range. During 2002, Iran produced about 3.5 million bbl/d of oil. Iran's current sustainable crude oil production capacity is estimated at around 3.75 million bbl/d, which is around 250,000 bbl/d above Iran's latest (February 1, 2003) OPEC production quota of 3.597 million bbl/d.

In 2002, Iran consumed an estimated 1 million bbl/d of oil and had net exports of around 2.5 million bbl/d In 2002, around 28 per cent of Iran's oil exports went to Europe, 20 per cent to Japan, and under 10 per cent to South Korea, with other large customers including China and India. Iran's domestic oil consumption is increasing rapidly (about 7 per cent per year) as the economy and population grow. Iran subsidizes the price of oil products heavily, to the tune of $3 billion or so per year, resulting in a large amount of waste and inefficiency in oil consumption. Iran also is forced to spend around $1 billion per year to import oil products (mainly gasoline) which it cannot produce locally. In early April 2003, as part of an effort to curtail the rise in gasoline subsidy expenditures, gasoline consumption and imports (both of which are growing rapidly), Iran raised gasoline prices by 30 per cent -35 per cent, to around 31-44 cents per gallon.

It is possible that with sufficient investment, Iran could increase its oil production capacity significantly. Iran produced 6 million bbl/d in 1974, but has not surpassed 3.8 million bbl/d on an annual basis since the 1978/79 Iranian revolution. During the 1980s, it is believed that Iran may have maintained production levels at some older fields only by using methods which have permanently damaged the fields. Also, Iran's oilfields are–according to Oil Minister Zanganeh–experiencing a depletion rate of 250,000-300,000 bbl/d per year, and are in need of upgrading and modernization. Despite these problems, Iran has ambitious plans to double national oil production–to around 5.6 million bbl/d by 2010 and 7.3 million bbl/d by 2020. The country is counting on foreign investment to accomplish this, possibly as high as $5 billion per year.

NIOC's onshore field development work is concentrated mainly on sustaining output levels from large, aging fields. Consequently, enhanced oil recovery (EOR) programs, including natural gas

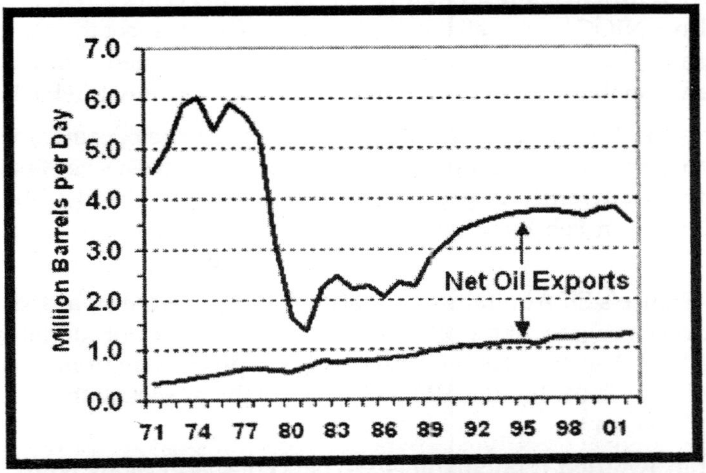

IRAN'S OIL PRODUCTION AND CONSUMPTION, 1971-2001

Note: 2002 values are estimated. Oil includes crude production and natural gas liquids.

injection, are underway at a number of fields, including Marun, Karanj, and the presently inactive Parsi fields. EOR programs will require sizeable amounts of natural gas, infrastructure development, and financing.

In October 1999, Iran announced that it had made its biggest oil discovery in 30 years, a giant onshore field called Azadegan located in the southwestern province of Khuzestan, a few miles east of the border with Iraq. Reportedly, the Azadegan field could contain oil reserves of up to 45 billion barrels, with potential production of 300,000-400,000 bbl/d. On November 1, 2000, agreement was reached between Japan and Iran for Japanese firms (Japex and Indonesia Petroleum, both majority-owned by the Japan National Oil Company–JNOC) to receive priority negotiating rights in developing Azadegan. In exchange, Japan is to loan Iran $3 billion; in April 2002, the second $1 billion installment on a $3 billion credit line was disbursed. In January 2001, the Majlis approved development of Azadegan by foreign investors using the so-called "buyback" model. In early March 2003, however, the Iranian official in charge of developing Azadegan said that Iran and Japan had not yet reached a final agreement on the project.

Since 1995, NIOC has made several sizable oil discoveries, including the 3-5-billion-barrel Darkhovin onshore oilfield, located near Abadan and containing low sulfur, 39° API crude oil. In late June 2001, Italy's ENI signed a nearly $1 billion, 5 1/2-year buyback deal to develop Darkhovin, with the added incentive of a limited risk/reward element (payment is to be linked to production capacity). ENI has a 60 per cent stake in the project, with NIOC holding the remaining 40 per cent. Ultimately, production at Darkhovin is expected to reach 160,000 bbl/d.

Foreign Investment/Buybacks

The Iranian constitution prohibits the granting of petroleum rights on a concessionary basis or direct equity stake. However, the 1987 Petroleum Law permits the establishment of contracts between the Ministry of Petroleum, state companies and "local and foreign national persons and legal entities." "Buyback" contracts, for instance, are arrangements in which the contractor funds all investments, receives remuneration from NIOC in the form of an allocated production share, then transfers operation of the field to NIOC after the contract is completed. This system has drawbacks for both sides: by offering a fixed rate of return (usually around 15 per cent -17 per cent), NIOC bears all the risk of low oil prices. If prices drop, NIOC has to sell more oil or natural gas to meet the compensation figure. At the same time, companies have no guarantee that they will be permitted to develop their discoveries, let alone operate them. Finally, companies do not like the short terms of buyback contracts.

The first major project under the buyback investment scheme became operational in October 1998, when the offshore Sirri A oil field (operated by Total and Malaysia's Petronas) began production at 7,000 bbl/d (Sirri A currently is producing around 20,000 bbl/d). The neighboring Sirri E field began production in February 1999, with production at the two fields expected to reach 120,000 bbl/d.

In March 1999, France's Elf Aquitaine and Italy's Eni/Agip were awarded a $1 billion contract for a secondary recovery program at the offshore, 1.5-billion-barrel Doroud oil and natural gas field located near Kharg Island. The program is intended to boost production from around 136,000 bbl/d to as high as 205,000 bbl/d by 2004. TotalFinaElf is operator of the project, with a 55 per cent share, while Eni holds the other 45 per cent.

In April 1999, Iran awarded TotalFinaElf (46.75 per cent share), along with Canada's Bow Valley Energy (15 per cent share), a buyback contract to develop the offshore Balal field. The field, which contains some 80 million barrels of reserves, started producing at a 20,000-bbl/d rate in early 2003, and is expected to reach 40,000 bbl/d by early 2004. In February 2001, ENI-Agip acquired a 38.25 per cent share in Balal.

In November 2000, Norway's Statoil signed a series of agreements with NIOC to explore for oil in the Strait of Hormuz area. The two companies also will cooperate on developing a natural gas-to-liquids processing plant for four southern onshore fields, and possibly will develop the Salman offshore field at a cost of $850 million, with eventual production of 130,000 bbl/d. Iran appears to be accelerating its plans to boost production of natural gas liquids (NGL), as well as liquefied petroleum gas (LPG). NGL expansion plans, including a $500 million plan to build two NGL plans on the south coast of Iran, are aimed mainly at making ethane feedstock available for Iran's growing petrochemical industry.

A much-sought-after deal to develop the giant Bangestan field has been delayed several times after an expected award in 2001. Bangestan includes three oilfields (Anwaz, Mansuri, Ab-Teymour) which currently produce about 250,000 bbl/d of oil. In April 2003, Shell stated that it was frustrated with the slow pace of negotiations on Bangestan, including numerous changes to terms of the project.

In May 2002, Iran's Oil Ministry signed a $585 million buyback contract with local company PetroIran to develop the Foroozan and Esfandiar offshore oilfields. PetroIran is expected to increase production at the fields from around 40,000 bbl/d at present to 109,000 bbl/d within 3 years. The two oilfields straddle the border with Saudi Arabia's Lulu and Marjan fields.

In other news related to "buyback" deals, the Cheshmeh-Khosh field, which had been awarded to Spain's Cepsa for $300 million, is likely to be re-awarded to a consortium of Cepsa and OMV. The two companies are to raise crude production at the field from 30,000 bbl/d to 80,000 bbl/d within four years.

Recently, Iran appears to have had some second thoughts about buybacks (including charges of corruption, insufficient benefits to Iran, and also worries that buybacks are attracting too little investment), and reportedly is considering substantial changes in the system. In late May 2002, Canada's Sheer Energy became the first foreign company since ENI's Darkhovin deal to reach agreement ($80 million to develop the Masjed-I-Suleyman, or MIS, field) under the ENI terms. Sheer aims to boost MIS production from 4,500 bbl/d to 20,000 bbl/d. In general, however, the addition of a limited risk/reward element has not attracted the flood of foreign energy investment which Iran both needs and wants. As a result, Iran reportedly is considering a further modification to its "buy-back" model, possibly extending the length of such contracts from the current 5-7 years.

Offshore Developments

The Doroud 1&2, Salman, Abuzar, Foroozan, and Sirri fields comprise the bulk of Iran's offshore oil output. Iran plans extensive development of existing offshore fields and hopes to raise its offshore production capacity to 1.1 million bbl/d (from around 600,000 bbl/d). It is estimated that development of new offshore Persian Gulf and Caspian Sea oil fields will require investment of $8-$10 billion.

In late 2001 and early 2002, Shell brought part of the $1.1 billion Soroush-Nowrooz development online, with production of around 60,000 bbl/d. The two fields are located offshore, about 50 miles west of Kharg Island, and contain estimated recoverable reserves of around 1 billion barrels of mainly heavy oil. Although Soroush was shut down briefly in March 2003 at the outset of war with Iraq, output from the field is still expected to reach 100,000 bbl/d by the end of 2003. In early 2003, a consortium of three Japanese companies bought a 20 per cent share in the Soroush/Nowrooz development project.

Caspian Sea Region

Aside from acting as a transit center for other countries' oil and natural gas exports from the Caspian Sea, Iran has potentially significant Caspian reserves of its own, including up to 15 billion barrels of oil and 11 trillion cubic feet of natural gas. It is important to note, however, that almost none of this is "proven" to be recoverable (although preliminary seismic surveys conducted by Lasmo and Shell indicated 2.5 billion barrels of oil). Currently, Iran has no oil or natural gas production in the Caspian region, although in March 2001, NIOC signed a $226 million deal with Sweden's GVA Consultants and Iran's Sadra to build an oil rig in the Caspian Sea off Mazandaran province. This marks Iran's first exploration attempt in the Caspian Sea, whose legal status among regional states remains in dispute.

At the present time, Iran maintains the most isolated position among the Caspian Sea's littoral states on the division of the Sea. Iran insists that regional treaties signed in 1921 and 1940 between Iran and the former Soviet Union, which call for joint sharing of the Caspian's resources between the two countries, remain valid. Iran has rejected as invalid all unilateral and bilateral agreements on the utilization of the Sea. As such, Iran is insisting that either the Sea should be used in common, or its floor and water basin should be divided into equal (20 per cent) shares. Under this plan, the so-called "condominium" approach, the development of the Caspian Sea would be undertaken jointly by all of the littoral states. However, using the equidistant method of dividing the seabed on which Kazakhstan, Azerbaijan, and Russia have agreed, Iran would only receive about 12 per cent -13 per cent of the Sea. In March 2002, Iran's Oil Minister Zanganeh asserted that Iran would begin exploiting its fifth of the Sea within a short time, and would not permit "any other party to engage in oil exploration" in this area. In January 2003, Iranian Foreign Minister Kamal Kharrazi reiterated the country's claim to a 20 per cent share of the Caspian, and in early April 2003, Oil Minister Zanganeh said that Iran would start Caspian drilling within a year or two.

As of April 2003, no agreement has been reached among Caspian Sea region states on this matter. In March 2003, Iran and Turkmenistan noted "the need to achieve a consensus between the five [littoral] countries," while the two countries reportedly moved ahead in charting their common border in the Sea. In late April 2002, a meeting between the five Caspian littoral states ended without agreement on a new treaty. On May 20, 2002, Iran and Azerbaijan also failed to reach agreement on Caspian Sea division. On July 23, 2001, tensions flared in the Caspian Sea region when an Iranian gunboat intercepted two BP oil exploration vessels off Azerbaijan's coast. Following the incident, BP suspended exploration in the disputed block (which Iran calls Alborz).

Crude Swaps

In order to get around restrictions in dealing with Iran, several firms have proposed oil "swaps" involving the delivery of Caspian (Azeri, Kazakh, Turkmen) oil to refineries in northern Iran, while the same amount of Iranian oil is exported through Persian Gulf terminals. According to Iranian Oil Minister Bijan Namdar-Zangeneh, Iran is planning to retool its oil infrastructure to accommodate such swaps, including construction of a $400 million, 240-mile pipeline from the Caspian area via Iran's Caspian port of Neka to refineries in northern Iran and to Tehran. Eventually, this could lead to the transport of 370,000 bbl/d of Caspian crude. Iran also plans to boost capacity at its northern refineries at Arak, Tabriz, and Tehran to about 800,000 bbl/d in order to process this oil. Currently, however, despite capacity of around 50,000 bbl/d, only 15,000-20,000 bbl/d of Turkmen oil are being shipped to Neka, and then on to Tehran by the existing Neka-Tehran pipeline. An equivalent amount of Iranian oil is then made available for export via Kharg Island terminal on the Persian Gulf. Meanwhile, in November 2002, Russia's Lukoil began sending around 25,000 bbl/d of Russian Siberian Light crude from the Caspian port of Astrakhan to Neka.

Refining and Transportation

As of January 2003, Iran had nine operational refineries with a combined capacity of 1.47 million bbl/d. In order to meet burgeoning domestic demand for middle and light distillates, Iran has imported refined products since 1982, and is attempting to boost its refining capacity to 2 million bbl/d. Two planned grassroots refineries include a 225,000-bbl/d plant at Shah Bahar and a 120,000-bbl/d unit on Qeshm Island. The $3 billion Shah Bahar refinery project was approved by the government in late 1994 and would be built by private investors.

Iran exports crude oil via four main terminals–Kharg Island (by far the largest), Lavan Island, Sirri Island (reopened on April 13, 2003 for the first time since 1988, when it was damaged by an Iraqi air raid), and Ras Bahregan. Refined products are exported via the Abadan and Bandar Mahshahr terminals. Many Iranian oil export terminals were damaged during the Iran-Iraq War, but all have been rebuilt.

Natural Gas

Iran contains an estimated 812 trillion cubic feet (Tcf) in proven natural gas reserves–the world's second largest and surpassed only by those found in Russia. Around 62 per cent of Iranian natural gas reserves are located in non-associated fields, and have not been developed, meaning that Iran has huge potential for gas development. Major non-associated gas fields include: South Pars (280-500 Tcf of gas reserves), North Pars (50 Tcf), Kangan (29 Tcf), Nar (13 Tcf), Khangiran (11 Tcf), and several others.

Despite the fact that domestic natural gas demand is growing rapidly, Iran has the potential to be a large natural gas exporter due to its enormous reserves. In 2001, Iran produced about 2.2 Tcf of natural gas. Of this, around 10 per cent is flared, and approximately 30 per cent is reinjected–in part for enhanced oil recovery efforts. Still, the amount of gas that Iran flares and reinjects is indicative of the abundance of this resource in the country.

Currently, natural gas accounts for nearly half of Iran's total energy consumption, and the government plans billions of dollars worth of further investment in coming years to increase this share. The price of natural gas to consumers is state-controlled. In March 2003, Russia's Gazprom said that it might form a joint company with Iran to develop Iranian gas resources.

South Pars

Iran's largest non-associated natural gas field is South Pars, geologically an extension of Qatar's 380-Tcf North Field, most likely the largest non-associated gas field in the world. South Pars was first identified in 1988 and originally appraised at 128 Tcf in the early 1990s. Current estimates are that South Pars contains 280 Tcf or more (some estimates go as high as 500 Tcf) of natural gas, of which a large fraction will be recoverable, and over 17 billion barrels of liquids. Development of South Pars is Iran's largest energy project, and already has attracted billions of dollars in investment. In early March 2003, the chairman of Petropars stated that another $8 billion would be spent on South Pars development during the Iranian year starting March 21, 2003. South Pars development is proceeding, but has been delayed by various problems–technical (*i.e.*, high levels of mercaptans–foul-smelling sulfur compounds–in the South Pars gas), contractual (*i.e.*, controversy over "buy-back" arrangements), etc. Phase 1, for instance, which is being handled by Petropars (owned 60 per cent by NIOC), has been delayed several times and now is scheduled for completion in mid-2004 (around 3 years behind schedule), involves production of 900 million cubic feet per day (Mmcf/d) of natural gas and 40,000 bbl/d of condensate.

Natural gas from South Pars largely is slated to be shipped north via the planned 56-inch, 300-mile, $500 million, IGAT-3 pipeline (a section of which is now being built by Russian and local contractors), as well as a possible IGAT-4 line. Gas also will be reinjected to boost oil output at the mature Aghajari field (output peaked at 1 million bbl/d in 1974, but has since fallen to 200,000 bbl/d), and possibly the Ahwaz and Mansouri fields (which make up part of the huge Bangestan reservoir in the southwest Khuzestan region). South Pars natural gas also is intended for export, by pipeline and also possibly by liquefied natural gas (LNG) tanker. Sales from South Pars could earn Iran as much as $11 billion per year over 30 years, according to Iran's Oil Ministry.

On September 29, 1997, Total (now TotalFinaElf) signed a $2 billion "buy back" deal (along with Russia's Gazprom and Malaysia's Petronas) to explore South Pars and to help develop the field during Phases 2 and 3 of its development. TotalFinaElf has a 40 per cent share of the project, with the other two companies each having 30 per cent shares. NIOC estimates that South Pars has a natural gas production potential of up to 8 billion cubic feet per day (Bcf/d) from four individual reservoirs.

In February 2003, Oil Minister Zanganeh officially inaugurated Phases 2 and 3 of South Pars development, which began to come onstream in September 2002. Already, Phases 2 and 3 reportedly are producing around 2 Bcf per day of natural gas, and 85,000 bbl/d of condensates. Twin undersea pipelines will carry gas from South Pars to onshore facilities at Asaluyeh. In March 2002, Hyundai signed another contract, this one for $1 billion, to build four natural gas processing trains. The Asaluyeh facility comprises four natural gas processing trains, sulphur recovery units, condensate stabilization and storage units, and export compressors.

Phases 4 and 5, estimated to cost $1.9 billion each, are being handled by ENI and Petropars, and involve construction (by Agip and Petropars) of onshore treatment facilities at the port of Bandar Asaluyeh. These two phases are expected to come online by late 2004 or early 2005.

Phases 6-8, which are to produce a combined 3 Bcf/d of natural gas and 120,000 bbl/d of condensate at a cost of $2.6 billion, are being handled by Petropars and Norway's Statoil, which signed an agreement in October 2002. First stages of the project are scheduled to come online in late 2004, with gas being transported via pipeline to the Agha Jari oilfield for injection as part of enhanced oil recovery efforts. NIOC is to take over as operator when development is finished.

Phases 9 and 10 are expected to supply the domestic market, possibly by 2007, while phases 11 and 12 are slated for LNG export and condensate production, possibly by 2008. Consortia reportedly interested in all or parts of phases 9-12 (expected to cost $4 billion) include Iran LNG (BP/Reliance), Persian LNG (Shell/Repsol-YPF), Pars LNG (Total/Petronas), and BG/Enel. In September 2002, South Korea's LG signed a $1.6 billion deal with NIOC on phases 9 and 10. LG's share is 42 per cent, and the deal reportedly uses international bank project financing rather than a "buy-back" model. Meanwhile, Shell hopes to win Phase 13, which will be dedicated to LNG production. Phase 14 is slated for gas-to-liquids (GTL) development, with Statoil and Shell reportedly interested. Phases 13 and 14 could be completed by 2008.

Other Natural Gas Development

In addition to South Pars, Iran's long-term natural gas development plans may involve: the 48-Tcf North Pars field (a separate structure from South Pars); the 6.4-Tcf, non-associated Khuff (Dalan) reservoir of the Salman oil field (which straddles Iran's maritime border with Abu Dhabi, where it is known as the Abu Koosh field); the 800-Bcf Zireh field in Bushehr province; the 4-Tcf Homa field in southern Fars province; the 14-Tcf Tabnak natural gas field located in southern Iran; the onshore Nar-Kangan fields, the 13-Tcf Aghar and Dalan fields in Fars province, and the Sarkhoun and Mand fields.

Natural Gas Trade

With almost unlimited natural gas production potential, Iran is looking to export large volumes of gas. Besides Turkey, potential customers for Iranian gas exports include: Ukraine (Kiev reportedly is interested in building an Iran-Armenia-Georgia-Crimea-Ukraine line), Europe, India, Pakistan, Armenia, Azerbaijan, Taiwan, South Korea, and coastal China. Exports could be either via pipeline or by LNG tanker, with possible LNG export terminals at Asaluyeh or Kish Island. Iran reportedly is developing two LNG plants (down from four planned earlier), with potential completion by 2007. In March 2003, British Gas and NIOC reportedly were in advanced talks on developing a $1.4 billion LNG plant at Bandar Tombak on the Persian Gulf coast. The plant is to comprise two LNG trains, with total capacity of 4.5-5 million tons per year.

In late January 2002, Iran and Turkey officially inaugurated a much-delayed natural gas pipeline link between the two countries. This follows several years of delays due to economic, political, and technical factors. In 1996, Iran and Turkey had signed a $20 billion agreement that called for Iran to supply Turkey with more than 8 Tcf of natural gas over a period of 22 years beginning in late 1999. Officials in Turkey and Iran variously blamed U.S. sanctions, financing problems on the Turkish leg of the $1.9 billion pipeline, economic recession in Turkey, and delays by the Iranians in completing an important metering station for delaying the project. Exports of Iranian natural gas to Turkey could reach 350 Bcf per year by 2007. There are questions, however, whether Turkish demand will grow rapidly enough to absorb this volume of gas from Iran, in addition to gas slated to be supplied by Russia, Algeria, and Nigeria. In June 2002, for instance, Turkey halted Iranian gas imports, ostensibly due to "quality problems" but more likely due to lack of demand in Turkey. In mid-November 2002, Turkey announced that it was resuming gas imports from Iran.

In October 2002, the International Atomic Energy Agency (IAEA) predicted that "Iran will be a major global natural gas supplier in the future," especially to Europe. Iran reportedly is shooting for around 300 Bcf per year of natural gas exports to Europe via Turkey by 2007. Along these lines, Greece and Iran signed a $300 agreement in March 2002 which calls for extending the natural gas pipeline from Iran to Turkey into northern Greece. After that, gas could be transported to Europe via Bulgaria and possibly Romania (a memorandum of understanding–MOU–was signed on this possibility in January 2003), or via an undersea pipeline to Italy, where gas demand is expected to grow rapidly in coming years. A deep water option could be extremely expensive, however, making an overland route more likely.

In January 2003, Iran and Kuwait signed an MOU on Iranian gas exports to Kuwait by 2005. The gas is to be used for power generation.

Electricity

As of 2001, Iran had installed power generation capacity of about around 31 gigawatts (GW), of which three-quarters or more was natural gas-fired, with the remainder either hydroelectric (7 per cent) or oil-fired. As a result of significant state investment in this area, a number of new power plants (mainly hydroelectric and combined cycle) have come online in recent years in Iran, including the 2,000-MW Shahid Rai thermal power station in Qazvin; a 1,290-MW combined-cycle plant in Rasht;

a doubling of the Tabriz power plant's capacity to 1,500 MW; two, 200-MW, steam-powered units at the Martyr Montazeri plant; a 215-MW steam-powered unit at the Ramin Power Plant; a 107-MW combined cycle generator at Montazer Qa'em Power Plant, and three-fourths of the Shazand power plant near Arak in central Iran.

With power demand growing rapidly (7 per cent -8 per cent annually), Iran is adding significant generation capacity–both thermal and hydroelectric, with the goal of adding total generating capacity of 30 GW within 10 years (according to Iran's Energy Minister). Around 3 GW is expected to come online during the current Iranian year, which ends on March 19, 2004. Currently, the largest hydropower projects are the 3,000-megawatt (MW) Karun 3 plant, the 2,000-MW Godar-e Landar facility, a 1,000-MW station in Upper Gorvand, and the 400-MW Karkheh dam. New thermal projects include two 1,040-MW combined cycle plants in the South, an 1,100-MW combined cycle plant at Arak, and a 1,000-MW facility in Bandar Abbas. In February 2003, 1,272-MW combined-cycle plant came online in Kerman. In January 2003, plans were announced to build Iran's first geothermal plant, in the northwestern province of Ardebil. In early April 2002, the 1,000-MW, natural-gas-fired, combined-cycle Shahid Raja'i power plant came online in the northern Iranian province of Qazvin.

Nuclear

Currently, Iran has several small nuclear research reactors, in addition to a large-scale nuclear power plant under construction at the southern town of Bushehr. Iran claims that its nuclear power is for peaceful purposes and that it will help free up oil and natural gas resources for export, thus generating additional hard-currency revenues. The country has stated its aim of having 6,000 MW of nuclear power online by 2020, accounting for 10 per cent of the country's power generation capacity at that point.

In December 2002, Iran and Russia signed a protocol for peaceful cooperation in nuclear power. Russia has been assisting Iran on the Bushehr nuclear power facility, work on which first began in 1974 by West Germany, but was halted (80 per cent complete) following the 1978/1979 revolution. Significant amounts of money, possibly billions of dollars, had been spent on Bushehr to that point. Following the Iran-Iraq War (1980-1988), during which time Bushehr was bombed six times and seriously damaged, progress on the plant resumed when Russia signed an $800 million contract in 1995. The contract with Russia called for completion of a 1,000-MW, pressurized-light-water reactor, as well as the possible supply of two modern VVER-440 units. Since then, work has proceeded slowly, although reports in early March 2003 indicated that Bushehr was 70 per cent complete, and is expected to come online as early as March 2004. Russia also reportedly has begun construction of a second unit at Bushehr.

Although Iran is a signatory to the Nuclear Non-Proliferation Treaty and insists that its nuclear program is for peaceful purposes (*i.e.*, power generation), the United States strongly opposes the Bushehr project and has in the past provided Russia with information pointing to the existence of an Iranian nuclear weapons program. In May 2002, U.S. Energy Secretary Spencer Abraham met with Alexander Rumyantsev, head of Russia's nuclear agency, and discussed this issue, with Rumyantsev stating the Russian position that Bushehr "is not a source of proliferation of nuclear material." In late March 2003, U.S. Undersecretary of State for Arms Control, John Bolton, said, "In the aftermath of Iraq, dealing with the Iranian nuclear weapons program will be of equal importance as dealing with the North Korean nuclear weapons program."

BUSHEHR-2

0 40 80 120 160 200 meters Scale 1 pixel = ~1 meter

Environment

In the context of its oil-based economy, environmental issues in Iran only recently have become important. Ongoing air pollution in urban areas, which reached a crisis level in Tehran in December 1999, have highlighted the need to improve Iran's environmental record. The rush to develop oil and natural gas resources in the Caspian Sea makes oil pollution in the Caspian a real environmental threat.

Huge increases in energy consumption over the past 20 years have contributed greatly to pollution levels as Iran's carbon emissions have nearly tripled over the same time span. Large numbers of old, inefficient cars on the road lacking catalytic converters account for much of the country's air pollution. Energy prices are kept artificially low in Iran through heavy state subsidies, resulting in wasteful consumption patterns.

In addition, Iran's abundance of fossil fuel resources has tended to discourage the country's incentive to shift to cleaner alternative energy sources for its energy needs. As Iran continues to struggle with air pollution in the 21st century, however, the country likely will need to take a variety of tough measures in order to avert an environmental crisis.

Country Overview

Islamic Republic Proclaimed: April 1, 1979

Population (7/02E): 66.6 million

Location/Size: Middle East–between the Persian Gulf and the Caspian Sea / 636,296 square miles

Major Cities: Tehran (capital), Meshed, Isfahan, Tabriz, Shiraz, Ahwaz, Kermanshah, Qom, Ardebil, Qazvin

Languages: Persian and Persian dialects (58 per cent), Turkic and Turkic dialects (26 per cent), Kurdish (9 per cent), Luri (2 per cent), Baluch (1 per cent), Arabic (1 per cent), Turkish (1 per cent)

Ethnic Groups: Persian (51 per cent), Azerbaijani (24 per cent), Gilaki and Mazandarani (8 per cent), Kurd (7 per cent), Arab (3 per cent), Lur (2 per cent), Baluch (2 per cent), Turkmen (2 per cent), other (1 per cent)

Religion: Shi'a Muslim (89 per cent), Sunni Muslim (10 per cent), Zoroastrian, Jewish, Christian, and Baha'i (1 per cent)

Defense (8/98): Army (350,000), Revolutionary Guard (120,000), Navy (20,600), Air Force (40,000-45,000), army reserves (350,000)

Economic Overview

Currency: Rial (R)

Exchange Rates (4/8/03): R7,978 per $US

Gross Domestic Product (GDP, at market exchange rates) (2002E): $111.0 billion

Gross Domestic Product (GDP, at purchasing power parity exchange rates) (2002E): $510.4 billion

Real GDP Growth Rate (2002E): 4 per cent *(2003F):* 4.3 per cent

Inflation Rate (2003F): 10 per cent -15 per cent

Unemployment Rate (2003E): 16 per cent -21 per cent

Current Account Balance (2002E): $5.0 billion *(2003F):* $4.2 billion

Major Trading Partners (2000): Japan, Italy, Germany, China, France, United Arab Emirates

Merchandise Exports (2002E): $24.8 billion

Merchandise Imports (2002E): $19.3 billion

Merchandise Trade Surplus (2002E): $5.5 billion

Major Export Products: Oil and oil products (90 per cent), carpets, pistachios

Major Import Products: Industrial supplies (37 per cent), machinery (30 per cent), consumer goods (18 per cent)

Oil Export Revenues (2002E): $18.0 billion *(2003F):* $18.7 billion

Oil Export Revenues/Total Export Revenues (2001E): around 90 per cent

External Debt (2003E): $11.5 billion

Energy Overview

Atomic Energy Organization of Iran: Gholamreza Aqazadeh

Proven Oil Reserves (1/1/03E): 89.7 billion barrels

OPEC Crude Oil Production Quota (as of 2/1/03): 3.597 MMBD

Crude Oil Production Capacity (3/03E): 3.75 MMBD

Oil Production (2002E): 3.5 MMBD (of which, 3.4 MMBD was crude oil)

Oil Production (1/03E): 3.8 MMBD (of which 3.6 MMBD was crude oil)

Oil Consumption (2002E): 1.0 MMBD

Net Oil Exports (2002E): 2.5 MMBD

Crude Oil Refining Capacity (1/1/03E): 1.47 MMBD

Major Crude Oil Customers: OECD Europe, Japan, China, South Korea

Natural Gas Reserves (1/1/03E): 812 trillion cubic feet (Tcf)

Natural Gas Production (2001E): 2.17 Tcf

Natural Gas Consumption (2001E): 2.32 Tcf

Recoverable Coal Reserves (2000E): 1,885 million short tons (Mmst)

Coal Production (2001E): 1.5 Mmst

Coal Consumption (2001E): 2.3 Mmst

Net Coal Imports (2001E): 0.8 Mmst

Electric Generation Capacity (2001E): 30.6 gigawatts (around 93 per cent thermal)

Electricity Consumption (2001E): 115.9 billion kilowatthours

Environmental Overview

Total Energy Consumption (2001E): 5.18 quadrillion Btu* (1.3 per cent of world total energy consumption)

Energy-Related Carbon Emissions (2001E): 90.1 million metric tons of carbon (1.4 per cent of world total carbon emissions)

Per Capita Energy Consumption (2001E): 80.3 million Btu (vs U.S. value of 341.8 million Btu)

Per Capita Carbon Emissions (2001E): 1.4 metric tons of carbon (vs U.S. value of 5.5 metric tons of carbon)

Energy Intensity (2001E): 40,770 Btu/ $1995 (vs U.S. value of 10,736 Btu/ $1995)**

Carbon Intensity (2001E): 0.71 metric tons of carbon/thousand $1995 (vs U.S. value of 0.17 metric tons/thousand $1995)

Fuel Share of Energy Consumption (2001E): Oil (51 per cent), Natural Gas (47 per cent), Coal (1 per cent), Hydroelectric (1 per cent)

Fuel Share of Carbon Emissions (2001E): Oil (53 per cent), Natural Gas (45 per cent), Coal (1 per cent)

Status in Climate Change Negotiations: Non-Annex I country under the United Nations Framework Convention on Climate Change (ratified July 18th, 1996). Not a signatory to the Kyoto Protocol.

Major Environmental Issues: Air pollution, especially in urban areas, from vehicle emissions, refinery operations, and industrial effluents; deforestation; overgrazing; desertification; oil pollution in the Persian Gulf; inadequate supplies of potable water.

Major International Environmental Agreements: A party to Conventions on Biodiversity, Climate Change, Desertification, Endangered Species, Hazardous Wastes, Marine Dumping, Nuclear Test Ban, Ozone Layer Protection and Wetlands. Has signed, but not ratified, Environmental Modification, Law of the Sea and Marine Life Conservation.

The total energy consumption statistic includes petroleum, dry natural gas, coal, net hydro, nuclear, geothermal, solar, wind, wood and waste electric power. The renewable energy consumption statistic is based on International Energy Agency (IEA) data and includes hydropower, solar, wind, tide, geothermal, solid biomass and animal products, biomass gas and liquids, industrial and municipal wastes. Sectoral shares of energy consumption and carbon emissions are also based on IEA data.

Oil and Gas Industries

Organizations: The Ministry of Petroleum (MoP) has overall responsibility for the country's energy sector. The MoP has four subsidiaries which function autonomously for the most part, but ultimately report to the Ministry: 1) National Iranian Oil Company (NIOC)–oil and gas exploration and production, refining and oil transportation; 2) National Iranian Gas Company (NIGC)–manages gathering, treatment, processing, transmission, distribution, and exports of gas and gas liquids; 3) National Iranian Petrochemical Company (NPC)–handles petrochemical production, distribution, and exports; and 4) National Iranian Oil Refining and Distribution Company (NIORDC) handles oil refining and transportation, with some overlap to NIOC. Also, the National Iranian Tanker Company (NITC) controls the second largest fleet of tankers in OPEC.

Foreign Oil Company Involvement: BG, BHP, Bow Valley, BP, Edison, ENI, Gazprom, OMV, Petronas, Royal Dutch/Shell, Sheer Energy, Statoil, TotalFinaElf

Major Oil Fields: Agha Jari, Ahwaz (Bangestan), Azadegan, Bibi Hakimeh, Darkhovin, Doroud, Gachsaran, Mansouri (Bangestan), Marun, Masjid-e Soleiman, Parsi, Rag-e-Safid, Soroush/Nowruz

Major Refineries (capacity, bbl/d) (1/1/03E): Abadan (400,000), Isfahan (265,000), Bandar Abbas (232,000); Tehran (225,000), Arak (150,000), Tabriz (112,000), Shiraz (40,000), Kermanshah (30,000), Lavan Island (20,000)

Major Oil Terminals: Ganaveh, Kharg Island, Lavan Island, Sirri Island, Cyrus, Ras Bahregan, Larak Island

Gas Pipeline System: The 780-mile, 40/42-inch IGAT-1 trunkline transports associated gas from Khuzestan area oilfields to consumption centers in the north; the 56-inch, $1.5 billion, 880-mile IGAT-2 line transports non-associated gas from the Kangan and Nar fields on the Persian Gulf coast near Bandar Taheri; the $500 million, 56-inch, 300-mile IGAT-3 pipeline, which would run from South Pars to Tehran, is under construction. Evaluation also has begun on a possible IGAT-4 line from South Pars to industrial northern Iran.

44

Iraq

Iraq is estimated to hold more than 112 billion barrels of proven oil reserves, and possibly much more undiscovered oil in unexplored areas of the country. Iraq also is estimated to contain at least 110 trillion cubic feet of natural gas. The country is a focal point for regional and international security issues.

General Background

In the aftermath of war in March and April 2003, Iraq now finds itself in a period of uncertainty and transition from over three decades of Ba'ath party rule. The events of 2003 mark the latest upheaval which Iraq has faced in its recent history. Since the early 1980s, for instance, Iraq experienced two major wars (Iran-Iraq and the Kuwait war), plus more than a decade of economic sanctions. As a result, the country's economy, infrastructure, environment, health care system, and other social indicators all deteriorated sharply. Iraq also assumed a heavy debt burden, possibly as high as $116 billion if debts to Gulf states and Russia are counted, and even more if $200 billion in reparations payments stemming from Iraq's 1990 invasion of Kuwait are included. It is possible, however, that much of Iraq's debt will be written off in the end, and that reparations will be capped at a certain level, possibly $40 billion or so. On August 8, 2003, the U.N. Humanitarian coordinator for Iraq, Ramiro Lopez da Silva, estimated that Iraq's reconstruction bill for 2004 would be $20 billion.

Now, with the regime of former President Saddam Hussein no longer in power, the country is being governed temporarily by a "Coalition Provisional Authority (CPA)" led by the United States and the United Kingdom. On May 6, 2003, President Bush named Ambassador L. Paul Bremer as presidential envoy to, and senior CPA official in, Iraq. In addition, there is a 25-member "Governing Council" made up of leading Iraqis, which met for the first time on July 13, 2003. The Governing Council's goals are to help rebuild Iraq as well as to prepare for "full, free and fair democratic elections." Also heavily involved in rebuilding Iraq are the U.S.-led Office of Reconstruction and Humanitarian Assistance (ORHA), the U.S. Army Corps of Engineers (headed by Brig. Gen. Robert Crear), private contractors

like Halliburton (and its subsidiary, Kellogg Brown & Root–KBR), non-governmental organizations, and coalition military forces.

Iraq's economy appears to have recovered somewhat from its condition just after the war, but the Economist Intelligence Unit (EIU) still forecasts a contraction in Iraq's real gross domestic product (GDP) of 7.5 per cent in 2003, with inflation of 75 per cent. This follows more than a decade of economic stagnation and decline. In 2004, on the contrary, the EIU expects a sharp upturn in real Iraqi GDP growth, to around 20 per cent, assuming continued increases in Iraqi oil production (and export revenues). In mid-August 2003, at the first meeting of the Council for International Coordination (CIC), which is tasked to coordinate donor countries financing Iraqi reconstruction, Paul Bremer stated that "Iraq is a rich country that is temporarily poor."

In July 2003, World Bank President James Wolfensohn stated that Bank assistance for Iraq's reconstruction could come after "a constitution and an elected government" were in place. Estimates of total, long-term Iraqi reconstruction costs run to $100 billion or higher. As of late June 2003, the CPA reportedly had $5.4 billion available in order to rebuild Iraq's power grid, sewer systems, and other public services. Paul Bremer, head of the CPA, said that the 2004 budget would run a deficit of $4 billion.

In May 2003, the U.N. Security Council passed Resolution 1483, lifting sanctions on Iraq, phasing out the U.N. "Oil-for-Food" program over six months, and designating a U.N. "special representative" to assist Iraq in its reconstruction efforts. In addition, according to the U.S. State Department, the resolution "stresse[d] the right of the Iraqi people to freely determine their own political future and control their own natural resources" (including oil), "encourage[d] international support for Iraq's recovery," and "enliste[d] the support of international financial institutions" in this effort. One billion dollars of unallocated funds in the U.N. escrow account were to be transferred to an Iraqi "Development Fund" to provide for immediate reconstruction needs. In June 2003, Paul Bremer suggested that revenues from oil sales "could be distributed to Iraq's citizens as 'dividends', along the lines of the system used by the State of Alaska."

On May 27, 2003, the U.S. Treasury Department lifted most U.S. sanctions on Iraq, thereby implementing U.N. Security Council Resolution 1483. For over a decade after Iraq's invasion of Kuwait in 1990, the United States had maintained unilateral economic sanctions against Iraq. Executive Order #12722 (August 2, 1990) imposed a complete trade embargo, and Executive Order #12724 (August 9, 1990) imposed additional restrictions. Under U.S. sanctions, goods or services were not permitted to be imported from or exported to Iraq, with the exception of the U.N. "oil-for-food" program.

Oil

On May 4, 2003, ORHA announced the appointment of Thamir Ghadban as chief executive of the interim management team for the Iraqi oil sector, in essence the *de facto* Oil Minister. Phillip Carroll, former head of Royal Dutch/Shell in the United States, also was appointed as head of an advisory board for the sector. Thamir Ghadban had formerly been a top official in Iraq's South Oil Company (SOC). Other important players include Ali Hassan at the State Oil Marketing Organization (SOMO), Adel Qazaz at the North Oil Company (NOC), and Jabbar Hussein Luaibi at SOC. In early August 2003, the CPA put the cost of rehabilitating Iraq's oil sector to its pre-war state at $1.144 billion, and the time frame to do so at nine months.

According to the *Oil and Gas Journal*, Iraq contains 112 billion barrels of proven oil reserves, the third largest in the world (behind Saudi Arabia and Canada). Estimates of Iraq's oil reserves and resources vary widely, however, given that only 10 per cent or so of the country has been explored. Some analysts (the Baker Institute, Center for Global Energy Studies, the Federation of American Scientists, etc.) believe, for instance, that deep oil-bearing formations located mainly in the vast Western Desert region, for instance, could yield large additional oil resources (possibly another 100 billion barrels or more), but have not been explored. Other analysts, such as the US Geological Survey, are not as optimistic, with median estimates for additional oil reserves closer to 45 billion barrels.

Iraqi oil development began in 1901, with the first well (Chia Surkh-1) drilled. The Iraq National Oil Company (INOC) was formed in 1964, and with Iraqi oil nationalization between 1972 and 1975, INOC took over from the international oil companies previously running the country's oil industry. In 1987, INOC was merged with the Ministry of Oil.

Iraq's oil development and production costs are amongst the lowest in the world (perhaps $3-$5 billion for each million barrels per day), making it a highly attractive oil prospect. However, only 17 of 80 discovered fields have been developed, while few deep wells have been drilled compared to Iraq's neighbors. Overall, only about 2,300 wells reportedly have been drilled in Iraq (of which about 1,600 are actually producing oil), compared to around 1 million wells in Texas for instance. In addition, Iraq generally has not had access to the latest, state-of-the-art oil industry technology (*i.e.*, 3D seismic, directional or deep drilling, gas injection), sufficient spare parts, and investment in general throughout most of the 1990s. Instead, Iraq reportedly utilized sub-standard engineering techniques (*i.e.*, overpumping, water injection/"flooding"), obsolete technology, and systems in various states of decay (*i.e.*, corroded well casings) in order to sustain production. In the long run, reversal of all these practices and utilization of the most modern techniques, combined with development of both discovered fields as well as new ones, could result in Iraq's oil output increasing by several million barrels per day.

Iraqi oil reserves vary widely in quality, with API gravities in the 22° to 35° range. Iraq's main export crudes come from the country's two largest active fields: Rumaila and Kirkuk. The southern Rumaila field, which extends a short distance into Kuwaiti territory, has around 663 wells and produces three streams: Basra Regular; Basra Medium (normally 30° API, 2.6 per cent sulfur); and Basra Heavy (normally 22°-24° API, 3.4 per cent sulfur). Basra Blend normally averages around 32° API, 1.95 per cent sulfur, but reportedly is heavier and more sour currently at around 29-30° API and 2 per cent + sulfur content.

The northern Kirkuk field, first discovered in 1927, forms the basis for northern Iraqi oil production. Kirkuk has around 337 wells and normally produces 35° API, 1.97 per cent sulfur crude, although the API gravity and sulfur content both reportedly deteriorated sharply in the months just preceding the war. Kirkuk's gravity, for instance, had declined to around 32-33° API, while sulfur content had risen above 2 per cent. Declining crude oil qualities–and an increased "water cut" as well–was likely the result of overpumping as Iraq attempted to sell as much oil as possible. An additional export crude, known as "Fao Blend," is heavier and more sour, with a 27° API and 2.9 per cent sulfur. Pre-war, Bai Hassan, Jambur, Khabbaz, Ajil (formerly "Saddam"), and Ain Zalah-Butmah-Safaia were the other oil fields in northern Iraq. An estimated 60 per cent of Northern Oil Company's (NOC) facilities in northern and central Iraq were damaged during the Gulf War.

Another major Iraqi oil field is the 11-billion barrel East Baghdad field, which came online in April 1989. Prior to the war, this centrally-located field currently produced around 50,000 bbl/d of heavy, 23° API oil as well as 30 million cubic feet per day (Mmcf/d) of associated natural gas.

Production

As of mid-August 2003, Iraqi oil output was fluctuating on a daily basis, but generally was averaging just under 1 million barrels per day (bbl/d). Some Iraqi oil–perhaps 200,000-300,000 bbl/d–was being reinjected into oil reservoirs in the North due to constraints on both domestic processing

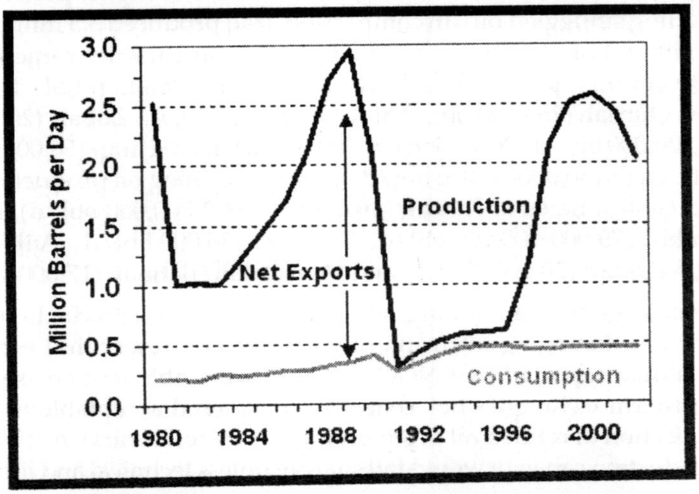

IRAQ'S OIL PRODUCTION AND CONSUMPTION, 1980-2000

Note: Production includes crude oil, condensate, natural gas liquids, ethanol and refinery gain.

ability as well as export outlets. According to the U.N. Joint Logistics Centre (JLC), as of mid-August 2003 "about 40 per cent of [northern Iraqi] production is transferred to the Baiji refinery, with the balance reinjected into the fields, ostensibly to maintain pressure. This is a most unusual practice but extraction of the surplus crude is necessary to produce much needed LPG. It means, however, that crude oil production is overstated by the volume reinjected (it not being available for refining or export, but counted as production). The reinjected crude may be lost forever."

On August 13, Iraq's main oil export pipeline from its main northern oilfield of Kirkuk to the Turkish port of Ceyhan reopened (see below for more details), but the line was shut down once again shortly thereafter due to sabotage on August 15 and 17. Iraq currently is aiming to increase its production to 2 million bbl/d by December 2003, and 2.8 million bbl/d by April 2004, but this goal may not be attained if problems continue in both the north and the south of the country.

Historically, Iraqi production peaked in December 1979 at 3.7 million bbl/d, and then in July 1990, just prior to its invasion of Kuwait, at 3.5 million bbl/d. From 1991, Iraqi oil output increased slowly, to 600,000 bbl/d in 1996. With Iraq's acceptance in late 1996 of U.N. Resolution 986, which allowed limited Iraqi oil exports in exchange for food and other supplies ("Oil-for-Food"), the country's oil output began increasing more rapidly, to 1.2 million bbl/d in 1997, 2.2 million bbl/d in 1998, and around 2.5 million bbl/d during 1999-2001.

During 2002, Iraqi oil production averaged 2.04 million bbl/d, down from about 2.45 million bbl/d in 2001 (and 2.69 million bbl/d in 2000), with large weekly and monthly fluctuations. Iraqi monthly oil output increased in the last few months of 2002 and in early 2003, peaking at around 2.58 million bbl/d in January 2003. Iraqi officials had hoped to increase the country's oil production capacity to 3.5 million bbl/d by the end of 2000, but did not accomplish this given technical problems with Iraqi oil fields, pipelines, and other oil infrastructure. Iraq also had hoped to expand production by 2010, to around 6 million bbl/d, through joint efforts by international oil companies (IOCs) and the Iraqi National Oil Company (INOC).

Prior to the latest war, oil industry experts generally assessed Iraq's sustainable production capacity at no higher than about 2.8-3.0 million bbl/d, with net export potential of around 2.3-2.5 million bbl/d (including smuggled oil). In comparison, Iraq produced 3.5 million bbl/d in July 1990. Approximately 2 million bbl/d of Iraq's production pre-war capacity came from oil fields in the southern part of the country, particularly North Rumaila (0.8 million bbl/d), South Rumaila (0.5 million bbl/d), West Qurnah (250,000 bbl/d at the end of 2002), Az Zubair (200,000-240,000 bbl/d), Misan/Buzurgan (100,000 bbl/d), Majnoon (50,000 bbl/d), Jabal Fauqi (50,000 bbl/d), Abu Ghurab (40,000 bbl/d), and Luhais (30,000-50,000 bbl/d). Iraq's remaining oil production capacity is located in the northern and central fields of Kirkuk (around 550,000-700,000 bbl/d), Bai Hassan (100,000-150,000 bbl/d), Jambur (75,000-100,000 bbl/d), Khabbaz (30,000 bbl/d), Ajil (formerly "Saddam," 25,000 bbl/d), East Baghdad (20,000 bbl/d), and 'Ayn Zalah/Batmah (17,000-20,000 bbl/d).

Among other challenges in maintaining, let alone increasing, oil production capacity, were Iraq's battle with "water cut" (damaging intrusion of water into oil reservoirs) especially in the south. Saybolt International had reported that NOC and SOC were able to increase their oil production through use of short-term techniques not generally considered acceptable in the oil industry (*i.e.*, "water flooding," injection of refined oil products into crude reservoirs). A U.N. report in June 2001 said that Iraqi oil production capacity would fall sharply unless technical and infrastructure problems were addressed.

Exports

Iraqi oil sales and exports currently are being handled by the State Oil Marketing Organization (SOMO), under CPA supervision. SOMO operations were seriously disrupted by war and turmoil during 2003, but the organization has now been reconstituted and has resumed many of its operations. On June 5, 2003, SOMO issued its first oil sales tender since the war started, for 8 million barrels of Kirkuk crude stored in tanks at Ceyhan and 2 million barrels stored at Basra. Dozens of companies placed bids for the oil, with winners including ChevronTexaco, Cepsa, ENI, Repsol, Total, and Tupras. Bids for the Kirkuk oil reportedly ranged around $2.70-$3.30 per barrel below dated Brent (f.o.b. Ceyhan). On June 22, a tanker arrived at Ceyhan to load the first oil since March 20, 2003, when the 600,000-barrel tanker "Caithness" completed loading one day after the outbreak of war. On July 3, SOMO issued its second spot tender, for 8 million barrels of Basra Light. At the time, SOMO stated that it was not yet ready to resume term sales, which generally require a certain level of stability, certainty, and predictability.

In late July 2003, however, SOMO signed its first term contracts since the war, for Basra Light oil from Iraq's southern fields. Exports of about 650,000 bbl/d were anticipated through the rest of 2003. Major purchasers included BP, ChevronTexaco, ConocoPhillips, ExxonMobil, Marathon Oil, Mitsubishi, Brazil's Petrobras, Repsol, Shell, Chinese trader Sinochem, and Vitol. However, with difficulties at the southern port of Mina al-Bakr, Iraq has only been averaging exports of 545,000 bbl/d or less since the restart of term contracts.

Pre-War State of Iraq's Oil Sector

In December 2002, the Council on Foreign Relations and the Baker Institute released a report on Iraq's oil sector. Among other things, the report concluded that: 1) Iraq's oil sector infrastructure is in bad shape at the moment, being held together by "band-aids," and with a production decline rate of 100,000 bbl/d per year; 2) increasing Iraqi oil production will require "massive repairs and reconstruction.costing several billions of dollars and taking months if not years;" 3) costs of repairing existing oil export installations alone would be around $5 billion, while restoring Iraqi oil production to pre-1990 levels would cost an additional $5 billion, plus $3 billion per year in annual operating costs; 4) outside funds and large-scale investment by international oil companies will be needed; 5) existing oil contracts will need to be clarified and resolved in order to rebuild Iraq's oil industry, with any "prolonged legal conflicts over contracts" possibly "delay[ing] the development of important fields in Iraq;" and 6) any "sudden or prolonged shut-down" of Iraq's oil industry could result in long-term reservoir damage; 7) Iraq's oil facilities could easily be damaged during any domestic unrest or military operations (in early February 2003, the Patriotic Union of Kurdistan claimed that Iraqi soldiers were mining oil wells in the north of the country in anticipation of war); and 8) given all this, a "bonanza" of oil is not expected in the near future.

According to the *Middle East Economic Survey (MEES)*, problems at Iraqi oil fields include: years of poor oil reservoir management; corrosion problems at various oil facilities; deterioration of water injection facilities; lack of spare parts, materials, equipment, etc.; damage to oil storage and pumping facilities; and more. *MEES* estimates that Iraq could reach production capacity of 4.2 million bbl/d within three years at a cost of $3.5 billion, and 4.5-6.0 million bbl/d within seven years.

Status of Oil Development Deals with Foreign Companies

Prior to the toppling of Iraq's Ba'athist regime, the country reportedly had signed several multi-billion dollar deals with foreign oil companies mainly from China, France, and Russia. Deutsche

Bank estimates $38 billion total on new fields–"greenfield" development–with potential production capacity of 4.7 million bbl/d if all the deals come to fruition (which Deutsche Bank believes is highly unlikely). The former Iraqi government reportedly had been growing increasingly frustrated at the failure of these companies actually to begin work on the ground, and was threatening to no longer sign deals unless firms agreed to do so without delay. Iraqi upstream oil contracts generally required that companies start work immediately, but U.N. sanctions dissuaded companies from doing so for the most part.

Now, following the toppling of Saddam Hussein's regime, the legal status of these agreements is up in the air, increasing the uncertainty level for companies interested in doing business with Iraq. Besides legal issues, companies are also looking for a relatively stable security situation, a functioning government, and other conditions to be in place before they move heavily into the country. In May 2003, Philip Carroll stated that contracts signed under the previous regime would be assessed to determine whether "they were made in the best interests of the Iraqi people."

Russia, which is owed billions of dollars by Iraq for past arms deliveries, has a strong interest in Iraqi oil development. This includes a $3.7 billion, 23-year deal to rehabilitate Iraqi oilfields, particularly the 11-15 billion barrel West Qurna field (located west of Basra near the Rumaila field). West Qurna is believed to have production potential of 800,000-1 million bbl/d. In a surprising and somewhat puzzling development, in mid-December 2002 the Iraqi Oil Ministry announced that it was severing its contract with the Lukoil consortium on West Qurna due to "fail[ure] to comply" with contract stipulations. Specifically, the Iraqis cited Lukoil's failure to invest a required $200 million over three years. Two other, smaller, stakes in West Qurna by Russian companies Zarubezhneft and Mashinoimport reportedly were left intact.

In addition, three exploration and production deals were signed between Iraq and Russian companies (Soyuzneftegaz, Stroytransgas-Oil, and Tatneft, to develop the 100,000-bbl/d Rafidain field, the Western Desert's Block 4, and the Western Desert's Block 9, respectively). Despite all this, Russia's Foreign Ministry said that it viewed the Iraqi decision on Lukoil and West Qurna "with regret." In mid-February 2003, following a month of talks between the two sides aimed at reversing Iraq's decision, the Iraqis announced that its decision to cancel the Lukoil deal was "finished and the contract has been scrapped." In May 2003, Lukoil said it would fight to keep the contract, and Russia's Deputy Foreign Minister said that Russia would seek compensation if contracts signed under the Saddam Hussein regime now were not honored.

In May 2003, another Russian company, Tatneft, set up a joint venture with Germany's MRH in order to win work in Iraq's oil sector. According to Tatneft's President, the company had been close to reaching a deal on exploring Block 9 in Iraq's Western Desert region prior to the war. In October 2001, a joint Russian-Belarus oil company, Slavneft, had signed a $52 million service contract with Iraq on the 2-billion-barrel, Suba-Luhais field in southern Iraq. Full development of Suba-Luhais could result in production of 100,000 bbl/d (35° API) at a cost of $300 million over three years.

In early April 2001, Russia's Zarubezhneft received U.N. approval to drill 45 wells in the Ajil (formerly"Saddam") field, plus Kirkuk and Bai Hassan, as part of an effort to reduce water incursion into the fields. The Ajil field contains 3 billion barrels of oil and 5 trillion cubic feet (Tcf) of associated gas. Iraq had been seeking foreign assistance for a second-phase Ajil development, which would raise oil production capacity to 50,000 bbl/d, as well as 300 Mmcf/d of gas.

The largest of Iraq's oilfields slated for post-sanctions development is Majnoon, discovered by Braspetro of Brazil in 1975, and containing reserves of 11-30 billion barrels of 28°-35° API oil. Majnoon is located 30 miles north of Basra on the Iranian border. In the 1990s, French company Elf Aquitaine (now merged with Total as TotalFinaElf) negotiated on a possible $4 billion deal with Iraq on development rights for Majnoon. In 1999, however, TotalFinaElf declined to sign a 23-year production sharing agreement (PSA) with Iraq on Majnoon. Following this, the field reportedly was brought onstream (under a "national effort" program begun in 1999) in May 2002 at 50,000 bbl/d. Future development on Majnoon ultimately could lead to production of 450,000 bbl/d within two years or so at an estimated (according to Deutsche Bank) cost of $4 billion. Eventually, Majnoon could produce significantly more oil than that, possibly as high as 3 million bbl/d.

In July 2001, angered by France's perceived support for the U.S. "smart sanctions" plan, Iraq announced that it would no longer give French companies priority in awarding oil contracts, and would reconsider existing contracts as well. Iraq also announced that it was inclined to favor Russia, which has been supporting Iraq at the U.N. Security Council, on awarding rights to Majnoon and another large southern oil field, Bin Umar. As of February 2003, Russian company Zarubezhneft reportedly was negotiating a contract to develop Bin Umar. The status of TotalFinaElf, which had previously expressed interest in the field, was not clear. In February 2003, TotalFinaElf's chief executive, Thierry Desmarest, said "We will fight in order to have the best chances for participating in the reconstruction of the country's oil industry."

In early June 2003, China's National Petroleum Company (CNPC) refuted a comment by Thamir Ghadban that CNPC's contract on the 90,000-bbl/d al-Ahdab development was now "void by mutual agreement." CNPC agreed in 1997 to spend $1.3 billion on Al-Ahdab, located in southern Iraq, but no progress was made while sanctions remained in place.

The 2.5-5 billion-barrel Halfaya project is the final large field development in southern Iraq. Prior to the war, several companies (BHP, CNPC, Agip) reportedly had shown interest in Halfaya, which ultimately could yield 200,000-300,000 bbl/d in output at a possible cost of $2 billion. Smaller fields with under 2 billion barrels in reserves also had received interest from foreign oil companies. These fields included Nasiriya (Eni, Repsol), Tuba (ONGC, Sonatrach, Pertamina), Ratawi (Shell, Petronas, CanOxy), Gharaf (Mashinoimport, Rosneftegasexport), Amara (PetroVietnam), Noor (Syria), and more.

In May 2003, Thamir Ghadban stated that three exploration agreements for blocks in Iraq's Western Desert were still valid. These included Indonesia's Pertamina on Block 3, Russia's Stroitransgas on Block 4, and Indian's Oil and Natural Gas Corp. for Block 8. In January 2003, Stroitransgas signed a $33.5 million contract for exploration on Block 4, and in July 2003, it indicated its interest in winning post-war business in Iraq.

Upstream Oil Sector: Current Status

During the war, approximately 7 Iraqi oil wells were set on fire, out of around 1,500 total wells. On April 14, 2003, the last of these fires were extinguished at the South Rumaila field. This was considered a significant accomplishment, given pre-war speculation that Iraq might set many of its oilfields ablaze as it did with Kuwaiti oilfields in 1991. However, in spite of the fact that little damage was done to Iraq's oil fields during the war itself, looting and sabotage after the war ended was highly destructive. On June 6, 2003, Phillip Carroll, Chairman of the Advisory Board to Iraq's Oil Ministry, indicated that there was an organized campaign of sabotage against Iraq's oil industry, and that "their techniques appear to be very professional and aim at causing harm to significant and important installations." On July 24, 2003, the CPA launched "Operation Power Crude" to protect key infrastructure (fuel pipelines, power facilities, etc.) against sabotage and looting.

On April 22, the first oil production since the start of the war began at the Rumaila field, with the restart of an important gas/oil separation plant (GOSP). Starting in mid-May 2003, the U.S. Army Corps of Engineers–which has the lead in restoring Iraq's oil output to pre-war levels–began a major effort to ramp up production in Iraq. At that time, Iraqi output was only about 230,000 bbl/d: 120,000 bbl/d from southern fields; and 110,000 bbl/d from northern ones. Plans at that time called for Iraq to reach over 1 million bbl/d in output by June 1. This proved overly optimistic. In actuality, Iraqi oil output did not hit 1 million bbl/d until late July

Iraq's southern oil industry was decimated in the 1990/1991 Gulf War, with production capacity falling to 75,000 bbl/d in mid-1991. That war resulted in destruction of gathering centers and compression/degassing stations at Rumaila, storage facilities, the 1.6-million bbl/d (nameplate capacity) Mina al-Bakr export terminal, and pumping stations along the 1.4-million bbl/d (pre-war capacity) Iraqi Strategic (North-South) Pipeline. Seven other sizable fields remain damaged or partially mothballed. These include Zubair, Luhais, Suba, Buzurgan, Abu Ghirab, and Fauqi. Generally speaking, oilfield development plans were put on hold following Iraq's invasion of Kuwait, with Iraqi efforts focused on maintaining production at existing fields.

Oil Export Pipelines

Under optimal conditions, and including routes through both Syria and Saudi Arabia that are now closed, Iraq's oil export infrastructure could handle throughput of more than 6 million bbl/d (2.8 via the Gulf, 1.65 via Saudi Arabia, 1.6 via Turkey, and perhaps 300,000 bbl/d or so via Jordan and Syria). However, Iraq's export facilities (pipelines, ports, pumping stations, etc.) were seriously disrupted by the Iran-Iraq War (1980-1988), the 1990/1991 Gulf War, the most recent war in March/ April 2003, and periodic looting and sabotage since then.

The 600-mile, Kirkuk-Ceyhan (Turkey) dual pipeline is Iraq's largest crude oil export line. One, 40-inch line has a fully-operational capacity of 1.1 million bbl/d, but reportedly could handle only around 900,000 bbl/d pre-war. The second, parallel, 46-inch line has an optimal capacity of 500,000 bbl/d and was designed to carry Basra Regular exports, but at last report was inoperable. Combined, the two parallel lines have an optimal capacity of 1.5-1.6 million bbl/d. On August 13, 2003, officials at the Turkish port of Ceyhan said today that Iraq had begun pumping fresh crude oil through the Kirkuk-Ceyhan pipeline for the first time since war broke out in late March 2003. However, the pipeline was operating far below capacity, at perhaps 300,000-400,000 bbl/d, with significant repairs still required. Also, the line was damaged by a bridge ("Al Fatah") that collapsed on it after being bombed by U.S. planes during the war. This will require major repairs, including the drilling of a new tunnel under the Tigris River and the laying of a new pipeline. In addition, the IT-1 pumping station on the Kirkuk-Ceyhan line was damaged by looters, but reportedly is operable manually. The IT-2 pumping station on the same line reportedly was looted and destroyed.

On August 16, 2003, two blasts on the Kirkuk-Ceyhan line once again shut down Iraqi oil flows to Turkey. Officials estimated that it would take 10 days to two weeks in order to repair the line, and also that the shutdown was costing Iraq $7 million per day in lost oil export revenues.

At least since 2001 until March 2003, Iraq and Syria were utilizing the 50-year-old Banias oil pipeline in violation of U.N. sanctions. The pipeline, from Iraq's northern Kirkuk oil fields to Syria's Mediterranean port of Banias (and Tripoli, Lebanon), reportedly was being used to transport as much as 200,000 bbl/d of Iraqi oil, mainly from southern Iraq, to Syrian refineries at Homs and Banias. The oil was sold at a significant price discount and freed up additional Syrian oil for export. Iraq and Syria also had talked of building a new, parallel pipeline as a replacement for the Banias line. In March

2003, flows on the pipeline were halted, although the US Defense Department denied that its forces had targeted the line.

During the Iran-Iraq War, Iraq also built a pipeline through Saudi Arabia (called IPSA) to the Red Sea port of Mu'ajiz, just north of Yanbu. IPSA has a design capacity of 1.65 million bbl/d, but was closed after Iraq invaded Kuwait in August 1990. In June 2001, Saudi Arabia expropriated the IPSA line, despite Iraqi protests. In June 2003, Thamir Ghadban said that he hoped Iraq would be able to use the IPSA line again.

In order to optimize export capabilities (*i.e.*, to allow oil shipments to the north or south), Iraq constructed a reversible, 1.4-million bbl/d "Strategic Pipeline" in 1975. This pipeline consists of two parallel 700,000-bbl/d lines. The North-South system allows for export of northern Kirkuk crude from the Persian Gulf and for southern Rumaila crudes to be shipped through Turkey. During the 1990/1991 Gulf War, the Strategic Pipeline was disabled after the K-3 pumping station at Haditha as well as four additional southern pumping stations were destroyed. In June 2003, the NOC estimated that it would take "a long time," possibly until the end of 2003, to repair the K-3 pumping station and resume operations on the Strategic Pipeline. The whole system also reportedly is in need of modernization.

In April 2003, there was some discussion of "reopening" the old oil pipeline from Mosul in northern Iraq to Haifa, Israel. The line, which was built in the 1930s, carried 100,000 bbl/d at its peak, but has been closed since Israel's establishment in 1948. Today, however the Mosul-Haifa pipeline is in extremely poor condition (the Iraqi section is completely rusted and the Jordanian section was sold as scrap metal several years ago), and reportedly would require hundreds of millions of dollars to repair/rebuild, even if this were politically feasible. Along those lines, Jordan strongly denied any interest in rebuilding this pipeline at the present time, stating that "the pipeline no longer exists in Jordanian territory."

Jordan and Iraq had agreed in 1998 to build a pipeline for the transport of Iraqi oil to Jordan's 100,000-bbl/d Zarqa refinery, and renewed this commitment in their most recent oil supply agreement. This would eliminate the necessity of transporting oil over 600 miles of highway from Haditha, Iraq, using a fleet of 1,500 tanker trucks, as was done for several years prior to 2003. Eventually, the line was seen as transporting as much as 300,000 bbl/d of Iraqi crude through Jordan, including oil for export. In December 2002, the Jordanian government was evaluating bids on the project from four competing contractors. However, it now seems unlikely that the project will move forward in the near future. With Iraqi oil supplies to Jordan halted since the war started, Jordan has been receiving oil from Kuwait, Saudi Arabia, and the UAE at a discounted price.

Oil Terminals

In the Persian Gulf, Iraq has three tanker terminals: at Mina al-Bakr, Khor al-Amaya, and Khor az-Zubair (which mainly handles dry goods and minimal oil volumes, plus natural gas liquids and liquefied petroleum gas). Mina al-Bakr is Iraq's largest oil terminal, with four 400,000-bbl/d capacity berths capable of handling very large crude carriers (VLCCs). Gulf War damage to Mina al-Bakr appears to have been repaired in large part and the terminal reportedly was handling around 1 million bbl/d in early 2003. A full return to Mina al-Bakr's nameplate capacity (1.6 million bbl/d) would require extensive infrastructure repairs. Mina al-Bakr also is constrained by a shortage of storage and oil processing facilities, most of which were destroyed in the Gulf War.

As of mid-August 2003, Mina al-Bakr was experiencing sporadic problems with power supplies (in part the result of wide-scale looting of copper power transmission cables), slowing the port's rate of tanker loading and throwing overall oil export targets into question. Reportedly, the power problems had slowed the time needed to load a VLCC from 2 days to 4-6 days. Mina al-Bakr's nameplate loading capacity is 85,000 barrels per hour (bph), but the loading rate now reportedly is only around 55,000 bph, about two-thirds of capacity.

Iraq's Khor al-Amaya terminal was heavily damaged during the Iran-Iraq War (and completely destroyed during Operation Desert Storm in 1991) and has been out of commission since then. As of March 2001, reports indicated that Iraq had largely completed repairing two berths at Khor al-Amaya, allowing for capacity of 600,000 bbl/d. Upon full completion of repairs, Iraq projects Khor al-Amaya's capacity will rise to 1.2 million bbl/d, and will help prevent delays at Mina al-Bakr while repairs are conducted there.

Refining

Iraq's refining capacity as of January 2003 was believed to be over 417,000 bbl/d, compared to a pre-Gulf War, nameplate capacity of 700,000 bbl/d. Before the latest war, it was believed that Iraq needed to refine 560,000 bbl/d in order to produce 400,000 bbl/d of needed products for domestic consumption. Currently, around 250,000 bbl/d of Iraqi heavy oils reportedly are being burned for power generation. In late April 2003, the Basra refinery restarted at 70,000 bbl/d, or half of its total capacity. As of mid-August, the 140,000-bbl/d plant was experiencing periodic stoppages due to electric power problems, reducing its output by around 40 per cent -50 per cent.

Overall, Iraq has 10 refineries and topping units. The largest are the 150,000-bbl/d Baiji North, 140,000-bbl/d (or higher) Basra, and 100,000-bbl/d Daura plants. During the Gulf War, both Baiji in northern Iraq as well as the refineries at Basra, Daura, and Nasiriyah were severely damaged. Prior to the war in March/April 2003, a lack of light-end products, low quality gasoline, and rising pollution levels because of a lack of water treatment facilities were some of the major problems faced by Iraq's refining sector. Following the war, significant investment will now be needed to perform refinery upgrades (Iraq had identified dozens of such projects prior to the war) and possibly to build a new $1 billion, 290,000-bbl/d "Central" refinery near Babylon.

At the present time, problems with Iraq's refineries are forcing the country to barter fuel oil for gasoline and liquid petroleum gas (LPG) from neighboring countries (mainly Kuwait, Jordan, and Turkey). In addition to supply shortfalls, distribution of LPG and other petroleum products, plus low "buffer stock levels," each remain major challenges that needs to be resolved. According to the JLC, "the supply of fuels in Iraq continue[d] to be poor," with an "LPG shortage in particular.affecting the most vulnerable elements of the population." In addition, as of mid-August 2003, the JLC was reporting that "gasoline was also in short supply.especially in the north, but at levels that the populace has become used to" and that "major shortages of LPG and diesel continue." However, the overall fuel supply situation in Iraq seems to be improving slowly if not steadily as of mid-August 2003. At the same time, the situation is being made somewhat worse by smuggling of refined products out of the country.

Natural Gas

Iraq contains 110 trillion cubic feet (Tcf) of proven natural gas reserves, along with roughly 150 Tcf in probable reserves. About 70 per cent of Iraq's natural gas reserves are associated (*i.e.*, natural gas produced in conjunction with oil), with the rest made up of non-associated gas (20 per cent) and dome gas (10 per cent). Until 1990, all of Iraq's natural gas production was from associated fields. In 2001, Iraq produced 97 billion cubic feet (Bcf) of natural gas, down drastically from peak output levels of 700

Bcf in 1979. Iraq plans to increase its natural gas output in order to reduce dependence on oil consumption and possibly for export at some point. Prior to the recent war, Iraq had even been developing plans to build a liquefied natural gas terminal.

Within two years after the lifting of U.N. sanctions, Iraq had hoped to produce 550 Bcf, and within a decade, Iraq had aimed to be producing about 4.2 Tcf of natural gas annually. Since most of Iraq's natural gas is associated with oil, progress on increasing the country's oil output will directly affect the gas sector as well. Associated gas often is simply flared off. Significant volumes of gas also are used for power generation and reinjection for enhanced oil recovery efforts.

Main sources of associated natural gas are the Kirkuk, Ain Zalah, Butma, and Bai Hassan oil fields in northern Iraq, as well as the North and South Rumaila and Zubair fields in the south. The Southern Area Gas Project was completed in 1985, but was not brought online until February 1990. It has nine gathering stations and a larger processing capacity of 1.5 billion cubic feet per day. Natural gas gathered from the North and South Rumaila and Zubair fields is carried via pipeline to a 575-Mmcf/d natural gas liquids (NGL) fractionation plant in Zubair and a 100-Mmcf/d processing plant in Basra. At Khor al-Zubair, a 17.5-million-cubic-foot LPG storage tank farm and loading terminals were added to the southern gas system in 1990. Natural gas also used to be pumped from Rumaila into northern Kuwait via a 40-inch, 105-mile pipeline. The gas was used to supply Kuwaiti power stations and LPG plants, but was halted following Iraq's invasion of Kuwait in August 1990.

Iraq's only non-associated natural gas production is from the al-Anfal field (200 Mmcf/d of output) in northern Iraq. Al-Anfal production, which began in May 1990, is piped to the Jambur gas processing station near the Kirkuk field, located 20 miles away. Al-Anfal's gas resources are estimated at 4.5 Tcf, of which 1.8 Tcf is proven. In December 2001, Russia's Gazprom reportedly was negotiating possible development of al-Anfal. In November 2001, a large non-associated natural gas field reportedly was discovered in the Akas region of western Iraq, near the border with Syria, and containing an estimated 2.1 Tcf of natural gas reserves. It is not clear whether the field is associated or non-associated.

Besides al-Anfal, Iraq has four large non-associated natural gas fields (Chemchamal, Jaria Pika, Khashm al Ahmar, Mansuriya) located in Kirkuk and Diyala provinces. In February 2000, Iraq's Oil Ministry named Agip and Gaz de France as leaders on a $2.3 billion PSA (production sharing agreement) project to develop these fields, which reportedly have total recoverable reserves of more than 10 Tcf.

Currently, Iraq has a major natural gas pipeline with the capacity to supply around 240 MMcf/d to Baghdad from the West Qurna field. The 48-inch line was commissioned in November 1988, with phases II and III of the project never completed due to war and sanctions. The last two phases of the pipeline project were meant to supply Turkey. Iraq's Northern Gas System, which came online in 1983, was damaged during the Gulf War as well as by the Kurdish rebellion of March 1991. The system supplied LPG to Baghdad and other Iraqi cities, as well as dry gas and sulphur to power stations and industrial plants. Iraq also has a Southern Gas System, which came online in 1985.

Electricity

As of late July 2003, indications were that Iraq had no more than perhaps 3,600 MW of power generating capacity, well below the amount needed to satisfy peak summer demand. Baghdad alone is estimated to require 2,400 MW of power during the summer's extreme heat for refrigeration and air conditioning, but was receiving perhaps half that amount. The Doura plant, which supplies the capital, was only running at 30 per cent capacity as of late July, while power lines between the Beiji facility, which also serves Baghdad, had been cut or looted. Overall, according to Paul Bremer, Iraq requires an extra 2,000 MW of generating capacity in order to meet demand, at a cost of perhaps $2 billion. In the meantime, the CPA has introduced a rationing system for the entire country, except for Basra, with three hours on and three hours off. Key facilities like hospitals, oil facilities, water and sewage plants were to receive power 24 hours a day under the plan.

Around 85 per cent -90 per cent of Iraq's national power grid (and 20 power stations) was damaged or destroyed in the 1990-1991 Gulf War. Existing generating capacity of 9,000 megawatts (MW) in December 1990 was reduced to only 340 MW by March 1991. In early 1991, transmission and distribution infrastructure also was destroyed, including the 10 substations serving Baghdad and about 30 per cent of the country's 400-kilovolt (kV) transmission network. In early 1992, Iraq stated that it had restarted 75 per cent of the national grid, including the 1,320-MW Baiji and Mosul thermal plants as well as the Saddam Dam. The U.N. Iraq Program estimated in November 2002 that Iraq's generating capacity was 4,300-4,400 MW. The U.N. Iraq Program further stated that, by the summer of 2004, Iraq's generating capacity could reach 5,900 MW, with several power stations (Al-Quds, Beji, Himreen,Yousfiya, Rumaila–all gas-fired) under construction and several others (Dibs, Hart, Najaf, Nassriya–gas and thermal) awaiting approval and/or funds.

Prior to the war, Iraq reportedly had signed contracts for renovating two generation units at the Harithah power plant, and another to rebuild the Yusufiyah plant, which stopped operating in 1990. Iraq's Electricity Authority reportedly also had signed several other contracts with Chinese, Swiss, French, and Russian companies, to build 3,000 MW of additional power generating capacity. In December 2000, it was reported that a Chinese company had completed work on the Abdullah power plant north of Baghdad. In October 2001, it was reported that Russia's Mosenergomontazh was working to modernize Iraq's Southern Heat and Power Plant in Najibia, Basra province. The project aims to add 200 MW of generating capacity to Iraq's grid. In August 2002, the Najaf governate in southern Iraq announced that two new power plants, with a combined capacity of 20 MW, had come online.

Country Overview

Independence: October 3, 1932

Population (2002E): 23 million

Location/Size: Middle East/168,709 square miles, slightly more than twice the size of Idaho.

Major Cities: Baghdad (capital), Basra, Mosul, Karbala, Kirkuk

Languages: Arabic, Kurdish

Ethnic Groups: Arab 75-80 per cent, Kurdish 15-20 per cent, Turkmen, Assyrian, or other 5 per cent

Religions: 97 per cent Muslim (Shi'a 60-65 per cent, Sunni 32-37 per cent), Christian or other (3 per cent)

Economic Overview

Currency: Iraqi Dinar (ID)

Unofficial Exchange Rate (7/03E): US$1=ID 1,500, compared to around US$1=ID3,000 just after the war

Gross Domestic Product (at market exchange rates) (2002E): $28.6 billion (down from $54 billion in 1988)

Gross Domestic Product (at purchasing power parity rates) (2002E): $15.5 billion (around one-third of 1989's economic output)

Real GDP Growth Rate (Global Insight: Base Case Scenario) (2002E): 5.5 per cent *(2003F):* -22.0 per cent *(2004F):* 37.0 per cent

Inflation Rate (Global Insight: Base Case Scenario) (consumer prices) (2002E): 24.6 per cent *(2003F):* 7.6 per cent *(2004F):* 6.9 per cent

Major Export Products (2002): Crude oil and oil products (regulated by the United Nations)

Major Import Products (2002): Food, medicine, consumer goods (regulated by the United Nations)

Merchandise Exports (2002E): $13.0 billion

Merchandise Imports (2002E): $7.8 billion

Merchandise Trade Balance (2002E): $5.2 billion

Current Account Balance (2002E): $2.3 billion

Oil Export Revenues (2002E): $12.3 billion (includes $3 billion or so in smuggling)

Oil Export Revenues/Total Export Revenues (2002E): 95 per cent or more

External Debt (2003E): estimates range from over $100 billion to more than $300 billion

Energy Overview

Proven Oil Reserves (1/1/03E): 112.5 billion barrels (around 75 billion barrels of which has not yet been developed; "probable" and "possible" reserves are as high as 220 billion barrels)

Oil Production (2002E): 2.04 million barrels per day (bbl/d), of which 2.02 million bbl/d was crude oil

Oil Production (January-May 2003E): 1.35 million barrels per day (bbl/d), ranging from a high of 2.58 million bbl/d in January to a low of 64,000 bbl/d in April; as of mid-August, production was believed to be above 1 million bbl/d)

Pre-war Oil Production Capacity, Maximum Sustainable. 2.8-3.0 million bbl/d (declining by about 100,000 bbl/d per year)

Current Oil Production Capacity, Maximum Sustainable (8/03E): 1.0 million bbl/d

Oil Export Routes: Kirkuk-Ceyhan pipeline; Mina al-Bakr port; to Jordan and Turkey via truck; reportedly to Syria via the Kirkuk-Banias pipeline; smuggling by boat along the Gulf coast

Oil Consumption (2002E): 460,000 barrels per day (bbl/d) *(8/03E):* 300,000-350,000 bbl/d

Net Oil Exports (2002E): 1.58 million bbl/d *(8/03E):* 650,000-700,000 bbl/d

U.S. Oil Imports from Iraq (2002E): 459,000 bbl/d (down from 795,000 bbl/d during 2001)

Crude Oil Refining Capacity (1/1/03E): 417,500 bbl/d (according to the *Oil and Gas Journal*)

Natural Gas Reserves (1/1/03E): 109.8 trillion cubic feet (Tcf)

Natural Gas Production/Consumption (2001E): 97 billion cubic feet (Bcf)

Electricity Generation Capacity (2002E): 4.3-4.4 gigawatts (90 per cent thermal) *(8/03E):* x.x gigawatts

Electricity Production (2001E): 36.0 billion kilowatthours

Environmental Overview

Total Energy Consumption (2001E): 1.08 quadrillion Btu* (0.3 per cent of world total energy consumption)

Energy-Related Carbon Emissions (2001E): 20.0 million metric tons of carbon (0.3 per cent of world total carbon emissions)

Per Capita Energy Consumption (2001E): 45.6 million Btu (vs U.S. value of 341.8 million Btu)

Per Capita Carbon Emissions (2001E): 0.85 metric tons of carbon (vs U.S. value of 5.5 metric tons of carbon)

Energy Intensity (2000E): 13,172 Btu/ $1995 (vs U.S. value of 11,014 Btu/ $1995)**

Carbon Intensity (2000E): 0.24 metric tons of carbon/thousand $1995 (vs U.S. value of 0.17 metric tons/thousand $1995)

Fuel Share of Energy Consumption (2001E): Oil (90 per cent), Natural Gas (9 per cent); Hydroelectric (<1 per cent)

Fuel Share of Carbon Emissions (2001E): Oil (90 per cent), Natural Gas (10 per cent)

Status in Climate Change Negotiations: Iraq is not a signatory to the United Nations Framework Convention on Climate Change or to the Kyoto Protocol.

Major Environmental Issues: Under Saddam Hussein, government water control projects drained most of the inhabited marsh areas east of An Nasiriyah by drying up or diverting the feeder streams and rivers. A once sizable population of "Marsh Arabs," who have inhabited these areas for thousands of years, were displaced, while the destruction of the natural habitat harmed the area's wildlife populations. Other problems include inadequate supplies of potable water, development of Tigris-Euphrates Rivers system contingent upon agreements with upstream riparian Turkey, air and water pollution, soil degradation (salination) and erosion, and desertification.

Major International Environmental Agreements: A party to the Law of the Sea and the Nuclear Test Ban.

The total energy consumption statistic includes petroleum, dry natural gas, coal, net hydro, nuclear, geothermal, solar, wind, wood and waste electric power.

Oil and Gas Industries

Major Oil Fields (proven reserves–billion barrels, 2002E): Majnoon (12-30), West Qurna (11.3-15.0), East Baghdad (11+), Kirkuk (10+), Rumaila (10+), Bin Umar (6+), Rattawi (3.1), Halfaya (2.5-4.6), Nassiriya (2-2.6), Suba-Luhais (2.2), Tuba (1.5), Khurmala (1.0), Gharaf (1.0-1.1), Rafidain (0.7), Amara (0.5)

Oil Refineries (crude refining capacity bbl/d, 2003E): Baiji (150,000), Basra (140,000), Daura (100,000), Khanakin (12,000), Haditha (7,000), Muftiah (4,500), Qayarah Mosul (2,000)

Major Ports: Mina al-Bakr, Khor al-Amaya, Khor al- Zubair, Umm Qasr

Major Pipelines (nameplate capacity): Kirkuk-Ceyhan (Dortyol) Pipeline–around 1.5-1.6 million bbl/d (currently partly operational); *Iraq-Saudi Arabia Pipeline* (IPSA1, 2)–possibly 1.65 million bbl/d (closed by Saudi Arabia in 1990); *Banias/Tripoli Pipeline*–possibly 0.3 million bbl/d (currently closed); *Iraq Strategic Pipeline*–less than 1.4 million bbl/d (reversible, internal transportation only).

45

Israel

Although Israel does not produce significant amounts of energy, it is located strategically for regional energy transit and also is an important variable in the Middle East security situation.

General Background

Israel currently is facing significant political and economic challenges, and its short-term economic outlook looks difficult at best. A combination of sluggish global economic conditions, a slump in the world high-technology sector (which is very important in Israel), lower foreign investment levels (only about $20 million in 2002) and Israeli consumer demand, the two-year Palestinian uprising, lack of progress towards peace, and sharply decreased tourism are all hurting the Israeli economy. Prime Minister Ariel Sharon, who was reelected in January 2003, has said that tackling Israel's economic problems will be the top priority of his new coalition government, sworn in by the parliament (Knesset) on February 27, 2003. Finance Minister Benjamin Netanyahu is focused on an emergency economic plan for the country, in coordination with Bank of Israel governor David Klein. Meanwhile, the United States has indicated that it may provide a significant increase in economic and military aid to Israel, partly in response to the war with Iraq. Israel has requested a $12 billion package of additional aid and loan guarantees from the United States.

Most importantly, the severe flareup in Palestinian-Israeli violence since September 2000 has cost Israel hundreds of lives and battered its economy, which had experienced a high-tech-driven boom and 6 per cent annual real economic growth for much of the 1990s. For 2003, the Israeli economy is expected to remain roughly flat, following declines of 0.9 per cent and 1.0 per cent in 2001 and 2002, respectively. Israel's economy also continues to be adversely affected by slowdowns in the global high-technology sector and in the world economy in general. Israel's unemployment rate has risen sharply since 1991, topping 10 per cent in the fourth quarter of 2002, and projected at 11.5 per cent for 2003.

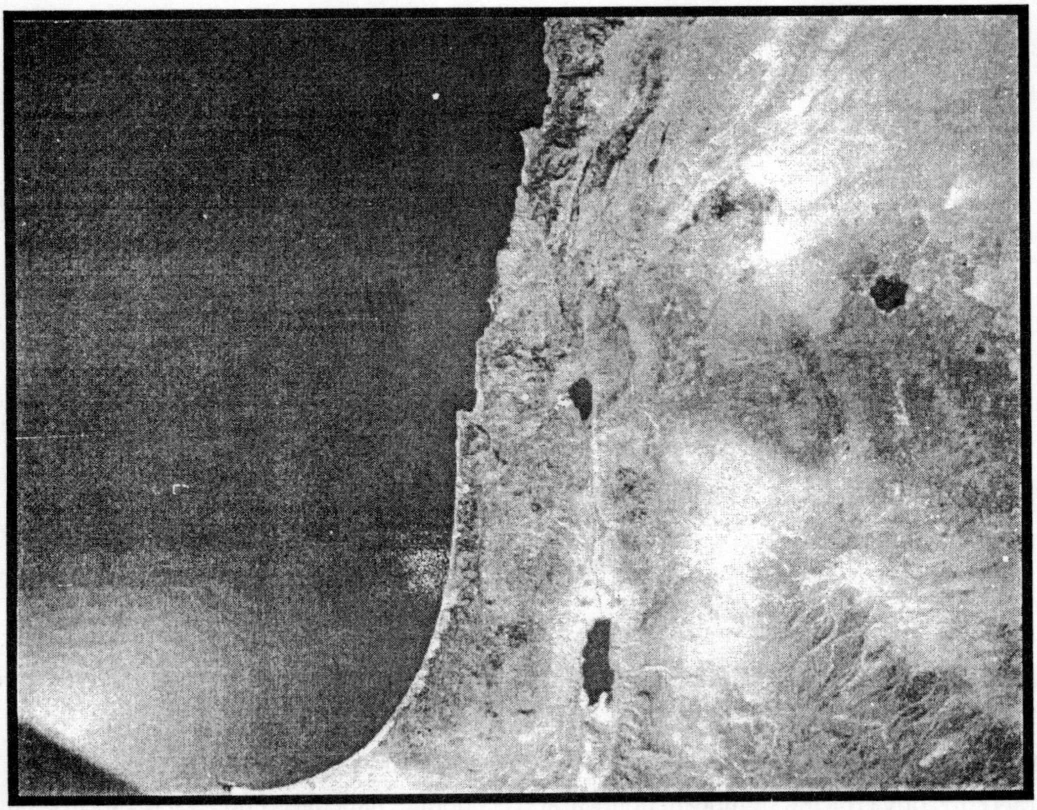

In response to a widening budget deficit (to around 4 per cent of gross domestic product–GDP–in 2002) caused by increased military spending, reduced tax receipts, and increased welfare spending due to the weak economy, Israel is operating under an austerity budget that has reduced subsidies and welfare payments and has raised taxes. Still, Israel's government is looking at even further spending cuts (possibly 11 billion shekels–around $2 billion) due to revenue shortfalls and a goal of reducing the budget deficit to 3.0 per cent of GDP in 2003. This budget austerity is not likely to help pull the economy out of its recession. On the other hand, the Bank of Israel cut its key lending rate in April by 0.2 per cent, to 8.7 per cent, despite concerns over higher inflation (5.7 per cent in 2002, up from 1.1 per cent in 2001).

Prior to the current round of Palestinian-Israeli violence, Israel had undertaken important structural reforms (such as privatization and reduced controls on foreign currency exchanges and profit remittances by foreign companies), although public spending still accounts for more than half of the country's GDP and the top marginal income tax rates exceed 60 per cent. Israel also had been making great progress in attracting tourism and foreign investment, but both of these are now down sharply –around two-thirds compared to 2000. Finally, Israel had been making halting progress towards privatizing government-owned companies, including banks, the state telecommunications company (Bezeq), and others. Given the ongoing Palestinian-Israeli turmoil, privatization and economic reform in general are now essentially on hold.

Energy

Until recently, with a significant offshore natural gas discovery, Israel has had essentially no commercial fossil fuel resources of its own, and has been forced to depend almost exclusively on imports to meet its energy needs. Israel has attempted to diversify its supply sources and to utilize alternatives like solar and wind energy. Traditionally, Israel has relied on expensive, long-term contracts with nations like Mexico (oil), Norway (oil), the United Kingdom (oil), Australia (coal), South Africa (coal), and Colombia (coal) for its energy supplies. Israel also has pursued other, cheaper sources of energy, like Egyptian gas, and hopes to expand natural gas significantly as a percentage of its energy mix in coming years.

Although the Israeli government in principle favors privatization of state-owned companies, the energy sector remains largely nationalized and state-regulated, ostensibly for national security reasons. In fact, little progress on energy sector privatization has been made since the late 1980s, when Paz Oil Company (the largest of three main oil-marketing companies in Israel) and Naphtha Israel Petroleum (an oil and gas exploration firm) were sold to private investors. Meanwhile, other energy companies such as the Oil Refineries Company, which operates Israel's two refineries (at Haifa and Ashdod), and the Oil Products Pipeline Company, which operates Israel's oil pipelines, remain state-owned, with no definite plans to privatize them in the near future (although the current Israeli government appears to favor such privatization, at least in principle). In early 1996, the Israel Electric Company's (IEC)'s monopoly was extended for another 10 years.

In February 2000, Israel and the United States signed an energy cooperation agreement. The agreement includes cooperation in the fields of gas, coal, solar power technology, and electric power generation. In addition, the two countries signed a letter of intent with Israel's Atomic Energy Commission to expand cooperation on nuclear non-proliferation and arms control issues. In May 2001, the Bush Administration announced that it would honor this agreement.

Oil

Israel produces almost no oil and imports nearly all its oil needs (around 237,000 barrels per day–bbl/d–in 2002). Traditionally, major oil import sources have included Egypt, the North Sea, West Africa, and Mexico. In recent years, however, Israel has stepped up its imports from Russia and the Caspian region (Kazakhstan, Turkmenistan, etc.) and now reportedly gets a majority of its oil from the former Soviet Union. In December 2002, Israel's Oil Refineries Ltd. (ORL) reportedly was negotiating with Mexico for annual supplies of around 3.7 million barrels (10,000 bbl/d). In late November 2002, ORL also signed a deal to purchase around 10,000 bbl/d from Angola at a cost of $100 million per year.

Although oil exploration in Israel has not proven successful in the past (current output is less than 1,000 bbl/d), drilling is being stepped up. Israel's Petroleum Commission has estimated that the country could contain 5 billion barrels of oil reserves, most likely located underneath gas reserves, and that offshore gas potentially could supply Israel's short-term energy needs. Geologically, Israel appears to be connected to the oil-rich Paleozoic petroleum system stretching from Saudi Arabia through Iraq to Syria.

Overall, around 420 oil wells have been drilled in Israel since the 1940s, with little success. In late September 2000, a contract was signed between U.S.-based Ness Energy International and Lapidoth Israel Oil Prospectors Corp. to commence further work on the Har Sedom 1 well. In 1994, Enserch Corp. of Dallas signed an agreement with two Israeli companies to examine a 1,500 square mile area on the Mediterranean coast. Isramco (a private company which absorbed the Israel National Oil Company when it was privatized in 1997), Delek, and Naphtha Israel Petroleum Corp. are partners in the Gevim 1 oil well being drilled near Sderot in the Negev desert. Isramco has stated that it is optimistic that the Gevim field will yield significant amounts of oil. Meanwhile, oil was discovered near the Dead Sea town of Arad in August 1996, and is currently flowing at the rate of about 600 barrels per day. In October 2001, Ness Energy announced that it was exploring for oil in the Masada Lease near the Dead Sea. Finally, Givat Olam Oil has drilled several wells–Meged 1-4–based on the company founder's reading of Deuteronomy 33. To date, no oil has been discovered by Givat Olam.

Israel has sizeable deposits of oil shale, perhaps 600 million tons recoverable, with average production of about 9,000 bbl/d. Most of Israel's shale oil resources are located in the Rotem basin region of the northern Negev desert near the Dead Sea. Oil shale is sedimentary rock containing organic material from which liquid fuel may be extracted, at a rate of perhaps 15-17 gallons of oil per ton of shale.

Downstream

Israel's oil sector remains highly centralized, but since 1988, the government has been moving to change this. Among other things, the process has ended the old cost-plus basis system, ended price controls for end users of petroleum products, and created more competitive conditions in general. Still, Israel has a ways to go in deregulating and privatizing the country's oil refining and distribution sectors, with ORL remaining a monopoly, for instance.

Israel has two major refineries, both run by ORL, located at Haifa (130,000 bbl/d) and Ashdod (90,000 bbl/d). The $1.3 billion, 100,000-bbl/d, Egyptian-Israeli joint venture MIDOR (Middle East Oil Refinery Ltd.) refinery in Alexandria, Egypt began operations in April 2001. The ultra-modern, environmentally-advanced facility includes a 25,000-bbl/d hydro cracker. In early June 2001, Israel's Merhav confirmed that it had sold its 20 per cent share in MIDOR to the National Bank of Egypt, making the refinery 100 per cent Egyptian owned. Merhav's sale effectively marked the end of what had been the largest Arab-Israeli joint venture to date. Although Merhav claimed that the sale was made solely for business reasons, most outside observers and analysts believed that political considerations were a major factor.

Although Israel itself produces almost no oil, a comprehensive settlement of the Arab-Israeli conflict could affect Middle East oil flows significantly. Israel's geographic location between the Arabian peninsula and the Mediterranean Sea offers the potential for an alternative oil export route for Persian Gulf oil to the West. At present, these oil exports must travel either by ship (through the Suez Canal or around the cape of Africa), by pipeline from Iraq to Turkey (design capacity 1.5-1.6 MMBD), or via the Sumed (Suez-Mediterranean) Pipeline (capacity 2.5 MMBD).

Utilization of the Trans-Arabian Pipeline (Tapline) could offer another potentially economic alternative. The Tapline was originally constructed in the 1940s with a capacity of 500,000 bbl/d, and intended as the main means of exporting Saudi oil to the West (via Jordan to the port of Haifa, then part of Palestine, now a major Israeli port city). The establishment of the state of Israel resulted in diversion of the Tapline's terminus from Haifa to Sidon, Lebanon (through Syria and Lebanon). Partly as a result of turmoil in Lebanon, and partly for economic reasons, oil exports via the Tapline were halted in 1975. In 1983, the Tapline's Lebanese section was closed altogether. Since then, the Tapline has been used exclusively to supply oil to Jordan, although Saudi Arabia terminated this arrangement to display displeasure with perceived Jordanian support for Iraq in the 1990/1 Gulf War. Despite these problems, the Tapline remains a potential export route for Persian Gulf oil exports to Europe and the United

States. At least one analysis indicates that the transportation at cost of exporting oil via the Tapline through Haifa to Europe would cost as much as 40 per cent less than shipping by tanker through the Suez Canal.

In April 2003, there was some discussion of "reopening" the old oil pipeline from Mosul in northern Iraq to Haifa. The line, which was built in the 1930s, carried 100,000 bbl/d at its peak, but has been closed since Israel's establishment in 1948. Today, however the Mosul-Haifa pipeline is in extremely poor condition (the Iraqi section is completely rusted and the Jordanian section was sold as scrap metal several years ago), and reportedly would require hundreds of millions of dollars to repair/ rebuild, even if this were politically feasible. Along those lines, Jordan has strongly denied any interest in rebuilding this pipeline at the present time, stating that "the pipeline no longer exists in Jordanian territory."

In the meantime, Israel has one main operational oil pipeline, known as the "Tipline," built in 1968 to ship Iranian oil from the Red Sea port of Eilat to Haifa (via the Mediterranean port of Ashkelon). As of March 2003, the Eilat-Ashkelon Pipeline Company (EAPC) reportedly was working to reverse flows on the 1.2-million-bbl/d line, so that oil would run from the Mediterranean to Eilat. Russia's Tyumen Oil Company reportedly was interested in the possibility of pumping Russian crude leaving Black Sea ports through the Israeli line to Eilat, where it could be loaded onto tankers for shipment to markets in Asia.

Natural Gas

Israel hopes to increase the share of natural gas in its fuel mix (especially for electricity generation, currently dominated by coal-fired plants) for energy security, economic, and environmental reasons, and has been looking at various options in recent years. One possibility is gas imports from Egypt's Nile Delta and offshore regions, either overland across the Sinai peninsula, or via underwater pipeline to the Israeli coast. Another strong possibility, which has arisen only in the past year or so, is using Israel's own, newly-discovered offshore gas resources.

As far as Egyptian gas is concerned, the East Mediterranean Gas (EMG) Company (a consortium of EGPC, Merhav of Israel, and Egyptian businessman Hussein Salem) has been set up to pursue this option. ENI, a major gas producer in Egypt, is nearly through construction on a $150-million gas pipeline from offshore fields north of Port Said, Egypt through the Sinai to El-Arish, near the border with the Gaza Strip, and only about 30 miles from Israel. Using Egyptian gas for power generation in the Palestinian Authority reportedly would cost 3.5 cents per kilowatt-hour, about half the price charged by the IEC. Currently, Gaza is almost totally dependent on the IEC for its electricity needs. Although Egypt and Israel announced a deal in early 2001 whereby Egypt would supply $3 billion worth of gas to Israel through 2012, in recent months the likelihood of Egyptian gas exports to Israel has decreased due to political factors, specifically the continuing Israeli-Palestinian conflict. There are also national security concerns in Israel over becoming too dependent on energy imports, especially from any one source. Egypt is now talking about exporting its gas directly to Jordan via 'Aqaba, although without Israel, the major potential regional market for Egyptian gas exports, the economics of this appear questionable.

Over the past three years, in an important development for a country which has never had significant domestic energy resources, several energy companies (Israel's Yam Thetis group, Isramco, BG, and U.S.-based Samedan) have discovered significant amounts of natural gas off the coast of Israel (and even more off the Gaza Strip). Initial estimates of 3-5 trillion cubic feet (Tcf) in proven reserves would be enough potentially to supply Israeli demand for years, even without natural gas imports, although this now seems optimistic. Israel's new offshore gas reserves belong mainly to two groups: 1) the Yam Thetis group (comprising the Avner Oil, Delek Drilling, and Noble Affiliates' Samedan subsidiary); and 2) a BG partnership with Isramco and others. In August 2000, Isramco/BG announced that it had discovered a large gas field 12 miles offshore at its Nir-1 well. The field reportedly contains gas reserves of 274 Bcf, and represents the third gas field discovered offshore Israel during 2000 (the largest two being Mary and Noa, with combined reserves of nearly 1.5 Tcf). In early September 2001, Isramco announced that BG was abandoning the Tommy, Orly, Shira and Aya concessions after analyzing geological and geophysical findings.

All else being equal, Israel appears to prefer utilizing domestic gas resources as opposed to imports from Egypt. Meanwhile, Israel's Yam Thetis gas consortium reportedly has been lobbying against Egyptian gas imports. In a related development, in June 2001, the Israeli Supreme Court ruled that Yam Thetis did not have a vested right to build a gas pipeline from its offshore reserves to IEC coastal power facilities, leaving open the option for possible competitors. In early May 2002, however, Noble Energy announced that it had obtained a license from the Israeli government to construct a pipeline from Mary-B to Ashdod. Field development and the pipeline are scheduled for completion by late 2003. Flow rates of 600 million cubic feet per day of natural gas are possible.

Several groups have been competing to supply gas to the IEC, including Yam Thetis and Isramco/ BG, plus EMG and local group Sdot Yam. In January 2001, IEC said that it would award EMG a significant share of its gas contracts, in large part because EMG's prices, at around $2.50 per 1,000 cubic feet, reportedly were 20 per cent -30 per cent below other offers. In May 2001, EMG reportedly was in an advanced stage of negotiations with IEC on supplying around 60 Bcf of gas per year for 10-15 years. However, as of June 2001, gas supplies from EMG were in doubt due to reports that Egypt was considering backing out of any gas deal with Israel. In March 2001, IEC announced that remaining gas supplies would come from Yam Thetis. In February 2001, Egypt's oil minister denied that any agreement had been reached on Egyptian gas sales to Israel. In early May 2002, IEC announced that it could begin taking delivery of natural gas from Yam Thetis as early as the second half of 2003. The gas would be delivered to IEC's Ashdod power station, which is being converted from fuel oil to natural gas. IEC also agreed to purchase half of its natural gas requirements from the Yam Thetis consortium, at a cost of $1.5-$2.5 billion over 11 years.

The other half remains uncertain, particularly since Belgium's Tractebel indicated in April 2002 that it was withdrawing from a $400 million project to construct a natural gas distribution grid in Israel, ostensibly due to security concerns. Tractebel, which had held a 60 per cent stake in the project consortium, had been the only bidder on the project, which was awarded in December 2001, and its expertise was considered crucial to the project. Following Tractebel's withdrawal, Israel's National Infrastructure requested that Israeli companies Paz Oil and Africa-Israel Investments, each of which held a 20 per cent stake in the Tractebel consortium, come up with a replacement for Tractebel by mid-May. In July 2002, BG decided not to join, dealing a major blow to the project, and triggering cancellation of the tender. Israel's gas law requires that an experienced foreign company hold at least a 10 per cent stake in the project. In August 2002, an Israeli consortium of Paz Oil, Africa-Israel Investments, and Batemen Engineering proposed adding a new foreign partner, Itera, which is affiliated with Russia's Gazprom. In September 2002, in yet another twist to the saga, IEC was authorized to build the entire grid, but in April 2003, Israel's cabinet decided to reverse course and prevent IEC from building the grid. The cabinet reportedly was concerned that IEC should not be a monopoly in both electricity and in natural gas.

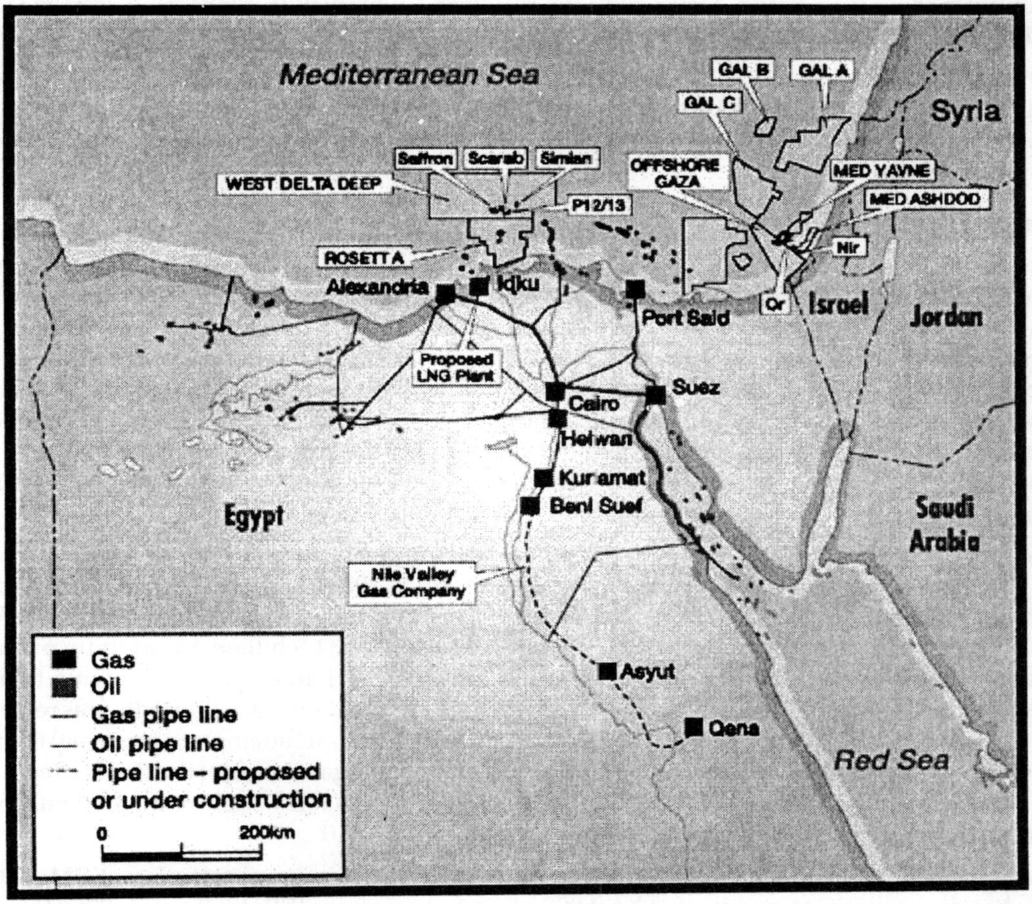

Meanwhile, gas has been discovered not only on Israel's side of the border, but also in areas that appear to lie in Palestinian territorial waters off the Gaza Strip. BG, which first struck gas in this area with its Gaza Marine-1 well in August 1999, has signed a 25-year contract to explore for gas and set up a gas network in the Palestinian Authority. In December 2000, BG successfully completed drilling a second gas well offshore Gaza. The drilling confirmed findings from the Marine-1 well, which had flowed at 37 million cubic feet per day, indicating possible reserves of around 1.4 Tcf. BG reportedly plans to invest $400 million in its offshore Gaza gas finds, which could be used to supply Israel, along with other sources.

Coal

Israel meets approximately 32 per cent of its energy demand requirements from coal (primarily for electric power generation). The National Coal Supply Corporation (NCSC) is the majority government-owned (74 per cent) company, established in 1981, solely responsible for securing the country's coal imports. In March 2001, the Israeli government approved the sale of NCSC to the IEC, which already owns 26 per cent of the company. Israel's coal supplies are all imported. In 2002, about 45 per cent of these imports came from South Africa, with the rest from Colombia (25 per cent), Australia (13 per cent), Indonesia (12 per cent), and others like Poland and China (5 per cent).

Overall, Israel is expected to import around 12.5 million short tons (Mmst) of coal in 2003. Growth in coal demand (and imports) is being driven mainly by rapid growth in electricity demand. A new coal terminal opened at Ashkelon in 2000 to handle coal imports for Israel's two coal-fired power plants located there.

Electricity

According to the IEC (Israel's monopoly national utility), Israel had about 9.1 gigawatts (GW) of installed electric generating capacity (at 20 power stations, including 7 major thermal plants) as of 2001, with nearly 70 per cent accounted for by coal-fired plants, 25 per cent by fuel oil-fired units, and the remainder by gasoil and independent power producers (IPPs). Israel also is a world leader in solar technology and relies heavily on solar energy for water heating (around 80 per cent of Israeli homes have solar water heaters). The 1,645-mile, IEC transmission grid is a closed loop system connecting power stations to major load centers throughout Israel and to the Palestinian Authority. The system includes EHV-400 KV transmission and 161 KV sub-transmissions systems, and serves 2.1 million customers, of which 1.8 million are residential.

Prior to the country's recent economic difficulties, electric power consumption had been increasing rapidly (around 4 per cent -5 per cent annually; 13 per cent in 2000), and the IEC had estimated that this growing power demand would require an increase in production capacity to nearly 15 GW by 2010. To meet this increased demand, IEC is aiming to raise \$1.2-\$1.3 billion a year in financing for generation, transmission, and distribution systems. However, in January 2003, following financial difficulties which resulted in a downgrade of the company's credit rating, IEC reportedly was reviewing its development strategy. More rapid privatization is one option, as Israel attempts to gradually liberalize its power sector. In April 2003, Israel's government approved a reorganization of the IEC in preparation for its privatization over the next few years. The current goal is that no single company will control more than 50 per cent of Israeli production or transmission by 2010.

As part of an effort to increase privatization of the country's power sector, Israel's Ministry of Energy has directed IEC to purchase at least 900 MW of power from IPPs by the year 2005 (of which possibly 150 MW are expected to come from solar and wind facilities, with the rest mainly natural gas-fueled). Israel's goal is for 10 per cent, or possibly as high as 20 per cent, of all electricity to be produced by IPPs. In June 1997, IEC announced the first tender for a large-scale private power plant in Israel–a 375-MW, dual-fired, combined-cycle plant to be built at Ramat Hovav (by a consortium of PSEG Global and the OPC energy company) in the Negev Desert. In February 2003, OPC reportedly decided to drop out due to difficulties in financing the \$250 million project. A second IPP in which OPC is involved, a 370-MW plant at Mishor Rotem, reportedly has not been affected. Also, in February 2003, EAPC and a group of private investors announced plans to build a \$450 million, 450-MW private power plant in Ashkelon, and to sell power at a lower price than the IEC.

The IEC is converting its oil and diesel-fired generators to natural gas, and hopes to generate 40 per cent of its electricity from gas by 2006. Natural gas would serve at least three goals: increased diversity in energy sources; benefits to the environment; and reductions in IEC's electric generation costs. In December 2002, IEC announced plans to acquire three new gas generating turbines, at a cost of \$230 million, from Siemens AG of Germany and Alstom of France. The turbines will boost IEC's generating capacity by 650 MW.

Israel's fourth coal plant, the Rutenberg facility at the Mediterranean Sea port of Ashkelon, was inaugurated on June 29, 2000. Rutenberg is the first power plant in Israel to have sophisticated anti-pollution scrubbers (Israel intends to install scrubbers at its two coal-fired plants in Hadera as well). In total, Rutenberg is to have four 550-MW generators, for a total production capacity of 2,200 MW. Meanwhile, a new coal dock at Ashkelon began operating in late 2000, saving IEC around $40 million a year in coal transportation costs. Previously, coal was shipped to another port, Ashdod, unloaded, and sent by rail to Ashkelon.

In December 2002, the Israeli government granted final approval for construction of the country's fifth coal-fired plant at Ashkelon. The 1,200-MW plant, which was originally approved in 2001, could enter service in 2009 (following a recent two-year postponement) at a cost of $1.3 billion. When finished, the plant will consume around 3 million tons of coal per year.

Besides fossil fuels, Israel also has looked to expand the utilization of other indigenous options, such as solar power. In early 2002, the IEC approved construction of Israel's first solar power station, a 100-MW plant to be built in the Dimona area of the Negev Desert for $250 million. In September 2002 the IEC issued a report which concluded that solar power had become uneconomical, but in January 2003, reports indicated that Israel would still move ahead with the plant.

At the present time, Israel has no nuclear power plants, although the country operates a reactor at Dimona, in the Negev Desert 25 miles west of the Jordanian border, as well as a smaller research facility at Nahal Sorek south of Tel Aviv. In December 2002, Israel's Infrastructure Ministry announced that it was proceeding with plans to study construction of a 1,200-MW nuclear plant at Shivta, in the Negev Desert near the border with Egypt. The Ministry has set 2020 as a target date for the plant.

One area of potential regional cooperation involves integration of individual national power transmission grids into a regional power network. Such a network would, among other benefits, allow power companies to take advantage of differences in peak demand periods, reduce the need for (and the costs associated with) installation and maintenance of reserve generating capacity, and provide outlets for surplus generating capacity (mainly from Israel to Jordan). Israel and Jordan held talks in October 1999 regarding possible cooperation on a shale-oil-fired plant as stipulated in the two countries' peace treaty. The two countries also have talked about linking their power grids and have discussed several proposed joint power stations, including a $1 billion, 1,000-MW plant to be located on the two countries' border, a 100-MW wind farm, a 150-MW solar thermal plant in the southern Arava desert near Eilat, and an 800-MW plant in Jordan that would supply power to Israel. In addition, IEC has developed plans for potential joint wind power development with Syria in the Golan Heights region should a peace treaty be signed. IEC estimates that up to 10 per cent of future electric supplies could come from outside the country.

IEC plans to spend about $1 billion over the next ten years to help reduce emissions from its power plants. New coal plants are to be equipped with flue gas desulphurization and combustion systems, and most of IEC's existing gas turbines have been retrofitted with low nitrogen combustion systems. Most of the coal ash waste produced by IEC's three coal-fired power plants is sold to the cement industry.

Country Overview

Independence: May 14, 1948 (from League of Nations)

Population (7/02E): 6.0 million

Location/Size: Eastern Mediterranean/8,019 sq. miles, slightly smaller than New Jersey

Major Cities: Jerusalem (capital), Tel Aviv, Haifa

Languages: Hebrew (official), Arabic, English

Ethnic Groups (1996E): Jewish (80 per cent), Arab and other (20 per cent)

Religions (1996E): Jewish (80 per cent), Muslim (15 per cent), Christian (2 per cent), Druze and other (3 per cent)

Defense (1998E): 175,000 regular forces (army: 134,000, navy: 9,000 air force: 32,000); 430,000 reserves

Economic Overview

Currency: New Israeli Shekel (NIS)

Market Exchange Rate (4/22/03): US$1 = NIS 4.60

Gross Domestic Product (GDP) (2002E): $102.7 billion

Real GDP Growth Rate (2001E): -0.9 per cent *(2002E):* -1.0 per cent *(2003F):* 0.4 per cent

Per Capita GDP (2002E): $15,622

Inflation Rate (consumer prices, 2001E): 1.1 per cent *(2002E):* 5.7 per cent *(2003F):* 4.4 per cent

Major Trading Partners: USA, European Community

Merchandise Exports (2002E): $27.4 billion

Merchandise Imports (2002E): $31.2 billion

Major Export Products: Machinery and equipment, cut diamonds, chemicals, textiles and apparel, agricultural products

Major Import Products: Military equipment, investment goods, rough diamonds, oil, consumer goods

Current Account Balance (2001E): -$2.3 billion *(2002E):* -$2.1 billion

Number of Tourists (2000E): 2.65 million *(2001E):* 1.22 million *(2002E):* 0.86 million

Unemployment Rate (2001E): 9.4 per cent *(2002E):* 10.3 per cent *(2003F):* 11.5 per cent

Total External Debt (2002E): $66.2 billion

Direct Investment Inflow (2002E): $1.5 billion (down from $4.5 billion in 2000)

Foreign Exchange Reserves (excluding gold) (3/03E): $23.5 billion

Energy Overview

Proven Oil Reserves (1/1/03E): 3.8 million barrels

Oil Production (2002E): 200 barrels per day (bbl/d)

Oil Consumption (2002E): 273,000 bbl/d

Net Oil Imports (2002E): 273,000 bbl/d

Crude Oil Refining Capacity (1/1/03E): 220,000 bbl/d

Coal Consumption (2001E): 10.9 million short tons (all of which is imported)

Natural Gas Reserves (1/1/03E): 1,375 billion cubic feet (Bcf) (Note: This figure includes the major recent offshore gas find)

Natural Gas Consumption/Production (2001E): 0.35 Bcf

Electric Generation Capacity (2000E): 9.1 gigawatts (70 per cent coal-fired; 25 per cent fuel oil; 5 per cent gasoil and IPPs)

Electricity Generation (2001E): 42.2 billion kilowatthours

Environmental Overview

Total Energy Consumption (2001E): 0.8 quadrillion Btu* (0.2 per cent of world total energy consumption)

Energy-Related Carbon Emissions (2001E): 16.3 million metric tons of carbon (0.2 per cent of world total carbon emissions)

Per Capita Energy Consumption (2001E): 122.8 million Btu (vs U.S. value of 341.8 million Btu)

Per Capita Carbon Emissions (2001E): 2.5 metric tons of carbon (vs U.S. value of 5.5 metric tons of carbon)

Energy Intensity (2001E): 7,381 Btu/ $1995 (vs U.S. value of 10,736 Btu/ $1995)**

Carbon Intensity (2001E): 0.15 metric tons of carbon/thousand $1995 (vs U.S. value of 0.17 metric tons/thousand $1995)

Fuel Share of Energy Consumption (2001E): Oil (68 per cent), Coal(32 per cent)

Fuel Share of Carbon Emissions (2001E): Oil (60.6 per cent), Coal (39.4 per cent)

Number of People per Motor Vehicle (1998): 3.8 (vs U.S. value of 1.3)

Status in Climate Change Negotiations: Non-Annex I country under the United Nations Framework Convention on Climate Change (ratified June 4th, 1996). Signatory to the Kyoto Protocol (signed December 16th, 1998- not yet ratified).

Major Environmental Issues: Limited arable land and natural fresh water resources pose serious constraints; desertification; air pollution from industrial and vehicle emissions; groundwater pollution from industrial and domestic waste, chemical fertilizers, and pesticides.

Major International Environmental Agreements: A party to Conventions on Biodiversity, Climate Change, Desertification, Endangered Species, Hazardous Wastes, Nuclear Test Ban, Ozone Layer Protection, Ship Pollution and Wetlands. Has signed, but not ratified, Marine Life Conservation.

The total energy consumption statistic includes petroleum, dry natural gas, coal, net hydro, nuclear, geothermal, solar, wind, wood and waste electric power. The renewable energy consumption statistic is based on International Energy Agency (IEA) data and includes hydropower, solar, wind, tide, geothermal, solid biomass and animal products, biomass gas and liquids, industrial and municipal wastes. Sectoral shares of energy consumption and carbon emissions are also based on IEA data.

Energy Industries

Organization: Isramco–private company, responsible for exploration and production; *Oil Refineries Limited*–Part privatized, runs Israel's 2 refineries at Haifa and Ashdod; *Paz Oil, Delek, and Sonol*–Israel's three largest oil retailers; *National Coal Supply Corporation*–government-owned company responsible for Israel's coal supply; *Israel Electric Corporation Ltd.*–state company responsible for Israel's electric power supply.

Major Ports: Ashdod, Haifa

Major Oil and Gas Fields: N.A.

Major Pipelines: Tipline–800,000 bbl/d (Eilat-Ashkelon-Haifa); Tapline–closed (Ras Tanura–Haifa)

Major Refineries (crude refining capacity): Haifa (130,000 bbl/d); Ashdod (90,000 bbl/d)

46

Italy

Italy is almost entirely dependent on imports to meet its energy needs. The country's heavy reliance on foreign oil and gas sources, such as Libya and Algeria, has made energy security and diversification of energy sources top concerns.

General Background

Italy is one of the world's largest economies, a founding member of both the European Union (EU) and the North Atlantic Treaty Organization (NATO), and belongs to Group of Eight (G-8) industrialized nations. It adopted the European single currency, the euro, on January 1, 1999.

Italy's 59th government since 1945 came to power in June 2001, headed by Prime Minister Silvio Berlusconi, leader of the Forza Italia party. The center-right government is a coalition of five parties called "House of Freedoms" which achieved majorities in both houses of Parliament, improving the new government's chances to implement reforms. Prior to Berlusconi's government, Italy had achieved a major economic policy objective: the reduction of its budget deficit to under 3 per cent of gross domestic product (GDP) in order to comply with the EU's Stability and Growth Pact, a cornerstone of the EU's monetary policy.

The current government will face some difficulties in maintaining this budget due to slow economic progress. The economic slowdown has meant that tax revenue has decreased, making achievement of reducing the government's budget deficit to 1.5 per cent of GDP in 2003 more difficult. Furthermore, this slowdown has complicated the realization of the government's 2001 election promises–tax cuts, privatization, deregulation of the labor market, and reform of the pension system. However, the 2003 budget plans to lower income taxes for employees earning below 25,000 euro a year, which addresses one of Prime Minister Berlusconi's campaign promises. Meanwhile, a new report issued from the European Commission characterized Italy as among the slowest EU members to make progress in reforming its economy. The Italian economy was weak throughout 2002, with an estimated real GDP growth of only 0.4 per cent for the year. Real GDP is forecast to grow 1.4 per cent in 2003.

The Italian energy sector has been undergoing considerable restructuring in recent years. EU Directives on electricity (96/92/EC) and on natural gas (98/30/EC) have established common rules for creation of internal markets and required privatization of Italy's dominant energy monopolies. In November 2002, European ministers set deadlines for complete energy liberalization across the EU. All business users will be free to choose energy suppliers by July 2004, and this will be extended to all domestic users in 2007.

In general, the Italian government has been working to liberalize its energy sector since the beginning of the 1990s. In 1995, the government created an independent body, the Electricity and Gas Authority, to regulate these areas. The government also issued the Bersani Decree on April 1, 1999, which seeks to accelerate the process of liberalizing Italy's electricity sector in accordance with the criteria laid down by EU Directive 96/92/EC. The Decree stipulated that, beginning on January 1, 2003, no single entity is authorized to produce or import, directly or indirectly, more than 50 per cent of the total electricity produced or imported in Italy. This Decree has required Enel, the state-owned electricity company, to divest 15,000 MW of its generation capacity by the 2003 deadline. In addition, Eni, the partially state-held oil and gas conglomerate, along with its main subsidiaries, Agip (hydrocarbons exploration and production) and Snam (gas supplies and distribution) all have been partially privatized. The Italian government sold off shares of Eni between 1995 and 1998, and now holds 35 per cent of the company. Liberalization of the Italian energy markets has resulted in increased competition over customers between the traditionally state-owned companies and other key players.

With limited domestic energy sources, Italy is highly dependent on energy imports. Historically, the country has relied heavily on imported oil, much of it from North Africa (mainly Libya). After declining for a number of years, oil consumption has remained static recently (although Italy remains the third largest oil consumer in Western Europe). Italy has, however, increased its consumption of natural gas. Natural gas is a much cleaner fossil fuel that helps Italy to meet domestic, European, and broader international requirements for a cleaner environment. North Africa and Russia are large exporters of natural gas to Italy. There have been concerns that this reliance on North African sources could have potentially negative implications for Italian energy security.

Oil

Italy holds 621 million barrels in proven oil reserves. In 2002, the country produced an estimated 149,000 barrels per day (bbl/d) of oil, a 2.7 per cent increase over 2001, while consumption remained the same at about 1.87 million bbl/d. Italy is more than 90 per cent reliant on imports and is Europe's third largest oil importer. Libya is Italy's main source of oil imports with other major import sources (in order of magnitude) including Iran, Saudi Arabia, and Algeria. About 50 per cent of Italy's gross oil imports are from the Middle East and North Africa. It is estimated that in 2000 oil's share of Italy's energy consumption fell to just under 50 per cent for the first time in over 20 years.

Italy is in the process of decreasing its reliance on oil, especially for heating and electricity generation. Heating oil consumption in 2000 was about one-third of that of 1981 and fuel oil consumption has fallen 38 per cent since 1995. Natural gas consumption is expected to rise as oil consumption falls in coming years.

Exploration and Production

Italy's oilfields are located in the north of the country, onshore and offshore along the Adriatic sea coast and Sicily. Production at two large Eni-operated fields, Villafortuna and Aquila, has declined in recent years. In order to offset dependence on foreign sources, Italy is in the process of increasing domestic oil production. Eni is the operator in a joint venture with Enterprise Oil to develop 600 million barrels of oil equivalent (including oil and associated natural gas) at Val d'Agri, in the southern Apennine region, considered to be Western Europe's most promising onshore development area. In

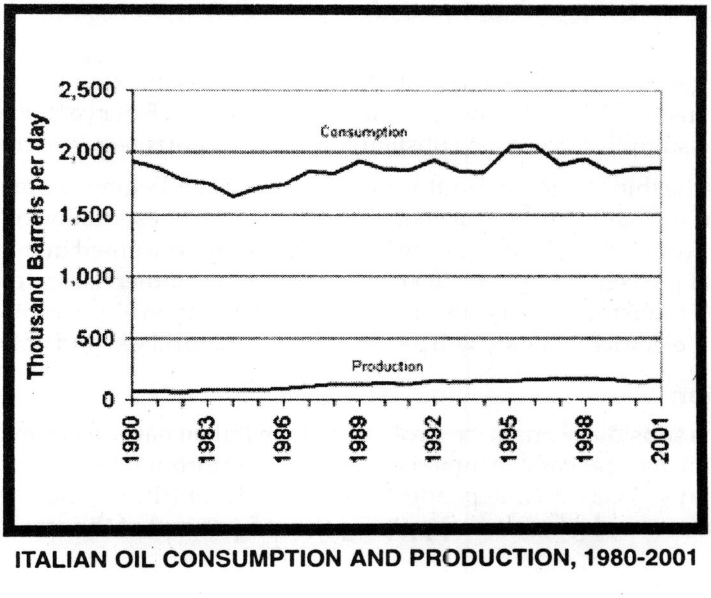

ITALIAN OIL CONSUMPTION AND PRODUCTION, 1980-2001

December 2001, Eni completed a 85-mile (136-kilometer), 150,000-bbl/d capacity pipeline connecting the fields to the Taranto refinery, which has allowed daily production in Val d'Agri to reach 45,000 bbl/d. Eni aims to have the fields producing 104,000 bbl/d by late 2003 and eventually 120,000 bbl/d.

In the Tempa Rossa field, neighboring the Val d'Agri fields, TotalFinaElf, which assumed operational ownership after purchasing Eni's 25 per cent interest, is developing over 400 million barrels of oil equivalent in a joint venture with Enterprise Oil (25 per cent) and ExxonMobil (25 per cent). Tempa Rossa has much heavier crude than Val d'Agri, and TotalFinaElf has finished 5 of the 7 planned wells at Tempa Rossa (as opposed to 42 at Val d'Agri). Tempa Rossa will reach a production plateau of 50,000 barrels of oil per day. The oil will be transported by pipeline to the Taranto refinery.

Downstream

To ensure access to foreign oil, the Italian government has promoted Italy as a refining center since the 1970s. There are large oil refining facilities along the Mediterranean coast and on Mediterranean islands, capable of processing a wide range of crude oils from North Africa and the Persian Gulf. As a result, Italy now has Europe's largest surplus refining capacity. However, refinery throughput was only 76 per cent of capacity in 2000, and Italy is importing almost as much refined product as it exports. Eni operates six of the 17 major refineries in Italy. While underutilized, the Italian refineries remain attractive because of their ability to handle large oil tankers and process many different fuel types, and also because of their catalytic and hydro cracking capabilities. ExxonMobil has a facility at Augusta on the Sicilian coast.

Italy has more gasoline stations than any other country in Europe. The government maintains that reducing the number of stations will help reduce the cost of gasoline. In June 2000, in further efforts to address steadily increasing gasoline costs that went above $4 per gallon in the summer of 2000, Italy's antitrust commission found seven oil companies guilty of price-fixing at the retail level. The companies were fined about $300 million, and in November 2000 the courts upheld the ruling on appeal but reduced the fine to about $200 million. In 2001, Eni sold 261 gasoline stations to Tamoil and 257 stations to other retailers as part of Eni's plan to concentrate more on upstream activities.

Natural Gas

Italy has natural gas reserves of 8 trillion cubic feet (Tcf). In 2001, the country produced 550 billion cubic feet (Bcf) and consumed 2.5 Tcf, relying on imports for nearly 75 per cent of total consumption. Italy's natural gas consumption has risen substantially since imports began in the early 1970s.

Italy currently is the third largest natural gas market in Europe, behind Germany and the United Kingdom. Natural gas use has increased substantially in recent years, especially for power generation. In 2001, natural gas represented about 34 per cent of total energy consumed in Italy, and this share is expected to grow to 44 per cent by 2010. In 2000, residential and commercial sectors accounted for 35 per cent of the market, industrial 30 per cent, and electricity generation 32 per cent. It is estimated that by 2010, 48 per cent of electricity in Italy will be generated by natural gas and only 7 per cent by oil.

Sector Organization

Eni along with its subsidiary Snam, control most of the Italian natural gas market. Eni supplied about 87 per cent of the natural gas consumed by Italy in 2000 through its domestic production and contracted imports. Edison Gas, an independent power producer (IPP), is the only other significant company in the Italian natural gas sector, supplying about 5 per cent of the market. Edison's shares

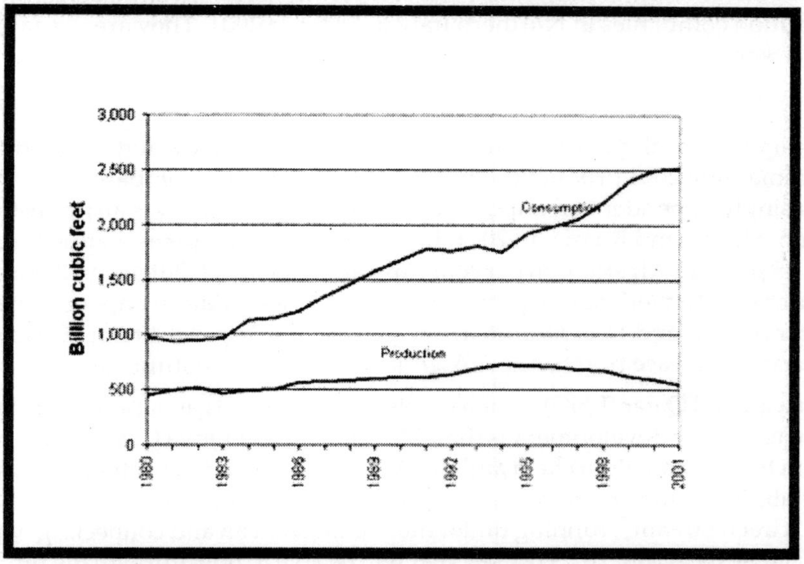

ITALIAN NATURAL GAS CONSUMPTION AND PRODUCTION, 1980-2001

are increasing rapidly, however, as natural gas supplies from a 20-year contract with Eni-Gazexport joint venture Promgas to import 70.6 Bcf per year began in 2002. Along with constructing a re-gasification plant in Porte Levante, Edison also is planning to have 15 per cent of Italy's natural gas market by 2007. Italian electricity conglomerate Enel also imports gas directly for its own generation activities and is aiming to increase its share of the natural gas market to about 20 per cent by 2007. Italy has had very high rates of return for natural gas suppliers, transporters, and distributors, when compared with other European countries. Italian consumers pay among the highest natural gas prices in Europe, though there is significant regional variability in prices.

As of January 1, 2003, Italian consumers gained the freedom to choose their natural gas supplier. Liberalization in Italy's gas market has ignited competition among large key players in the sector, such as Gaz de France, BG, BP, Shell and ExxonMobil, as well as hundreds of domestic distribution groups. This process began in 1998 with the EU Natural Gas Directive, directing member states to open their natural gas markets to competition. In May 2000, the Italian government unveiled a decree, authorizing that no single company can supply more than 50 per cent of the natural gas sold to final users by 2003. No company will be able to send more than 75 per cent of natural gas put into the transmission system beginning in 2002, and this will be reduced to 61 per cent by 2010. The legislation also requires corporate and accounting separation of natural gas storage and transport activities. Snam will retain control of Italy's 29,000-kilometer (18,125-mile) pipeline natural gas grid, but parent company Eni must split Snam's pipeline transport activities from commercial and sales activities. In late November 2001, the new company controlling the natural gas grid, Snam Rete Gas Italia, had an initial public offering (IPO) of 40.24 per cent of its shares, which was heavily oversubscribed. The launching of Eni's new gas distribution company, Italgas Piu, was also in November 2001. Through Stoccaggi Gas Italia, Eni also operates a system where it stores and modulates natural gas. Eni's storage system is made up by 9 fields, 8 of which are located in Northern Italy (one of them is not yet operational) and one in Central Italy. Enel received permission from antitrust authorities to acquire five gas distribution companies in Northern Italy in October 2001. They are: Arda Gas, Gead, Adda Gas, Geico, and Sein.

Imports

Most of Italy's natural gas fields are located in the Po Valley and offshore in the Adriatic. Preliminary estimates of 2001 production show a 7 per cent decline compared to 2000. Hence, Italy is growing increasingly dependent on imports. Diversification of supply is an important issue, as Italy relies heavily on Algeria and Russia. In 2001, Italian natural gas sources were estimated to be 22 per cent domestic, 45 per cent Algerian, 36 per cent Russian, 13 per cent Dutch and 2 per cent Norwegian. Snam imports from the Netherlands, Algeria, Russia and Norway; Edison imports from Russia, Norway, Libya and Qatar (LNG); and Enel imports from Algeria and Nigeria. Snam, Edison, and Enel have contracts in place to increase purchases of Algerian and Russian natural gas.

Libya, Norway, and Qatar (LNG) are major alternative natural gas suppliers for the Italian market. The biggest project under development is the $4.5 billion "West Libya Gas Project", which will draw natural gas from the offshore Bouri field and the Wafa field, located in southern Libya. The gas will be treated in Melitah, 80 kilometers from Tripoli, where it will be transported via the 600-kilometer (372-mile) pipeline "Green Stream", running under the Mediterranean and connecting with the TransMed pipeline in Gela in Sicily. About 282.5 Bcf per year will be able to flow through the pipeline. The project was finalized in July 1999 as a joint venture between Libya's state-held National Oil Company and Eni, and an Eni contract to buy 141 Bcf per year from the joint venture was signed in early 2000. In November 2000, Gaz de France agreed to buy another 70.6 Bcf from Eni. Start-up is planned for 2004. Snam began importing 6 billion cubic meters per year (212 Bcf) of Norwegian natural gas through existing pipelines in October 2001, and will continue to do so for 24 years.

At the end of 2002, the Algerian oil group Sonatrach along with Edison, Enelpower, Germany's BASF, Wintershall and Eos Energia formed a joint venture called Galsi. The joint venture will study the feasibility of building a 1,500 km $2 billion pipeline from Algeria's eastern terminal Skikda to Sardinia. A potential starting date for operation is 2008. Russia's Gazprom also has plans to begin constructing a pipeline via Bulgaria, with an operating capacity of 353 billion cubic feet per year, but it is unlikely that the project will begin in 2003.

Liquefied Natural Gas

Although LNG makes up a small percentage of Italy's natural gas imports (9 per cent in 2000), it is becoming increasingly important. In 1992, Enel signed a contract under which Nigerian LNG will be delivered to France and swapped out to Enel for 22 years, beginning in 1999. In June 2001, Edison signed a 25-year agreement to buy 3.5 million metric tons of LNG per year (equivalent to 173 Bcf) from the fourth natural gas train of Rasgas in Qatar when the train is completed in 2005. Delivery will be made to a $430-million floating re-gasification terminal which Edison and ExxonMobil are constructing 11 miles offshore Marina di Rovigo in the northern Adriatic. It will be completed in 2005 with a capacity of 161 billion cubic feet (Bcf) per year. Enel has announced that it also has an agreement with Qatargas and Spain's Repsol-YPF to import LNG. In 2001, the two European companies signed a memorandum to carry out a feasibility study on constructing a gas train with a 4.8 million metric tons (equivalent to 233 Bcf) per year capacity.

In order to increase LNG imports, a number of companies are in the process of building re-gasification terminals. Presently, Italy has one LNG terminal–Eni's 122.5 billion cubic feet (Bcf) per year in Panigaglia. BG Group (British Gas) recently won approval for its proposed LNG recovery terminal at Brindisi in southern Italy. BG is aiming to have the terminal on stream at the end of 2006 with a capacity of 3 million metric tons (equivalent to 146 Bcf) per year. Italy has five other LNG terminal proposals under consideration. Enel has proposed three sites located in Toranto, Vado Ligure and Muggia. The Italian company Falck is proposing two sites along the southern Calabrian coast. Edison is proposing an additional terminal in Tuscany.

Coal

Coal consumption in Italy is dominated by power generation, which is increasing (estimated to reach 22 per cent by 2010), and coke production for steel, which is decreasing. Coal has played a small role in the Italian energy sector, and Italy produces almost no coal domestically. In 2001, nearly 8 per cent of Italy's primary energy demand was met with coal. The power sector is expected to increase its coal consumption in coming years, as Eni works to decrease reliance on imported oil, though coal will not be as important as natural gas. Clean coal technology will figure prominently in this increased coal usage, as EU environmental stipulations, Kyoto targets, and Italian public opinion demand that Italy's energy sector become increasingly clean.

Increased coal usage will be supplied by a combination of increased domestic production and increased imports. Main exporters of steam coal to Italy are South Africa, Indonesia, Colombia, United States, Poland, China and Australia.

Electricity

Italy has electric generation capacity of 69 million kilowatts. In 2001, the country generated 203.4 billion kilowatt hours (bkwh) and consumed 289.1 bkwh. Generation is mostly from thermal sources (78 per cent)–oil, gas, and coal. The mix of thermal power is shifting away from oil and toward natural gas, such that natural gas should be the dominant fuel source for electricity generation by the end of the decade. Non-hydro renewable electricity generation (mostly solar and geothermal) almost doubled in the 1990s, and over 2 per cent of Italian electricity is now produced from renewable sources.

Italy's extensive electricity network is linked to its neighbors. Electricity imports come mostly from France, Switzerland and a small percentage from Slovenia. In the summer of 2002, Italy and Greece completed the construction of a new 163-kilometer (102-mile), 400-kilovolt underwater cable to link Italy and Greece. The cable will allow Greece to transfer electricity to the EU, as well as serve as a back-up source for Italy.

Sector Organization

Enel is Italy's dominant electricity company, which the government founded in 1962. Enel, which was completely state-owned until November 1999, produced about 70 per cent of Italy's electricity usage in 2000. EU Directive 96/92/EC, issued in 1996, requires member countries to open their electricity markets to competition and also requires that no single company generate more than 50 per cent of any member country's electricity by 2003. This has resulted in several important changes in Enel, including partial privatization and the sale of some of its assets to reduce market share. The Italian government issued the Bersani Decree in 1999, which implemented the EU Directive, beginning the liberalization of the electricity market in Italy. The November 1999 Enel privatization stock sale was Europe's largest initial public offering (IPO). The government floated 32 per cent of the company, which sold for 18 billion euros ($15 billion), on the Milan and New York stock exchanges.

The Decree also required Enel to divest 15,000 MW of its total 56,000 MW installed capacity by January 1, 2003. In order to comply with the Decree, Enel divested three electricity generation companies: 5,438 MW Elettrogen in July 2001 to Endesa, 7,009 MW Eurogen in March 2002 to Edipower, a consortium led by Edison, and 2,611 MW Interpower to Electrabel and Energia in November 2002. Furthermore, no company will be allowed to acquire or hold stakes in more than one of the three companies, and no buyer will be able to be more than 30 per cent government-held. This last requirement was to prevent Electricite de France (EdF) from acquiring these companies. In 2001, EdF and Fiat took control of Edison. Edison is 76 per cent controlled by Italenergia, an energy consortium in which EdF has an 18 per cent stake. After EdF's acquisition of Edison, the Italian government passed a decree to reduce EdF's voting rights in Italenergia to 2 per cent in the hope of blocking the French state-owned company from taking it over. This has resulted in a dispute over EdF between Italy and the European Union. Because liberalization of the energy sector has proceeded at a slower pace in France, EdF has remained a state-owned company. EdF's purchase of privatized assets would in effect transfer them from Italy to France, so Italy has attempted to restrict this sale. The European Commission ruled in June 2001, that capital flows may not be restricted merely because of varying degrees of liberalization. However, the initial privatization sale may be restricted, but such restrictions can only be in place for a limited period, after which the privatized companies can be resold to state-owned companies.

In January 2003, the European Commission issued statements of plans to take legal action against Italy for establishing the so-called "golden share laws", which allow the state to influence former state-owned companies. Enel and EdF have also been conducting their own negotiations to settle the dispute. EdF has offered Enel a stake in four nuclear plants in France in exchange for recovery its voting rights in Edison.

Another facet of liberalization is that Enel must also sell its distribution networks in Italy's large urban centers. It has already sold off many local grids, including the Rome network to ACEA for 568 million euros in April 2001 and the Turin and Milano networks to AEM. However, local electricity companies have complained that Enel is blocking some of their access and several lawsuits in this regard. In July 2002, the Italian government issued a bill that would require Enel to reduce its distribution network to less than 10 per cent of market share within three years (This also applies to Eni's gas pipeline network). The bill was approved by the cabinet and has been transferred to parliament for ratification. Enel has 100 per cent of the unlisted company Terna, which completely owns Italy's high voltage grid.

In early November 2000, the European Commission approved a merger that gave Italian conglomerate Compart SpA sole control of Falck SpA, forming the then third-largest electricity generation company in Italy. The deal also gave Compart control of Falck's subsidiary, Sondel SpA. Enel, while selling off domestic assets, has made some foreign acquisitions. In December 2000, for instance, Enel purchased CHI Energy of the United States for $170 million. In September 2001, Enel purchased Nueva Viesgo of Spain, making Enel the fifth-largest generator in Spain. Siemens and Fiat Engineering are to build a 390-MW, combined cycle power plant in Torino worth 190 million euros. Eni has also been expanding its electricity generation capacity in Italy. In May 2002, it began the construction of a 1,000-MW power plant in northern Italy. More recently, Eni announced that it plans to build five 2,000-MW power plants by 2005.

Nuclear

Italy has four nuclear power plants–Latina in central Italy, Caorsa in the northeast, Garigliano also in central Italy and Trino Vercellese in the northwest. Since 1987, none of the stations have been in operation after a public vote decided to abandon the use of nuclear power. The plants are currently being dismantled. The Italian government is in the process of identifying a single site to store all radioactive waste located within the country in order to increase security.

More recently, the Italian government has been reconsidering its policy towards nuclear energy. In December 2002, the Italian parliament considered a draft law which would permit Italian utilities to buy stakes in nuclear power plants outside of Italy. The document also included prospects of returning to nuclear energy generation in Italy. As mentioned in the previous section, Enel is already seeking to gain stakes in four nuclear power plants in France.

Environment

Environmental awareness has grown in Italy in recent years. Although Italy has relatively low per capita energy consumption and energy intensity levels in comparison to other OECD countries, air pollution remains a serious environmental challenge.

Because of Italy's heavy reliance on oil imports to meet its energy needs, energy security and diversification of energy sources are a top priority in Italy's energy strategy. Italy is well endowed with renewable energy resources, such as solar, biomass and geothermal which could be captured and utilized for energy. The government's goal of doubling the country's production of energy from renewable resources by 2012 will help enable Italy to meet its growing energy demand in the 21st century in a more sustainable manner.

Country Overview

Location/Size: Southern Europe/301,230 sq km (116,305 sq mi, slightly larger than Arizona)

Major Cities: Rome (capital), Milan, Naples, Turin, Palermo, Genoa

Languages: Italian, German (parts of Trentino-Alto Adige region are predominantly German speaking), French (small French-speaking minority in Valle d'Aosta region), Slovene (Slovene-speaking minority in the Trieste-Gorizia area)

Ethnic groups: Italian (includes small clusters of German-, French-, and Slovene-Italians in the north and Albanian-Italians and Greek-Italians in the south)

Religion: predominately Roman Catholic with established Protestant and Jewish communities and a growing Muslim immigrant community

Population (10/02): 57.9 million

Defense (8/98): Army, 165,600; Navy, 40,000; Air Force, 63,600; Paramilitary forces, 255,700; Conscripts, 134,100

Economic Overview

Currency: Euro–On January 1, 2002, the Euro became the single currency of the 12 member states. After that date, the Lira existed as legal tender until February 28, 2002.

Market Exchange Rate (03/05/03): US$1=.91870 Euro

Nominal Gross Domestic Product (GDP, 2002E): $1,777 billion *(2003F):* $1,302 billion

Real GDP Growth Rate (2002E): 0.4 per cent ; *(2003F):* 1.4 per cent

Unemployment Rate (2001E): 9.5 per cent ; *(2002E):* 9.1 per cent

Inflation Rate (consumer prices, 2002E): 2.4 per cent ; *(2003F):* 2.4 per cent

Major Export Products: Textiles, clothing, machinery, transportation equipment

Major Import Products: Crude oil, other fuels, machinery, transport equipment

Major Trading Partners: Germany, France, Netherlands, United States, United Kingdom

Energy Overview

Proven Oil Reserves (1/1/03E): 621 million barrels

Oil Production (2002E): 149,000 barrels per day (bbl/d), of which 84,700 bbl/d is crude oil

Oil Consumption (2002E): 1.87 million bbl/d

Net Oil Imports (2002E): 1.69 million bbl/d

Crude Oil Refining Capacity (12/1/02E): 2.30 million bbl/d

Natural Gas Reserves (12/1/02E): 8.0 trillion cubic feet (Tcf)

Natural Gas Production (2001E): 550 billion cubic feet

Natural Gas Consumption (2001E): 2.5 Tcf

Net Natural Gas Imports (2000E): 2.0 Tcf

Recoverable Coal Reserves (2000): 37 million short tons (Mmst)

Coal Production (2001E): 0.02 Mmst

Coal Consumption (2001E): 22.4 Mmst

Electric Generation Capacity (1/1/01): 69 million kilowatts

Electricity Generation (2001E): 203.4 billion kilowatthours

Electricity Consumption (2001E): 289.1 billion kilowatthours

Environmental Overview

Total Energy Consumption (2000E): 7.96 quadrillion Btu* (2 per cent of world total energy consumption)

Energy-Related Carbon Emissions (2001E): 121.5 million metric tons of carbon (1.9 per cent of world total carbon emissions)

Per Capita Energy Consumption (2000E): 138.3 million Btu (vs U.S. value of 351 million Btu)

Per Capita Carbon Emissions (2000E): 2.0 metric tons of carbon (vs U.S. value of 5.6 metric tons of carbon)

Energy Intensity (2000E): 6,603 Btu/ $1990 (vs U.S. value of 10,918 Btu/ $1990)**

Carbon Intensity (1999E): 0.1 metric tons of carbon/thousand $1990 (vs U.S. value of 0.17 metric tons/thousand $1990)

Fuel Share of Energy Consumption (2000E): Oil (48.7 per cent), Natural Gas (32 per cent), Coal (6.2 per cent), Hydro (5.7 per cent), and other Renewables (1.58 per cent)

Fuel Share of Carbon Emissions (2000E): Oil (57.5 per cent), Natural Gas (31.4 per cent), Coal (11.1 per cent)

Renewable Energy Consumption (1998E): 560 trillion Btu* (1 per cent increase from 1997)

Status in Climate Change Negotiations: Annex I country under the United Nations Framework Convention on Climate Change (ratified April 15th, 1994). Under the negotiated Kyoto Protocol (signed on April 29th, 1998, and ratified the Treaty on May 31st, 2003), Italy, as a member of the European Union, has agreed to reduce greenhouse gases 8 per cent below 1990 levels by the 2008-2012 commitment period.

Major Environmental Issues: Air pollution from industrial emissions such as sulfur dioxide; coastal and inland rivers polluted from industrial and agricultural effluents; acid rain damaging lakes; inadequate industrial waste treatment and disposal facilities.

Major International Environmental Agreements: A party to Conventions on Air Pollution, Air Pollution-Nitrogen Oxides, Air Pollution-Sulphur 85, Air Pollution-Sulphur 94, Air Pollution-Volatile Organic Compounds, Antarctic-Environmental Protocol, Antarctic Treaty, Biodiversity, Climate Change, Desertification, Endangered Species, Environmental Modification, Hazardous Wastes, Law of the Sea, Marine Dumping, Nuclear Test Ban, Ozone Layer Protection, Ship Pollution, Tropical Timber 83, Tropical Timber 94, Wetlands and Whaling. Has signed, but not ratified: Air Pollution-Convention on Persistent Organic Pollutants.

The total energy consumption statistic includes petroleum, dry natural gas, coal, net hydro, nuclear, geothermal, solar, wind, wood and waste electric power. The renewable energy consumption statistic is based on International Energy Agency (IEA) data and includes hydropower, solar, wind, tide, geothermal, solid biomass and animal products, biomass gas and liquids, industrial and municipal wastes.

Energy Industry

Oil and Gas Company: Ente Nazionale Idrocarburi (ENI); Chief Subsidiaries: Agip (hydrocarbons exploration and production), Snam (gas supplies and hydrocarbon transportation), ENIchem (petrochemicals)

Major Pipelines (gas): TransMed, Trans-Austria Gasleitung

Major Ports: Cagliari (Sardinia), Genoa, La Spezia, Livorno, Naples, Palermo, Trieste, Venice

National Electricity Company: Ente Nazionale per l'Energia Elettrica (ENEL, undergoing privatization)

47

Japan

Japan is the world's fourth largest energy consumer and second largest energy importer (after the United States). Over the past decade, Japan has been experiencing a period of slow economic growth, and has taken important steps towards economic deregulation and restructuring. As a result of slow economic growth, Japanese demand for energy has been stagnant in recent years. Oil consumption has stabilized after several years of decline, but is still 9 per cent below its 1996 peak level.

Recent Developments

Japan's economy has resumed modest growth in 2003 after stalling in 2001 and 2002 due to weak demand for Japanese exports. Japan's real gross domestic product (GDP) rose by only 0.1 per cent in 2002, and it is projected to rise by 1.1 per cent in 2003. The slight upturn largely reflects a surge in exports to China, at a time when exports to Japan's largest market, the United States, have declined. Unemployment has still risen to 5.5 per cent, which is historically a very high level for Japan.

Japan's prime minister Junichiro Koizumi, who took office in 2001, has pressed for structural reforms in Japan's economy. In one major change, Koizumi reversed the previous policy of increasing government spending to stimulate the country's economy, and has set a deficit ceiling of 30 trillion yen. Spending on public works projects, which had been funded as part of previous stimulus packages, has been scaled back significantly. The Bank of Japan, however, has adopted a more expansionary monetary policy, which has provided some stimulus to the economy.

Much of the growth projected in the second half of 2003 reflects an anticipated recovery in exports to the United States. Capital spending by Japanese businesses, however, and domestic demand for consumer products and services, are likely to recover more slowly.

Japan's economic stagnation since the early 1990s has led to a period of consolidation in the country's energy sector. Energy demand has been stable, and Japan's energy industries, particularly

the downsteam oil sector, have undergone a period of downsizing and consolidation. Japan remains important to the world energy sector, though, as one of the major exporters of energy-sector capital equipment, and engineering, construction, and project management services.

Energy

Japan lacks significant domestic sources of energy and must import substantial amounts of crude oil, natural gas, and other energy resources, including uranium for its nuclear power plants. In 2001, the country's dependence on imports for primary energy stood at more than 79 per cent. Oil provided Japan with 50 per cent of its total energy needs, coal 17 per cent, nuclear power 14 per cent, natural gas 14 per cent, hydroelectric power 4 per cent, and renewable sources 1.1 per cent. About half of Japan's energy is used by industry and about one-fourth by transportation, with nearly all the rest used by the residential, agricultural, and service sectors. Japan's energy intensity (energy use per unit of GDP) is among the lowest in the developed world.

Oil

Japan contains almost no oil reserves of its own (59 million barrels of proven oil reserves), but it was the world's second largest oil consumer (after the United States) in 2002. China is likely to surpass Japan in oil consumption for 2003. In 2002, Japan consumed an estimated 5.32 million barrels per day (bbl/d) of oil, down from 5.42 million bbl/d in 2001. Most (75 per cent -80 per cent) of this oil came from OPEC, particularly Persian Gulf countries like the United Arab Emirates, Saudi Arabia,

Kuwait, Qatar, and Iran. Japan has worked–with relatively little success–to diversify its oil import sources away from the Middle East. Another oil supplier to Japan is China, which, while it is a net oil importer, supplies light oil from its Daqing field for use in Japanese power plants. Until 1996, when Japan's oil consumption peaked at nearly 5.9 million bbl/d, Japanese oil consumption (and imports) had been growing steadily for years. After 1997, Japan's oil consumption declined as its economic slump caused demand by industrial and other users to decline.

Japanese oil companies have been active overseas since 1967, when the government established a state-run company to promote overseas oil exploration, the Japan National Oil Company (JNOC). Over the years, JNOC amassed numerous bad loans through extensive investment programs and loan guarantees to Japanese exploration firms. A study of JNOC for the Japanese Ministry of Economy Trade and Industry (METI), conducted by the American consulting firm Booz Allen and Hamilton, concluded that Japan's policy of subsidies for oil exploration left firms with little incentive to seek high rates of return on their investments. In November 2001, Prime Minister Koizumi called for the abolishment of JNOC, and the Japanese government has been planning its liquidation and considering what role it should have in financing overseas oil projects. Current plans, announced in March 2003, call for the final dissolution of JNOC to take place in March 2004. A new company, however, would be formed out of three of the more financially-sound units of JNOC. One of the other major subsidiaries of JNOC, Japex, is to be listed on the Tokyo Stock Exchange in late 2003.

The loss of drilling rights by Japan's Arabian Oil Company (AOC) in the Saudi Arabian portion of the Neutral Zone dealt a major blow to Japan's policy of seeking overseas equity in oil projects. AOC's rights to the concession, which produced 280,000 bbl/d, expired at the end of February 2000. Efforts to negotiate an extension with Saudi authorities failed when Japan refused to commit to investment in development projects desired by the Saudis. Saudi Aramco has taken over operation of the former AOC fields. AOC's concession for the Kuwaiti portion of the Neutral Zone expired in January 2003. In December 2002, however, Kuwait signed a service contract with AOC, similar to those held by other foreign oil companies in Kuwait, allowing the firm to maintain its operations there. While AOC does not hold an equity stake in the field, its offtake from operations there is essentially unchanged, despite the change in contract terms.

Japan has been trying to make up for the loss of the AOC concession in Saudi Arabia by increasing its investment in Iran. Iran announced in November 2000 that it would begin exclusive negotiations with Japan Petroleum Exploration Corporation (Japex) and Indonesia Petroleum (Inpex), both of which are majority-owned by JNOC, for development rights to the huge onshore Azadegan oilfield. Azadegan has been estimated to contain 6 billion barrels of recoverable reserves. The consortium submitted a preliminary development plan for Azadegan in mid-2001, but progress in negotiations has been slow and no final agreement has been reached. Recent disclosures in early 2003 regarding Iran's possible nuclear ambitions have added a further political complication to the negotiations. The Japanese consortium had concluded an agreement to work with Shell on the project once a final contract is signed with Iran.

Apart from its interests in the Persian Gulf, Japan also has been seeking equity stakes in the Caspian Sea region. In July 1998, Mitsui purchased a 15 per cent share, along with Azerbaijan's State Oil Company, of concessions in the Caspian Sea's Kur Dashi oil field. Oil reserves in the contract area are estimated at 500 million-1 billion barrels. In February 1999, JNOC announced that it would help finance the oil development project, the first since it revealed its major financial difficulties in June 1998. The Kur Dashi oil field is important to Japan's strategic goal of reducing its dependence on Middle Eastern oil imports. In December 1998, four other Japanese companies (Japex, Teikoku Oil Co.,

Inpex, and Itochu Corporation) signed a deal to purchase a different field (Atashgyakh-Mugandeniz-Yanan Tava) in the Caspian. JNOC's most significant purchase in 2000 was a 7 per cent stake in Kazakhstan's offshore Kashagan field in the Caspian Sea, through its subsidiary Inpex.

Another possible source for Japanese oil imports, which has recently received increased attention, is the Russian Far East. Japan has been promoting a proposed pipeline from oilfields near Anagarsk in Siberia to an export terminal on the Pacific coast at Nakhodka. An alternate proposed route would transport the oil from Anagarsk to Daqing in China, connecting to China's existing crude oil pipeline network. A Russian export terminal on the Pacific is a very appealing idea to Japan, since it could reduce Japan's dependence on imports from the Persian Gulf. The Japanese government has reportedly offered to provide financing for the project on very favorable terms, but the option of a pipeline to China is considered more likely to be implemented in the near future, due to the lower construction cost.

Refining/Downstream

As of January 2003, Japan had 4.8 million bbl/d of oil refining capacity at 33 refineries, down from 5.0 million bbl/d as recently as 2001. In recent years, as Japan's petroleum product consumption has stagnated, the country's refining industry has suffered from overcapacity. Japan also began to allow imports of petroleum products in the mid-1990s, putting additional pressure on Japanese refiners to cut costs and become internationally competitive.

In response to these pressures, Japan's refining industry went through a round of consolidations in 1999 and 2000. Nippon Oil and Mitsubishi Oil completed a merger in early 1999, forming Nippon Mitsubishi Oil. Nippon Mitsubishi then acquired Koa Oil from Caltex in September 1999, and in February 2002 merged Koa Oil with another subsidiary, Tohuku Oil. In October 1999, Nippon Mitsubishi announced a strategic alliance with another independent Japanese refiner, Cosmo Oil. The move, while not a merger, allows the two companies to coordinate distribution of refined products and to reduce costs through reduced duplication of some functions. Another similar strategic alliance was formed with Idemitsu Kosan in December 2002. As a result of the agreement, Idemitsu shut down the 80,000 bbl/d Hyogo refinery in April 2003 and will shut down the 110,000 bbl/d Okinawa refinery in early 2004.

A second alliance coalesced around Showa Shell, Royal Dutch Shell's Japanese subsidiary, in which it owns a 50 per cent stake. In January 1999, Showa Shell and Japan Energy announced a strategic alliance in petroleum product distribution and crude oil procurement.

The third major player in Japan's refining sector is ExxonMobil, through its Japanese subsidiary, Tonen General Sekiyu, which resulted from the merger of two ExxonMobil subsidiaries in February 2000. Tonen is the third largest of the alliances in terms of market share.

Fuji Oil, which has a 200,000 bbl/d refinery near Tokyo, was acquired by AOC in January 2003. This created a vertically-integrated, albeit small, company based on AOC's crude oil imports from Kuwait's half of the Neutral Zone.

While many regulatory restrictions on Japanese refiners have been removed over the last few years, the firms remains saddled with a requirement to maintain mandatory large petroleum stocks. This requirement permits Japan to maintain a strategic reserve without having to build a government-run storage facility like the U.S. Strategic Petroleum Reserve, but also imposes significant additional capital costs on refiners operating in Japan.

Natural Gas

Liquefied Natural Gas

Japan has about 1.4 trillion cubic feet (Tcf) in proven natural gas reserves, with possibly more under the seabed surrounding Japan. Because domestic natural gas production is minimal, about 97 per cent of Japan's natural gas is imported, all in the form of liquefied natural gas (LNG). Unlike oil, demand for natural gas is still rising about 3-4 per cent per year. Most of this LNG comes from Southeast Asia, with 32 per cent from Indonesia and 20 per cent from Malaysia. The United States also supplies a small quantity of LNG to Japan from a facility in Alaska, which accounts for slightly over 2 per cent of Japan's natural gas consumption. Most of the LNG is used either for electric power generation or as feedstock for petrochemical plants.

Three Japanese companies, Tokyo Gas, Osaka Gas, and Toho Gas signed a binding contract in February 2002 for the import of natural gas from Malaysia's MLNG Tiga project, covering deliveries beginning in 2004. The contract is noteworthy in that it includes much more flexible terms for the purchaser than most traditional LNG contracts, which commit the purchaser to a specific volume over 15 to 20 years. The three firms also renewed their baseload contracts with Malaysia's first two LNG export terminals, on terms more flexible than the original contracts. Tokyo Gas and Toho Gas signed a binding contract in October 2001 for LNG purchases from Australia's North West Shelf LNG project, to begin in 2004.

Japanese firms have been considering the possibility of imports, either by pipeline or as LNG, from large natural gas deposits on the Russian island of Sakhalin. ExxonMobil and Shell are backing rival development options–ExxonMobil a project (Sakhalin-1) which would export matural gas via a pipeline to Japan's main island of Honshu, and Shell a project (Sakhalin-2) which would feed into an an LNG export terminal with Japanese firms as its primary customers. Shell's Sakhalin-2 project has made faster progress, having signed contracts in the first half of 2003 with both Tokyo Electric Power (TEPCO) and Tokyo Gas. Deliveries are set to begin in early 2007. ExxonMobil has said it doesn't expect the Japanese market to have sufficient demand for its pipeline project to begin deliveries as planned in 2008, and it might consider export options targeted at South Korea or China.

Much of Japan's urban area is not served by a natural gas distribution system. With plans to increase the natural gas portion of its primary energy supply to 13 per cent by 2010, however, Japan is considering expansion of its internal natural gas pipeline system. Many analysts cite the absence of an effective natural gas distribution system as a key reason for Japan's high retail energy prices.

City gas consumption has increased by more than 70 per cent in the last decade due to a 25 per cent increase in natural gas customers and also to a large rise in consumption by industry. Japan's major natural gas companies include Tokyo Gas, Osaka Gas and Chubu Gas. The Japanese government has begun to gradually deregulate the natural gas industry, which is leading to the increased competition.

Coal

Japan has small coal reserves of 852 million short tons (Mmst), and the country ceased production in January 2002 with the closure of its last operating coal mine at Kushiro, on the northern island of Hokkaido. Japan's coal mines had been heavily subsidized in recent years, since they were not cost-competitive with other producers.

Japan is by far the world's largest importer of steam coal, mainly for power generation, paper plants, and cement production. Japan also is the world's largest importer of coking coal for its steel industry. Overall, Japan accounts for about 22 per cent of total world coal imports. Sources of imported steam coal are Australia, South Africa, the United States, and China. Japanese coking coal imports come mainly from Australia, Canada, the United States and Russia.

Prices paid by Japanese firms for coal are currently declining, as increased exports of coal from China put downward pressure on the market. Australia still supplies about 60 per cent of the Japanese market, but the trend has been toward a steadily increasing market share for China.

Electricity

Japan generated 1,037 billion kilowatthours (Bkwh) of electricity on 235 gigawatts of capacity in 2001. Of Japan's total generation in 2001, about 60 per cent came from thermal (oil, gas, and coal) plants, 30 per cent from nuclear reactors, 8 per cent from hydroelectric dams, and less than 2 per cent from geothermal, solar, and wind.

Due to the country's desire to enhance its energy security, Japan has increased its reliance on nuclear power generation. The last year, however, has been challenging for Japan's nuclear power industry. In August 2002, it emerged that maintenance inspection findings at some nuclear reactors owned by Tokyo Electric Power (TEPCO) had not been properly reported to government regulators. This led to the shutdown of all 17 of TEPCO's nuclear reactors over the following several months. Several new reactor projects, including some proposed by other utilities, have been put on hold while the issue is resolved. In the short term, this has led to increases in Japan's fuel oil and LNG consumption, as generating capacity using fossil fuels has been brought online to make up for the shortage of nuclear generating capacity. As of mid-July 2003, only 3 of the 17 TEPCO nuclear power plants had resumed operation. It remains unclear how long it will be before the rest of TEPCO's nuclear generating capacity is available again. The long-term effects on public confidence in the nuclear power industry also have caused concern. Japan's government has indicated that it is still comitted to increasing nuclear power's share of generating capacity in the future, once the current maintenance and documentation issues are resolved.

Japan's nuclear output nearly doubled between 1985 and 1996, as Japan attempted to move away from dependence on oil following the 1973 Arab oil embargo. During the past few years, public opposition to Japan's nuclear power program has increased in reaction to a series of accidents at Japanese nuclear plants, especially the accident at the Tokaimura uranium processing plant in September 1999 and the recent TEPCO reactor shutdowns.

By raising its reliance on nuclear-generated electricity, Japan is hoping to reduce its carbon dioxide emissions. Japan's current 10-year energy plan, approved in March 2002, calls for the expansion of nuclear generation by about 30 per cent by 2010. This is expected to entail the construction by between 9 and 12 new nuclear power plants. The Japanese government also plans to offer subsidies for nuclear power plant construction, to offset expected cost-cutting pressures on utilities due to deregulation which might lead to increased reliance on fossil fuels for electricity generation. Currently Japan ranks third worldwide in installed nuclear capacity, behind the United States and France. Japan currently has 51 reactors with an installed capacity of 45 GW, including those units currently shut down by TEPCO.

In August 1998, the Atomic Energy Commission approved the construction of a new light-water reactor, which will be built in Higashidori in Aomori prefecture in northern Japan. Also, in March 1999, the Japanese Nuclear Safety Commission approved plans for Hokuriku Electric Power Company to build a new nuclear power plant in the central town of Shika, which will be operational by 2006.

To enhance its energy security, Japan's government advocates uranium and plutonium recovery through reprocessing of spent fuel. The Power Reactor and Nuclear Fuel Development Corporation (PNC) operates a reprocessing plant with an annual capacity of 90 tons, but a larger reprocessing plant, Rokkasho-Mura, with a capacity of 800 tons per year, planned for completion in July 2005, is under construction. Reprocessing is expensive and costs can quickly rise with new safety requirements and the development of new technologies. Estimated in 1993 to cost about $8 billion, more recent estimates put the cost of the facility much higher. In the meantime, Japan is negotiating with the French firm COGEMA for the reprocessing of spent nuclear fuel in France. COGEMA may continue to reprocess some spent fuel even after the Rokkasho plant is completed. Japan also is interested in recycling recovered plutonium. In 1999, Japan began–in two prefectures–a controversial mixed-oxide utilization plan, which involves burning a highly toxic mix of plutonium and uranium on a commercial scale.

Despite its relatively high cost, natural gas, mainly imported as LNG, also is likely to experience considerable growth as a fuel for electricity generation. Renewables, chiefly hydropower and geothermal energy, also are expected to grow, and both coal and nuclear are projected to grow in absolute terms (although nuclear power's share of the market is expected to drop). An accelerating decline is projected for oil-fired generation, which is still more significant in Japan than in most other developed countries. In the short-term, Japan's economic slowdown had resulted in a sharp downturn in capital spending by utilities, which has delayed several new power plant projects.

Japan's electricity prices are by far the highest in the OECD, and Japan has begun a program of reforms designed to make its electric utility sector more efficient. Currently, Japan is served by 10 vertically integrated utilities which each have a specific geographic zone. The Japanese Diet passed a bill in May 1999 which amended the Electric Utilities Industry Law (EUIL) to allow a partial opening to competition. Beginning in March 2000, about 8,000 large industrial and commercial Japanese electric power consumers, comprising roughly one-third of the Japanese power market, have been able to chose their electricity suppliers. Regional utilities currently are obligated to allow power from other suppliers to transit their grids to these large consumers.

While a small percentage of Japan's electricity has been provided by independent power producers (IPPs) since 1995, the new deregulation is expected to generate much more investor interest in developing IPPs, though progress in this direction has been slower than expected due to weak demand.

Environment

In recent years Japan has begun to take a more active role in protecting the environment. As air pollution in dense urban areas has persisted, the Japanese government has taken a number of measures to mitigate it, such as encouraging the adoption of low-polluting mass transportation buses and garbage collection vehicles as alternatives to diesel.

The effects of the oil shocks in the 1970s forced Japan to streamline heavy industrial production and reorient its economy toward less energy intensive industries. As a result, while Japanese energy consumption is high, Japan is a world leader in terms of energy efficiency, and a newly revised energy savings law is aiming to reduce the per capita energy consumption in Japan even further.

Japan committed to reducing its total carbon emissions by 6 per cent under the 1997 Kyoto Protocol, but as Japan is the world's fourth largest producer of greenhouse gases, the task ahead is difficult. Japan's economy is heavily dependent on imported oil, but the government is placing increased emphasis on the diversification of its energy sources. In addition to nuclear power Japan is looking to increase its share of solar, hydro, and other carbon-free, non-polluting renewable energy sources. In the 21st century Japan will need to continue in its role as a leading developer of environmental technologies in order to protect its environment.

Country Overview

Population (2002E): 127.0 million

Location/Size: Eastern Asia–island chain between the North Pacific Ocean and the Sea of Japan/ 145,882 square miles (slightly smaller than California)

Major Cities: Tokyo (capital), Osaka (Kansai), Nagoya, Fukuoka/Kitakyushu, Sapporo

Languages: Japanese

Ethnic Groups: Japanese (99.4 per cent)

Religion: Shinto and Buddhist (84 per cent), other (16 per cent)

Defense (8/98): Army (151,800), Navy (43,800), Air Force (45,600), U.S. Forces in Japan (39,100)

Economic Overview

Currency: Yen

Exchange Rate (7/15/03): US$1 =117.9 Yen

Gross Domestic Product (GDP, at market exchange rate) (2002E): $4.0 trillion *(2003F):* $4.2 trillion

Real GDP Growth Rate (2002E): 0.1 per cent *(2003F):* 1.1 per cent

Inflation Rate (consumer prices)(2002E): -1.7 per cent *(2002F):* -2.1 per cent

Current Account Balance (2002E): $111.8 billion

Major Trading Partners: United States, Germany, Asian NIEs, China, OPEC

Merchandise Exports (2002E): $396.0 billion

Merchandise Imports (2002E): $302.8 billion

Merchandise Trade Surplus (2002E): $93.2 billion

Major Export Products: Machinery and transport equipment; chemical and other manufactured goods

Major Import Products: Chemical and other manufactured goods; machinery and transport equipment; mineral fuels; foodstuffs; crude material

Energy Overview

Proven Oil Reserves (1/1/03E): 59 million barrels

Oil Production (2002E): 81,105 barrels per day (bbl/d), of which 5,332 bbl/d was crude oil

Oil Consumption (2002E): 5.32 million bbl/d

Net Oil Imports (2002E): 5.24 million bbl/d

Crude Oil Refining Capacity (1/1/03E): 4.8 million bbl/d

Major Crude Oil Import Sources (2003E): United Arab Emirates, Saudi Arabia, Kuwait, Iran, Qatar, Indonesia

Natural Gas Reserves (1/1/03E): 1.4 trillion cubic feet (Tcf)

Natural Gas Production (2001E): 0.09 Tcf

Natural Gas Consumption (2001E): 2.84 Tcf

Net Natural Gas Imports (2001E): 2.75 Tcf

Coal Reserves (2001E): 852 million short tons (Mmst)

Coal Production (2001E): 3.5 million short tons (Mmst)

Coal Consumption (2001E): 165.9 Mmst

Net Coal Imports (2001E): 163.4 Mmst

Electric Generation Capacity (1/1/01E): 235 gigawatts

Electricity Production (2001E): 1,037 billion kilowatthours

Environmental Overview

Total Energy Consumption (2001E): 21.9 quadrillion Btu* (5.4 per cent of world total energy consumption)

Energy-Related Carbon Emissions (2001E): 315.8 million metric tons of carbon (4.8 per cent of world total carbon emissions)

Per Capita Energy Consumption (2001E): 172.2 million Btu (vs U.S. value of 341.8 million Btu)

Per Capita Carbon Emissions (2001E): 2.5 metric tons of carbon (vs U.S. value of 5.5 metric tons of carbon)

Energy Intensity (2001E): 3,879 Btu/ $1995 (vs U.S. value of 10,736 Btu/ $1995)**

Carbon Intensity (2001E): 0.06 metric tons of carbon/thousand $1995 (vs U.S. value of 0.17 metric tons/thousand $1995)**

Fuel Share of Energy Consumption (2001E): Oil (50.1 per cent), Coal (16.8 per cent), Natural Gas (13.5 per cent)

Fuel Share of Carbon Emissions (2001E): Oil (57.6 per cent), Coal (28.8 per cent), Natural Gas (13.5 per cent)

Status in Climate Change Negotiations: Annex I country under the United Nations Framework Convention on Climate Change (ratified May 28th, 1993). Under the negotiated Kyoto Protocol (signed on April 28, 1998; ratified June 4, 2002), Japan has agreed to reduce greenhouse gases 6 per cent below 1990 levels by the 2008-2012 commitment period.

Major Environmental Issues: Air pollution from power plant emissions results in acid rain; acidification of lakes and reservoirs degrading water quality and threatening aquatic life; Japan is one of the largest consumers of fish and tropical timber, contributing to the depletion of these resources in Asia and elsewhere.

Major International Environmental Agreements: A party to the Antarctic-Environmental Protocol, Antarctic Treaty, Biodiversity, Climate Change, Desertification, Endangered Species, Environmental Modification, Hazardous Wastes, Law of the Sea, Marine Dumping, Nuclear Test Ban, Ozone Layer Protection, Ship Pollution, Tropical Timber 83, Tropical Timber 94, Wetlands and Whaling.

The total energy consumption statistic includes petroleum, dry naturai gas, coal, net hydro, nuclear, geothermal, solar, wind, wood and waste electric power. The renewable energy consumption statistic is based on International Energy Agency (IEA) data and includes hydropower, solar, wind, tide, geothermal, solid biomass and animal products, biomass gas and liquids, industrial and municipal wastes.

Energy Industries

Organizations: Electric Power Development Co. (EPDC)–a quasi-governmental wholesale power company established in 1952 to help secure stable supplies of electricity. EPDC is scheduled for privatization within the next 5 years. Power Reactor and Nuclear Fuel Development Corp. (PNC, or Donen)–the government nuclear oversight organization; Japan National Oil Company–government owned

Major Oil Companies: Arabian Oil (Tokyo), Cosmo Oil, Tonen General Sekiyu, Indonesia Petroleum Ltd., Itochu, Japan Energy, Japan National Oil Corp. (JNOC), Japan Petroleum Exploration., Mitsui Oil Exploration, Nippon Mitsubishi Oil, Showa Shell Sekiyu, Sodeco, Sumitomo, Teikoku Oil

Major Electric Power Companies: Chubu Electric, Chugoku Electric, Hokkaido Electric, Hokuriku Electric, Japan Atomic Power, Kansai Electric, Kyushu Electric, PNC, Shikoku Electric, Tohoku Electric, Tokyo Electric

Major Refineries (capacity, bbl/d): Negishi (365,750), Ichihara–Chiba (228,000), Mizushima (190,000), Cosmo–Chiba (228,000), Showa Yokkaichi (222,000), Tonen Kawasaki (292,950)

Pipelines: Crude oil 52 miles; petroleum products 200 miles; natural gas 1,116 miles

48

Jordan

Jordan occupies a strategic location in the Middle East, and is an important crossroads for regional energy integration. Jordan also is a potential alternative transit center for oil and gas exports from the Persian Gulf region.

General Background

Despite tensions in the Middle East, Jordan's economy has been experiencing growth. Real gross domestic product (GDP) grew at 4.9 per cent in 2002, but is projected to drop back to 3.5 per cent in 2003. Much of the recent growth stems from expansion in the country's manufacturing sector. This has been driven in part by the U.S.-Jordan Free Trade Area agreement, which was ratified by the U.S. Senate in September 2001.

Exports of goods to neighbors Iraq and Saudi Arabia also are important to Jordan's economy, and growth in regional exports is helping to further Jordan's economic recovery. Remittances from Jordanian workers in the Persian Gulf countries also are very important to Jordan's balance of payments.

Jordan was admitted to membership in the World Trade Organization in April 2000, after agreeing to a package of trade and investment liberalization measures, as well as improvement in protections for foreign-owned intellectual property.

A privatization program has been undertaken to reduce the Jordanian government's stake in sectors of the economy previously dominated by state-controlled firms. The Jordanian government sold a 40 per cent stake in the Jordan Telecommunications Company to France Telecom in 2000, in the most significant privatization to date. Oil and electric power generation also are targeted for privatization, but progress in the energy sector has been relatively slow. As part of its preparations for admission to the WTO, Jordan in 2000 lifted most limits on foreign ownership of formerly state-owned companies.

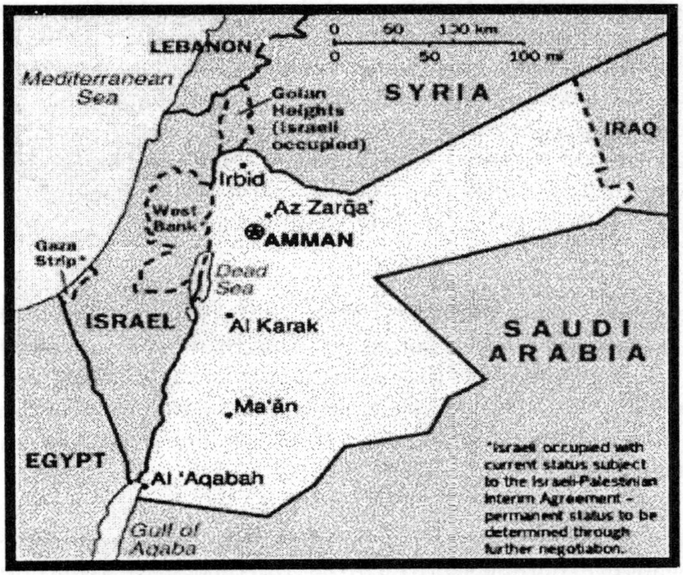

Oil

Jordan has no significant oil resources of its own, and relies on Iraqi oil for nearly all of its needs (around 106,000 barrels per day–bbl/d–in 2002). Jordan's oil imports from Iraq are permitted by the United Nations under a special dispensation from the general U.N. sanctions regime on Iraq.

In December 2002, Jordan's government renewed its agreement with Iraq on oil supplies. Under the agreement, Jordan receives half of its crude oil free of charge, and receives steeply discounted prices for the rest. In addition to the crude oil imports, Jordan also imports about 20,000 bbl/d of refined petroleum products from Iraq, also priced at a substantial discount. Jordan has reportedly held discussions with potential alternate suppliers in the Persian Gulf about replacement of Iraqi oil supplies if they were were disrupted, but has stressed that no final agreements for such supplies have been reached. Another measure taken is the acquisition of an oil tanker, with a capacity of 2 million barrels, which is berthed at the port of Aqaba. It has been used to build up a strategic reserve, and has reportedly now been filled to capacity.

Jordan has one refinery, at Zarqa, with a capacity of 90,400 bbl/d. An expansion of the facility to a capacity of 150,000 bbl/d is under consideration, but has not yet been implemented. Jordan and Iraq had agreed in 1998 to build a pipeline for the transport of Iraqi oil to Jordan's Zarqa refinery, and renewed this commitment in their most recent oil supply agreement. This would eliminate the necessity of transporting oil over 600 miles of highway from Iraq with a fleet of 1,500 tanker trucks. The Jordanian government received bids on the project in late 2002. However, it seems doubtful that the project will move forward in the near future.

Jordan does possess a significant quantity of oil shale resources, possibly as much as 40 billion tons. Canada's Suncor has conducted limited exploration digging in the Lajjun area, southwest of Amman, and has conducted discussions with the Jordanian government on the possible development of an oil shale extraction facility, but no development agreement has been signed.

Jordan's state Natural Resources Authority (NRA) has been promoting oil exploration within the country, which has been relatively unexplored until now. TransGlobal Corporation holds a concession for the Wadi Araba area in Western Jordan. Other small independent companies have conducted surveys of other areas as well, but without yet finding commercial quantities of oil. To help attract foreign investment, the Jordanian government has plans to privatize its oil sector. In October 1995, the country set up the state-owned National Petroleum Co. (NPC) to handle upstream oil and gas exploration and development. In mid-1999, NPC divested its oil-drilling operation, which now operates as Petra Drilling Company. NPC is still active in the natural gas sector.

Oil Transportation

A comprehensive settlement of the Arab-Israeli conflict could affect Middle East oil flows significantly. Jordan's geographic location between the Arabian peninsula and the Mediterranean coastal states of Israel and Lebanon offers the potential for alternative oil export routes for Persian Gulf oil to the West. At present, these oil exports must travel either by ship (through the Strait of Hormuz), by pipeline from Iraq to Turkey (capacity 0.8-1.6 million bbl/d), or via the Sumed (Suez-Mediterranean) Pipeline (capacity 2.4 million bbl/d).

Utilization of the Trans-Arabian Pipeline (Tapline) could offer another potentially economic alternative. The Tapline was originally constructed in the 1940s with a capacity of 500,000 bbl/d, and intended as the main means of exporting Saudi oil to the West (via Jordan to the port of Haifa, then part of Palestine, now a major Israeli port city). Following the establishment of the state of Israel, the Tapline's terminal was diverted from Haifa to Sidon, Lebanon (through Syria and Lebanon).

Partly as a result of turmoil in Lebanon, and partly for economic reasons, oil exports via the Tapline were halted in 1975. In 1983, the Tapline's Lebanese section was closed altogether. Afterwards, the Tapline was used exclusively to supply oil to Jordan, from 1983 until Saudi Arabia terminated this arrangement in 1990 after the Iraqi invasion of Kuwait. With the future of Iraqi oil supplies to Jordan open to question, the Tapline has once again begun to attract attention as a potential alternate source of oil for the country. According to press reports, Saudi Arabia has quietly started examining the condition of the pipeline and the feasibility of returning it to operation if necessary, but given the condition of the pipeline, that could not be done quickly.

Natural Gas

Jordan has modest reserves of natural gas, 230 billion cubic feet (Bcf), and has developed one gas field, at Risha in the eastern desert near Iraq. The current output of around 30 million cubic feet per day (Mmcf/d) from the Risha field is used to fuel one nearby power plant, which generates about 10 per cent of Jordan's electricity.

For several years, Jordan has been exploring the option of importing natural gas from Egypt. In 1999, a decision was made to delay imports until a more thorough evaluation of reserves at the Risha field was completed. When this review showed that quantities available were not sufficient to meet the country's needs, Jordan decided to reopen talks on imports from Egypt. In May 2001, a 30-year agreement was concluded with Egypt for gas sales to begin at a rate of 100 MMcf/d in 2003. Construction of the section of the pipeline in Egypt began in late 2001, starting from the existing pipeline terminus at El-Arish in Sinai. The Egyptian state-owned firm EGAS was awarded a contract by the government of Jordan in August 2002 for the construction of the Jordanian section of the pipeline, from Aqaba to power plants near Amman. Most of the construction work is complete, and the pipeline is slated to begin operation in May 2003.

Arab governments also have been discussing the potential of extending the Egypt-Jordan pipeline and increasing its capacity–dubbed the Arab Gas Pipeline project. The extensions under discussion have included a link to Syria and Lebanon, an extension to Turkey, and a subsea pipeline from Lebanon to Cyprus. At present, though, it seems unlikely that extensions to Turkey or Cyprus will be built in the near term.

Electricity

Almost all electricity production in Jordan currently is carried out by the National Electric Power Company (NEPCO), a state-owned utility, and its subsidiaries. The Zarqa power plant, with a capacity of 400 megawatts (MW), and the Aqaba power plant, with a capacity of 650 MW, are the country's two main power generation facilities. Electricity demand is growing–around 3.5 per cent in 2001–and the Jordanian government has been seeking ways to attract foreign capital to fund additional capacity. In December 2001, the Jordanian government awarded a contract to Rothschild and Sons of Britain to provide advice on the privatization scheme. Jordan's basic plan for the future of its electric utility system involves having NEPCO maintain ownership of transmission assets, but relying on private power generation and privatizing existing generation assets. NEPCO's distribution subsidiary, the Electricity Distribution Company (EDCO), also is to be privatized. Two other private distribution firms already exist–the Jordan Electric Power Company (JEPCO) in the Amman area and the Irbid District Electricity Company (IDECO) covering the area around Irbid.

In May 2000, Jordan awarded its first contract for an independent power producer (IPP) to Tractebel of Belgium. The plant will have a capacity of 450 MW, and will be located near Amman. It is to be fuelled with natural gas from Egypt when it is completed in 2004.

An area of potential regional cooperation involves integration of individual national power transmission grids into a regional power network. Such a network would, among other benefits, allow power companies to take advantage of differences in peak demand periods, reduce the need for (and the costs associated with) installation and maintenance of reserve power generating capacity, and provide outlets for surplus generating capacity (mainly from Israel to Jordan).

The first step in this enterprise, the linking of the Egyptian and Jordanian power grids via an underwater cable between Aqaba and Taba (across the Gulf of Aqaba in Sinai), was completed in October 1998 and formally inaugurated in April 1999. Syria and Jordan also have linked their electric grids. Israel, for the time being, has been excluded from the grid linking projects, but has continued discussions with Jordan on the subject and may join the network once political hurdles related to the Arab-Israeli peace process have been overcome.

Country Overview

Independence: May 25, 1946 (from the United Kingdom)

Population (2002E): 5.3 million

Location/Size: Middle East, northwest of Saudi Arabia/89,213 sq. kilometers, slightly smaller than Indiana

Major Cities: Amman (capital), Irbid, Al'Aqabah, Ma'an

Languages: Arabic (official), English

Ethnic Groups: Arab (98 per cent), Circassian (1 per cent), Armenian (1 per cent)

Religions: Sunni Muslim (96 per cent), Christian (4 per cent)

Defense (8/98): Army (90,000); Air Force (13,400); Navy (650); Reserves (35,000); Paramilitary Forces (30,000)

Economic Overview

Currency: Jordanian Dinar (JD)

Market Exchange Rate (3/18/03): US$1 = JD 0.71

Gross Domestic Product (GDP) (2002, market exchange rate): $9.5 billion

Real GDP Growth Rate (2002E): 4.9 per cent *(2003F):* 3.5 per cent

Consumer Price Inflation (2002E): 1.8 per cent *(2003F):* 1.5 per cent

Major Trading Partners: Iraq, USA, Germany, India, Italy, United Kingdom, Saudi Arabia

Merchandise Exports (2002E): $5.1 billion

Merchandise Imports (2002E): $6.1 billion

Merchandise Trade Balance (2002E): -$1.0 billion

Major Export Products: Chemicals; phosphates; potash; food; manufactures; textiles

Major Import Products: Food; machinery; chemicals; transport equipment; crude oil

Current Account Balance (2002E): -$10.0 million

Total External Debt (2002E): $8.9 billion

International Reserves (2002E): $4.1 billion

Energy Overview

Proven Oil Reserves (1/1/03E): 890,000 barrels

Oil Production (2002E): 40 bbl/d

Oil Consumption/Net Imports (2002E): 106,000 bbl/d

Main Oil Import Source: Iraq

Crude Oil Refining Capacity (1/1/03E): 90,400 bbl/d

Natural Gas Reserves (1/1/03E): 230 billion cubic feet (Bcf)

Natural Gas Production/Consumption (2001E): 10 Bcf

Electric Generation Capacity (1/1/02E): 1.47 gigawatts (99.6 per cent thermal; 0.4 per cent hydro)

Electricity Generation (2001E): 7.1 billion kilowatthours

Environmental Overview

Total Energy Consumption (2001E): 0.23 quadrillion Btu* (<0.1 per cent of world total energy consumption)

Energy-Related Carbon Emissions (2001E): 4.3 million metric tons of carbon (<0.1 per cent of world carbon emissions)

Per Capita Energy Consumption (2001E): 33.7 million Btu (vs U.S. value of 341.8 million Btu)

Per Capita Carbon Emissions (2001E): 0.63 metric tons of carbon (vs U.S. value of 5.5 metric tons of carbon)

Energy Intensity (2001E): 28,986 Btu/ $1995 (vs U.S. value of 11,014 Btu/ $1995)**

Carbon Intensity (2001E): 0.54 metric tons of carbon/thousand $1995 (vs U.S. value of 0.17 metric tons/thousand $1995)

Fuel Share of Energy Consumption (2001E): Oil (94.0 per cent), Natural Gas (4.6 per cent), Coal (0.0 per cent)

Fuel Share of Carbon Emissions (2001E): Oil (96.4 per cent), Natural Gas (3.6 per cent), Coal (0.0 per cent)

Status in Climate Change Negotiations: Non-Annex I country under the United Nations Framework Convention on Climate Change (ratified November 12th, 1993). Not a signatory to the Kyoto Protocol.

Major Environmental Issues: Limited natural fresh water resources; deforestation; overgrazing; soil erosion; desertification.

Major International Environmental Agreements: A party to Conventions on Biodiversity, Climate Change, Desertification, Endangered Species, Hazardous Wastes, Law of the Sea, Marine Dumping, Nuclear Test Ban, Ozone Layer Protection and Wetlands.

The total energy consumption statistic includes petroleum, dry natural gas, coal, net hydro, nuclear, geothermal, solar, wind, wood and waste electric power. The renewable energy consumption statistic is based on International Energy Agency (IEA) data and includes hydropower, solar, wind, tide, geothermal, solid biomass and animal products, biomass gas and liquids, industrial and municipal wastes. Sectoral shares of energy consumption and carbon emissions are also based on IEA data.

Energy Industries

Organization: Natural Resources Authority (NRA)–state body responsible for overall direction of Jordan's energy resources; National Petroleum Company; Petra Drilling Company; Jordan Petroleum Refinery Co.–operates Jordan's single refinery at Zarqa, near Amman; National Electric Power Corporation (NEPCO)

Major Port: Aqaba

Major Oil and Gas Fields: Risha (natural gas)

Major Pipelines: Tapline–closed (Ras Tanura–Haifa)

Major Refineries (crude capacity): Zarqa (90,400 bbl/d)

49

Kazakhstan

Kazakhstan is important to world energy markets because it has significant oil and natural gas reserves. As foreign investment pours into the country's oil and natural gas sectors, the landlocked Central Asian state is beginning to realize its enormous production potential. With sufficient export options, Kazakhstan could become a major world energy producer and exporter over the next decade.

General Background

Kazakhstan has the Caspian Sea region's largest recoverable crude oil reserves, and accounts for approximately two-thirds of the roughly 1.5 million barrels per day (bbl/d) currently being produced in the region (including regional oil producers Kazakhstan, Azerbaijan, and Turkmenistan). Accordingly, Kazakhstan has Central Asia's largest economy. Kazakhstan's real gross domestic product (GDP) grew by 9.5 per cent in 2002, to $24.5 billion, resulting in a per capita GDP of $1,640 (roughly comparable to Guatemala and Ecuador). This marked only the third consecutive year of significant economic growth in Kazakhstan since independence.

Economic growth in recent years has been propelled by Kazakhstan's growing petroleum industry, and as a result, the state's budget revenue is roughly 55 per cent dependent on the oil sector. Several economic research efforts conducted in 2002 and 2003 highlighted the growing danger of possible over-reliance on the oil sector, with some analysts predicting that without more investment into the country's non-oil sectors, Kazakhstan's economic capacity will be strained by 2007, thus stifling growth in the next decade. In an effort to reduce Kazakhstan's exposure to price fluctuations for energy and commodities exports, the government has created the National Fund of Kazakhstan. As of June 2003, the National Fund held $2.3 billion.

Kazakhstan's President, Nursultan Nazarbaev, has been involved in Kazakhstan's national politics since 1977 when he served as Secretary of the Central Committee of the Communist Party of Kazakhstan. In April 1990, he became interim president of the newly independent Republic of

Kazakhstan, and was later elected to the post in the country's first national elections, held in December 1991. Nazarbaev was re-elected in 1999 (after a 1995 referendum extended his term) and will be up for reelection again in 2006. The Kazakhstani executive branch was re-shuffled in June 2003 when then Prime Minister Imanghaliy Tasmaghambetov resigned from his position. A new Prime Minister, Daniyal Akmetov, has been appointed along with a new cabinet, many of which are reportedly holdovers from the previous administration.

Oil

Kazakhstan sits astride the northeast portion of the Caspian Sea and claims most of the Sea's biggest known oil fields. Kazakhstan's combined onshore and offshore proven hydrocarbon reserves have been estimated to be between 9 and 17.6 billion barrels (comparable to OPEC members Algeria on the low end and Qatar on the high end). Although only a minor world oil exporter in 2002, Kazakhstan is poised to become a more significant player in world oil markets over the next decade.

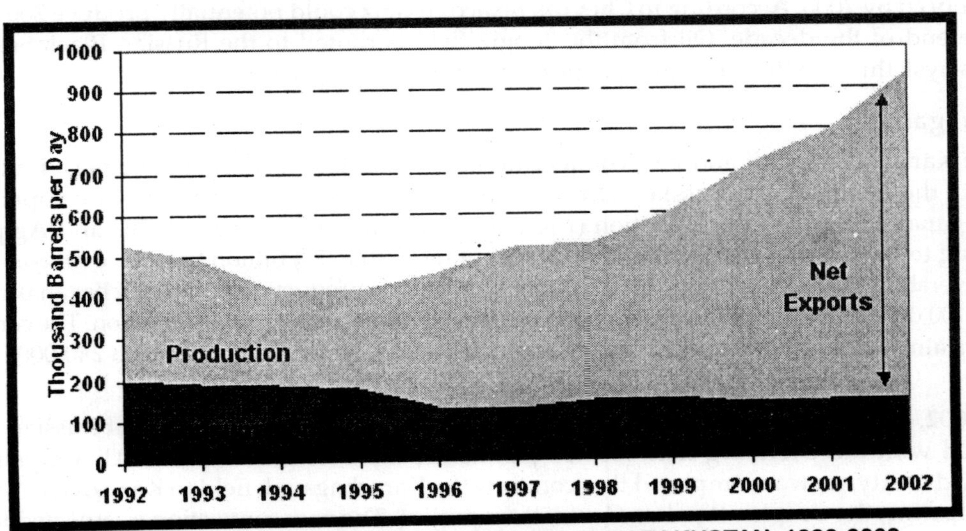

OIL PRODUCTION AND CONSUMPTION IN KAZAKHSTAN, 1992-2002

Kazakhstan produced approximately 939,000 barrels per day (bbl/d) of total liquids in 2002 and consumed only 140,000 bbl/d, resulting in net exports of 799,000 bbl/d. Markets for exported Kazakhstani oil are growing rapidly, with oil being delivered to world markets through the Black Sea (via Russia) and the Persian Gulf (via swaps with Iran), as well as some additional traffic northward to Russia via pipeline and rail.

Between 1999 and 2002, Kazakhstan's oil production grew year-on-year by approximately 16 per cent annually, resulting in a doubling (roughly) of oil production since independence (see graph). Conversely, other major economic indicators declined markedly during the decade since independence (including GDP, and the production and consumption of natural gas, coal, and electricity–see below). Increased oil production has been the result of an influx of foreign investment into Kazakhstan's oil sector. International projects have taken the form of joint ventures with Kazmunaigaz (formerly Kazakhoil), the national oil company, as well as production-sharing agreements (PSAs), and exploration/field concessions. In June 2003, the government of Kazakhstan announced a new Caspian Sea development program, which calls for new offshore blocks to be auctioned beginning in 2004. According to the plan, the first offers will be made to Kazmunaigaz which will then conduct tenders for partnerships. According to statements made in June 2003, the government of Kazakhstan expects that the country will produce 2.4 million bbl/d (120 million tons per year) by 2010, and 3.6 million bbl/d (180 million tons per year) by 2015. Most of this growth will come from three enormous fields: Tengiz, Karachaganak, and Kashagan.

Tengiz

The Tengiz field is located in the swamplands along the northeast shores of the Caspian Sea. Recoverable crude oil reserves have been estimated at six to nine billion barrels by consortium member ChevronTexaco. Tengiz has been developed by the Tengizchevroil (TCO) joint venture (ChevronTexaco 50 per cent, ExxonMobil 25 per cent, Kazmunaigaz 20 per cent, LukArco 5 per cent) since 1993, and in 2002, the consortium produced 285,000 barrels per day, or one third of Kazakhstan's daily production. In January 2003, after contentious negotiations with the government of Kazakhstan, the TCO consortium members initiated a $3 billion expansion project designed to boost production to approximately 450,000 bbl/d by 2006. According to ChevronTexaco, Tengiz could potentially produce 700,000 bbl/d by the end of the decade. Oil from the Tengiz field is routed to the Russian Black Sea port of Novorossiysk through the Caspian Pipeline Consortium (CPC) project.

Karachaganak

The Karachaganak oil and gas/condensate field is located onshore, in northern Kazakhstan, and near the border with Russia's Orenburg field. Karachaganak is being developed by the Karachaganak Integrated Organization (KIO), a consortium led by Britain's BG and Agip (Italy). According to British Gas, the field holds reserves of more than 2.4 billion barrels of oil and 16 Tcf of gas, recoverable over the 40-year life of the project. Condensate production from Karachaganak averaged roughly 100,000 bbl/d in 2002, representing 11 per cent of total Kazakhstani production. The consortium members aim to increase output from Kashagan to 140,000 bbl/d in 2003, and to 240,000 bbl/d by 2008.

In 2002, all of Karachaganak's liquid condensate production was processed at Russian facilities associated with the Orenburg field located just across the border. In April 2003, a pipeline spur southward to Atyrau was completed that connects the Karachaganak field to Kazakhstan's primary export pipeline, the Caspian Pipeline Consortium project. The new connection is expected to enable

increased exports from Karachaganak, and reduce the consortium members' dependence on Russian buyers. Initial shipments through the new pipeline are expected in August 2003. The consortium members are also working with the government of Kazakhstan to develop a plan for the construction of an on-site gas processing facility for the field. A feasibility study has been commissioned and is expected to be completed at the end of 2003.

Kashagan

The Kashagan field is located off the northern shore of the Caspian Sea, near the city of Atyrau. Although the field is still being appraised, in June 2002 the consortium operating the field, the Agip Kazakhstan North Caspian Operating Company–Agip KCO (formerly known as OKIOC), estimated the field's recoverable reserves at 7-9 billion barrels of oil equivalent, with further potential totaling 9 to 13 billion barrels using secondary recovery techniques (gas injection, for example). Assuming proven crude oil reserves in the neighborhood of 8 billion barrels, the Kashagan field alone would hold roughly the same amount of oil as Brazil, South America's second largest oil producer. Output at the first stage of development, planned for 2005, is expected to reach 100,000 bbl/d, with significant potential for further development. However, Kazakhstan needs to resolve two major issues–Caspian ownership rights and export routes–before it can reach its full oil-producing potential.

Oil Exports

In 2002, Kazakhstani oil was exported in three directions: northward (via the Russian pipeline system and rail network); westward (via the Caspian Pipeline Consortium Project and barge to Azerbaijan); and southward (via swaps with Iran). Connections to ports on the Black Sea and the

Persian Gulf have allowed some Kazakhstani oil (or proxy oil from Iran) to be traded on the world market. Efforts are underway to expand the country's export infrastructure over the next decade as Kazakhstan's oil production is expected to increase.

Caspian Pipeline Consortium (CPC)

The CPC connects Kazakhstan's Caspian Sea area oil deposits with Russia's Black Sea port of Novorossiysk. The CPC project, which was developed by the governments of Russia, Kazakhstan, and Oman in conjunction with a consortium of international oil companies, is actually an extension of the existing oil transit infrastructure surrounding the Caspian Sea. Newly constructed components of the line run from the Russian town of Komsomolskaya, straight westward to Novorossiysk. The pipeline is supplied with Kazakhstani oil through the Soviet-era links surrounding the Sea, which the consortium members have refurbished. The first crude oil was loaded onto a tanker in Novorossiisk on October 15, 2001, and the pipeline was officially opened on November 27, 2001. Initial capacity of the CPC pipeline is 560,000 bbl/d, with plans to increase capacity to 1.34 million bbl/d.

In 2002, CPC transited roughly 260,000 bbl/d, or roughly one-third of Kazakhstan's exports. Most of this oil came from the Tengiz field. Throughput at CPC is expected to increase in 2003 as two pipeline spurs (one from the Kenkiyak area completed in March 2003, and another from the Karachaganak field which is near completion) are expected to bring throughput to between 350,000 and 415,000 bbl/d by the end of the year.

Atyrau-Samara

Kazakhstan's other major oil export pipeline is a northbound link to the Russian distribution system, Atyrau-Samara. Before the CPC pipeline was completed at the end of 2001, almost all of Kazakhstan's exports were distributed through this system. In June 2002, Kazakhstan and Russia signed a 15-year oil transit agreement under which Kazakhstan will export at least 350,000 bbl/d of oil annually via the Russian pipeline system. As the CPC project grows with Kazakhstani production, absolute volumes though Atyrau-Samara are expected to grow, but this pipeline will become relatively less significant. For example, in 2001, Atyrau-Samara transited virtually all of Kazakhstan's exports, whereas in 2003, according to the Kazakhstani Ministry of Energy and Mines, Atyrau-Samara will account for less than half of Kazakhstan's export oil transit.

Besides the CPC and Atyrau-Samara, Kazakhstani exports are sent for swaps in Iran, via rail to Russia, and across the Caspian by barge. Other export routes reportedly under consideration are a possible line to China, and a subsea trans-caspian pipeline connecting to the Baku-Tiblisi-Ceyhan project.

Downstream/Refining

Kazakhstan has three major oil refineries supplying the northern region (at Pavlodar), western region (at Atyrau), and southern region (at Shymkent), with total refining capacity of 427,000 bbl/d. The refinery at Pavlodar is supplied mainly by a crude oil pipeline from western Siberia (since Russian reserves are well placed geographically to serve that refinery), the Atyrau refinery runs solely on domestic crude from northwest Kazakhstan, and the Shymkent refinery currently uses oil from Kazakh fields at Kumkol, Aktyubinsk, and Makatinsk, although it is linked by pipeline to Russia. Reconstruction of the Atyrau refinery is expected to begin in 2004 and will be carried out by the Marubeni Corporation (Japan).

Natural Gas

Despite Kazakhstan's sizeable proven natural gas reserves of 65 trillion cubic feet (Tcf) (comparable to Canada and Kuwait and ranking it in the top 20 countries in the world), the country is currently a net natural gas importer. In 2001, Kazakhstan produced 360 billion cubic feet (Bcf) and consumed 505 Bcf, resulting in net imports of 145 Bcf. Most of Kazakhstan's natural gas imports come from Uzbekistan and go to the south of the country.

Natural gas production in Kazakhstan has increased significantly since 1999. In August 1999, the Kazakh government passed a law requiring subsoil users (such as oil companies) to include natural gas utilization projects in their development plans. As a result, natural gas production in 2000 doubled compared to 1999, reaching 314 billion cubic feet, the highest level since independence. Production increased again in 2001, and was 16 per cent higher than in 2000. Preliminary estimates indicate that Kazakhstan's natural gas production increased even further in 2002. According to the 15-year strategy of the Kazakhstani Ministry for Energy and Mineral Resources, the country plans to increase its natural gas production to 1.2 Tcf by 2005, to 1.66 Tcf by 2010, and to 1.84 Tcf by 2015.

Most of Kazakhstan's natural gas reserves are located in the west of the country, with roughly 25 per cent of proven reserves situated in one field, Karachaganak. The Karachaganak oil and gas condensate field reportedly has proven natural gas reserves of 16 Tcf. Press reports indicate that in 2001 the field yielded 132 Bcf of natural gas, representing 37 per cent of the country's total production. Because of Kazakhstan's disjunct distribution network, Karachaganak's natural gas is exported northward to Russia's Orenburg processing plant (as opposed to being delivered to Kazakhstani consumers in the south–see below). Efforts are also underway to export Karachaganak's gas condensate and other liquids through the CPC pipeline system. The Karachaganak Integrated Organization, which is developing the field, has thus far focused its efforts primarily on extraction of the field's liquid condensate reserves. Several of the country's other oil fields, Tengiz and Kashagan for example, also contain associated natural gas.

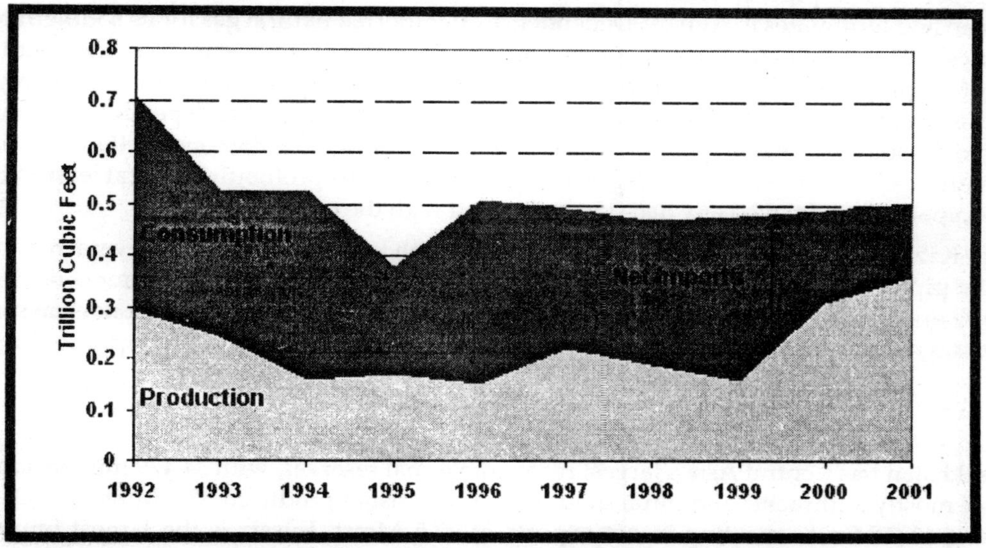

NATURAL GAS PRODUCTION AND CONSUMPTION IN KAZAKHSTAN, 1992-2001

Another important natural gas field, Amangeldy, is situated in the south of the country, near Zhambul. Exploratory drilling in 2001 indicated reserves of up to 1.8 Tcf. The field is being developed by Kazmunaigas in conjunction with Repsol-YPF (Spain), and the partners expect initial production of roughly 24 Bcf by the end of 2003. Because of the field's key location in the energy-strapped south, the government of Kazakhstan has expressed its hope that Amangeldy will allow the country to cease import dependence in the near future.

Natural Gas Distribution

Kazakhstan has two separate domestic natural gas distribution networks, one in the west which services the country's producing natural gas fields, and one in the south which mainly delivers imported natural gas to the southern consuming regions. The lack of internal pipelines connecting Kazakhstan's natural gas-producing areas to the country's industrial belt (between Almaty and Shymkent) has hampered the development of natural gas resources. Kazmunaigaz, the state oil and natural gas company operates Kazakhstan's main natural gas pipelines.

In the north, Kazakhstan is developing its ability to export its natural gas through Russia's natural gas pipeline system. In 2002, natural gas from the Karachaganak field was sent northward to Russia's Orenburg gas processing plant; however, efforts are currently underway to to expand that link and boost export capacity. Russia and Kazakhstan have agreed on preliminary export schedules for 2003 and 2004, which call for 212 Bcf and 247 Bcf respectively to be sent northward to the Orenburg plant for processing. Some of the gas will then be routed for marketing in the Russian system and some will be sent back to Kazakhstan. In July 2002, Kazmunaigaz, working in conjunction with Gazprom (Russia) under the joint venture KazRosGas, began a $500 million program to upgrade Kazakhstan's natural gas pipeline network. The program, which is being developed with the financial backing of several international banks, will increase Kazakhstan's export capacity in the north, beginning in 2005.

Southern Kazakhstan receives its natural gas supplies from Uzbekistan via the Tashkent-Bishkek-Almaty pipeline. This pipeline snakes through Uzbekistan before reaching Shymkent, then transits Kyrgyzstan and terminates in Almaty. Dependence on imported natural gas for its southern regions has at times been problematic as erratic pricing and supplies from Uzbekistan, combined with illegal tapping of the pipeline by Kyrgyzstan, has resulted in significant supply disruptions to Almaty in the middle of the heating season. As a result, Kazakhstan is determined to end its dependence on imported supplies for its southern regions. Although Kazakhstan has considered the construction of an internal north-south pipeline, thereby alleviating import dependency, the prohibitive cost (at least $1 billion) of such a pipeline has delayed any decision to proceed with the project.

Since Kazakh natural gas is a potential competitor with Russian natural gas, several new natural gas export pipelines from the Caspian Sea region also are in development or under consideration, potentially opening up new markets for Kazakh natural gas. In the meantime, Kazakhstan serves as an important natural gas transit center for Turkmen and Uzbek natural gas that is piped to Russia and beyond.

Coal

Kazakhstan has Central Asia's largest recoverable coal reserves, with 34,479 million short tons (Mmst) of mostly anthracitic and bituminous coal. Kazakhstan produced 88.5 Mmst in 2001, while consuming 61.0 Mmst, resulting in net exports of 27.5 Mmst. Russia is the largest importer of

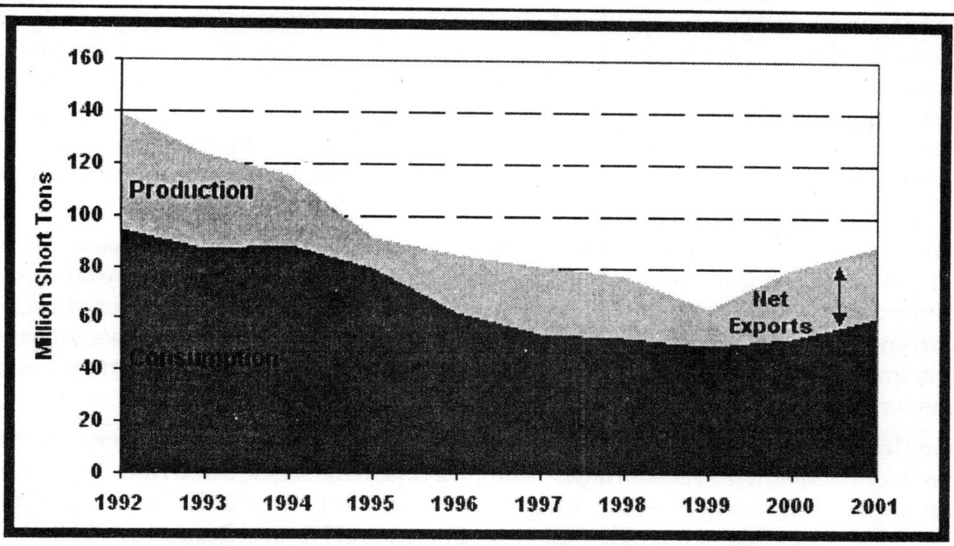

COAL PRODUCTION AND CONSUMPTION IN KAZAKHSTAN, 1992-2001

Kazakhstani coal, followed by Ukraine.

Coal production in Kazakhstan, which was the Soviet Union's third-largest producer behind Russia and Ukraine, has fallen by roughly 50 per cent since independence. EIA data show a modest upswing in coal production in 2000 and 2001, however, preliminary 2002 estimates indicate that output fell again last year. According to the Ministry of Energy and Natural Resources, the country plans to increase coal production in 2003 by approximately 5 per cent over 2002 levels, with a further boost of 25 per cent expected by 2010.

Kazakhstan's largest coal producer, Bogatyr Access Komir, which accounts for roughly 35 per cent of the country's coal output, is a subsidiary of Access Industries Incorporated (U.S.A.). Bogatyr Access Komir develops northern Kazakhstan's Bogatyr and Severny coal fields and is the country's largest exporter to Russia. Russian firms are also stake holders in the Kazakhstani coal industry and roughly 16 Mmst are transited annually from Kazakhstan northward via rail to power plants in southern Russia.

Since independence, Kazakhstani coal consumption has fallen from 94 Mmst in 1992 to 61 Mmst in 2001. As a percentage of total energy consumption, coal accounted for 41 per cent in 2001, down slightly from 46 per cent in 1992. The majority of Kazakhstan's electric generating plants are coal-fired, including the country's largest power generator, Ekibastuz No. 1, located in north-central Kazakhstan.

Electricity

Kazakhstan has 71 power plants, including five hydroelectric power stations, giving the country an overall installed generating capacity of 17 gigawatts (GW), 88 per cent of which is thermal, and 12 per cent of which is hydroelectric. Most of the country's power generation comes from coal-fired plants located in the northern coal producing regions. Kazakhstan's hydroelectric facilities are located primarily along the Irtysh river, which flows from China across northeast Kazakhstan.

The production and consumption of electricity in Kazakhstan fell significantly following independence. However, robust economic growth since 2000 has helped to boost production to 52.4 billion kilowatthours (Bkwh) in 2001, and consumption to 48.4 Bkwh, marking the first time since independence that Kazakhstan's electricity generation has significantly exceeded its consumption. Transmission issues, however, necessitate that Kazakhstan continue to import electricity in the southern part of the country, as the country's northern generating units are connected to a separate transmission grid.

The decade-long decline in Kazakhstani electricity consumption has come primarily at the expense of thermal power, while generation at hydroelectric facilities has remained constant. As a result, hydropower now accounts for almost 20 per cent of Kazakhstan's electricity consumption, more than twice its percentage in 1992, making the Irtysh River, which starts upstream in China's Altai mountains, increasingly important as a source of hydropower. Kazakhstan and China have held joint negotiations on management of the Irtysh River since 1999.

Although Kazakhstan technically generates enough electricity to meet its demand, the country has suffered from frequent power shortages since 1992 due to the sector's deteriorating infrastructure. Kazakhstan incurs large energy losses during transmission and distribution over its 285,000 miles of distribution lines. According to Kazakh Minister of Energy and Natural Resources Vladimir Shkolnik, an average of 15 per cent of the electricity generated in Kazakhstan is lost before it reaches consumers, owing to the widespread deterioration of Kazakhstan's power infrastructure.

Transmission and Distribution

Kazakhstan's electricity transmission and distribution system is divided into three networks, two in the north and one in the south, each of which is connected to external systems (Russia in the north, and the Unified Energy System of Central Asia in the south). However, these transmission systems are not yet sufficiently interconnected. The northern networks, which service the coal-fired

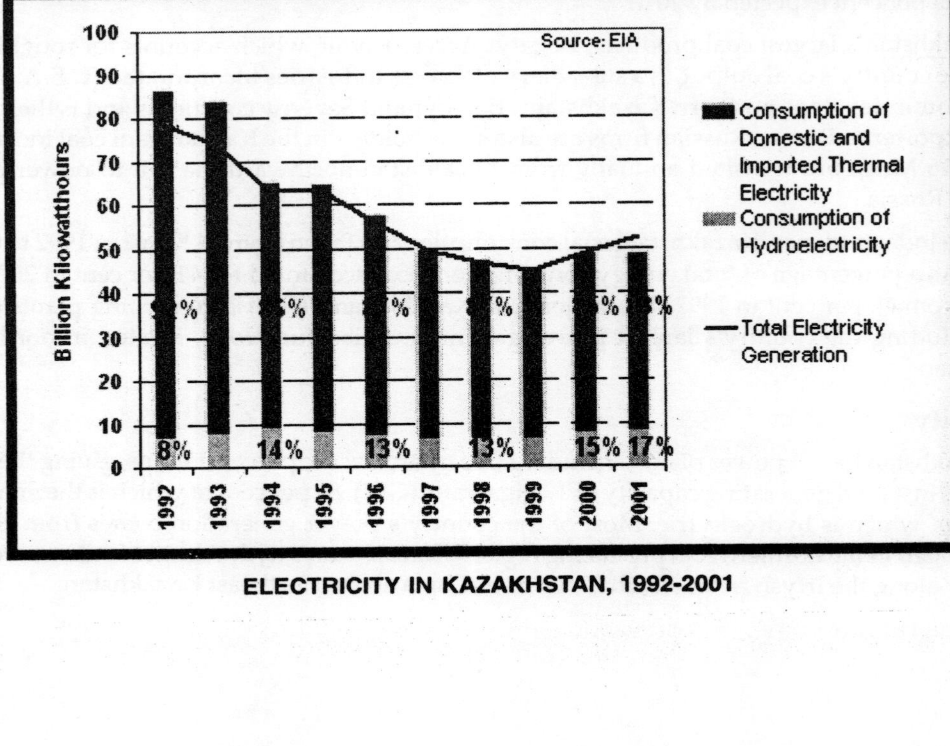

ELECTRICITY IN KAZAKHSTAN, 1992-2001

power plants that make up most of the country's installed capacity, have recently begun exporting electricity to Russia. In January 2003, the Ekibastuz Power Plant No. 2, located in the northern Pavlodar region, began exporting electricity northward. Conversely, the southern network, which is connected to the Unified Energy System of Central Asia, is forced to import electricity from neighboring Kyrgyzstan and Uzbekistan owing to its lack of installed generating capacity.

Since Kazakhstan's southern regions are largely dependent on expensive imported electricity supplies, the Kazakhstan Electricity Grid Operating Company (KEGOC) has been considering the construction of a second North-South power line to complement the existing, 600-MW-capacity line, making it possible to supply the country's southern regions fully with energy generated in Kazakhstan. The line would cost an estimated $300 million to build. The country also has other plans for modernizing its electricity sector. In 2003, KEGOC is reportedly planning to invest $150 million to upgrade the country's high-voltage transmission lines, upgrade automated substations and purchase new distribution equipment. Investments in 2003 are part of a bigger $258.4 million upgrading project which is being planned with financial assistance from the World Bank and European Bank for Reconstruction and Development.

Industry Organization

Kazakhstan has privatized all of its power plants, but the sale of regional electricity distribution companies has proceeded more slowly, and the majority of the distribution networks have not yet been privatized. KEGOC has granted management rights to several private companies, but KEGOC maintains control over high-voltage transmission lines, substations, and the central dispatching apparatus.

Non-payment of electricity bills, an inadequate collection system, and the lack of market-based transportation tariffs all have been obstacles to further large-scale investment in Kazakhstan's transmission and distribution sector. Under the former Soviet Union, Kazakhstan utilized a system of fixed electricity tariffs that were unrelated to production costs and investment needs. Kazakhstan's State Anti-Monopoly Committee is working to bring electricity tariffs in line with those in other countries and to allow the market to determine transmission tariffs. Effective July 1, 2001, KEGOC increased electricity transmission rates across the country by an average of 23.7 per cent.

Nuclear Power

Kazakhstan's sole nuclear power plant–the 90-MW Mangyshlak Nuclear Power Plant at Aqtau, which has been shut down since April 1999–was sold in April 2003 by the government of Kazakhstan to Kazatmomprom, the national nuclear power company. Kazatomprom, which has exclusive rights to the production and sale of Kazakhstani plutonium, plans to maintain and run the plant's thermal generators and water distribution facilities for regional consumption.

According to press reports, the Kazakhstani government is still considering the construction of a new nuclear plant in the east, near Lake Balkash. This project was first announced in 1998, but later shelved in September 2002 because of safety concerns, as well as public opposition to the project. In February, 2003, Kazakhstan's Minister of Energy and Mineral Resources, Vladimir Skholnik, suggested to the Kazakhstani Parliament that a new international tender be held for the construction of the Balkash nuclear power facility. In June 2003, Mr. Skholnik commented that the plant will be built within 15 years.

Country Overview

Independence: December 16, 1991; National holiday: Republic Day, October 25, 1990 (date on which Kazakhstan declared its sovereignty)

Population (7/02E): 16.7 million

Location: Central Asia, bordering the Caspian Sea, Russia, Turkmenistan, Uzbekistan, Kyrgyzstan, and China

Size: 1,052,100 sq. miles (slightly less than four times the size of Texas)

Major Cities: Almaty; Astana (capital, moved from Almaty in December 1998); Karaganda; Shymkent

Languages: Kazakh (Qazaq, state language) 40 per cent, Russian (official, used in everyday business) 66 per cent

Economic Overview

Currency: Tenge

Market Exchange Rate (7/01/2003): US $1=146.7 Tenge (KZT)

Nominal Gross Domestic Product (GDP) (2002E): $24.5 billion; *(2003E):* $28.5 billion

Real GDP Growth Rate (2002E): 9.5 per cent ; *(2003E):* 7.6 per cent

Inflation Rate (Change in Consumer Prices, Dec. 2001-Dec. 2002E): 6.0 per cent ; *(2003E):* 7.4 per cent

Unemployment Rate (2002E): 9.3 per cent

Current Account Balance (2002E): -$596 million; *(2003E):* -$396 billion

Major Trading Partners: Russia, U.S., Uzbekistan, China, Turkey, U.K., Germany, Ukraine,

Merchandise Exports (2002E): $10 billion; *(2003E):* $11.7 billion

Merchandise Imports (2002E): $7.6 billion; *(2003E):* $9.4 billion

Merchandise Trade Balance (2002E): $2.4 billion; *(2003E):* $2.3 billion

Major Exports: oil, ferrous and nonferrous metals, machinery, chemicals, grain, wool, meat, coal

Major Imports: machinery and parts, industrial materials, oil and gas, vehicles

Gross Foreign Debt (2002E): $12.95 billion

Energy Overview

Proven Oil Reserves (1/1/03): 9-17.6 billion barrels

Oil Production (2002E): 939,280 bbl/d, of which 818,359 bbl/d was crude

Oil Consumption (2002E): 140,000 bbl/d

Net Oil Exports (2002E): 799,280 bbl/d

Crude Oil Refining Capacity (1/1/03E): 427,000 bbl/d

Natural Gas Reserves (1/1/03E): 65 trillion cubic feet

Natural Gas Production (2001E): 360 billion cubic feet (Bcf)

Natural Gas Consumption (2001E): 505 Bcf

Net Natural Gas Imports (2001E): 145 Bcf

Coal Reserves (2002E): 34,479 million short tons (Mmst)

Coal Production (2001E): 88.5 million short tons (Mmst)

Coal Consumption (2001E): 61.0 Mmst

Electric Generation Capacity (2001E): 17 gigawatts (GW) (88 per cent thermal, 12 per cent hydroelectric)

Electricity Generation (2001E): 52.4 billion kilowatt-hours (Bkwh)

Electricity Consumption (2001E): 48.4 Bkwh

Environmental Overview

Total Energy Consumption (2001E): 1.73 quadrillion Btu* (0.43 per cent of world total energy consumption)

Energy-Related Carbon Emissions (2001E): 33.4 million metric tons of carbon (0.5 per cent of world total carbon emissions)

Per Capita Energy Consumption (2001E): 117.0 million Btu (vs. U.S. value of 341.8 million Btu)

Per Capita Carbon Emissions (2001E): 2.2 metric tons of carbon (vs. U.S. value of 5.5 metric tons of carbon)

Energy Intensity (2001E): 79,352 Btu/ $1995 (vs. U.S. value of 10,736 Btu/ $1995)**

Carbon Intensity (2001E): 1.53 metric tons of carbon/thousand $1995 (vs. U.S. value of 0.17 metric tons/thousand $1995)

Fuel Share of Energy Consumption (2001E): Coal (41.1 per cent), Natural Gas (30.5 per cent), Oil (23.7 per cent)

Fuel Share of Carbon Emissions (2001E): Coal (53.4 per cent), Natural Gas (28.8 per cent), Oil (23.8 per cent)

Status in Climate Change Negotiations: Non-Annex I country under the United Nations Framework Convention on Climate Change (ratified May 17th, 1995). Signatory to the Kyoto Protocol (March 12th, 1999).

Major Environmental Issues: Radioactive or toxic chemical sites associated with its former defense industries and test ranges are found throughout the country and pose health risks for humans and animals; industrial pollution is severe in some cities; because the two main rivers which flowed into the Aral Sea have been diverted for irrigation, it is drying up and leaving behind a harmful layer of chemical pesticides and natural salts; these substances are then picked up by the wind and blown into noxious dust storms; pollution in the Caspian Sea; soil pollution from overuse of agricultural chemicals and salination from poor infrastructure and wasteful irrigation practices

Major International Environmental Agreements: A party to Conventions on Air Pollution, Biodiversity, Climate Change, Desertification, Endangered Species, Ozone Layer Protection, Ship Pollution. *Signed, but not ratified:* Climate Change.

The total energy consumption statistic includes petroleum, dry natural gas, coal, net hydro, nuclear, geothermal, solar and wind electric power. The renewable energy consumption statistic is based on International Energy Agency (IEA) data and includes hydropower, solar, wind, tide, geothermal, solid biomass and animal products, biomass gas and liquids, industrial and municipal wastes.

Energy Industries

Organization: Kazmunaigaz (vertically-integrated state oil and natural gas company, created in February 2002 by combining state-run Kazakhoil (oil) and TransNefteGaz (oil and natural gas transport, made up of KazTransOil and KazTransGaz)); Kazakhstanugol Corporation (state coal company); Kazakhstan Electricity Grid Operating Company (KEGOC)

Major Oil and Gas Fields: Tengiz (mostly oil), Karachaganak (oil and gas/condensate), Kashagan (oil), Amangeldy (natural gas)

Major Oil Ports: Atyrau and Aqtau on the Caspian Sea

Oil Export Pipelines: Tengiz-Novorossiisk (Russia); Uzen-Atyrau-Samara (Russia); Kenkyak-Orsk (Russia) line that transports oil from the Aktyubinsk fields to the Orsk refinery

Major Oil Refineries (crude oil refining capacity): Pavlodar (162,666 bbl/d); Atyrau (104,427 bbl/d); Shymkent (160,000 bbl/d)

Major Power Plants (capacity): Ekibastuz No.1 (4,000 megawatts, MW), Yermak (2,400 MW), Zhambyl (1,230 MW)

50

Kuwait

Kuwait contains 96.5 billion barrels of proven oil reserves (including its share of the Neutral Zone), or roughly 9 per cent of the world's total oil reserves.

General Background

As one of the world's leading oil producing states, Kuwait's economy is heavily dependent on oil revenues. Current high oil prices are producing a surge in oil export revenues for Kuwait. For fiscal year 2002/03 (which runs through March 2003), Kuwait assumed oil prices of only $15 per barrel (for Kuwaiti oil), at which price the country would experience a budget deficit. However, as of early March 2003, oil prices were running significantly above (more than twice) $15 per barrel, meaning that a budget surplus was more likely. Kuwait's real gross domestic product (GDP) growth for 2002 was estimated at about 1.8 per cent, with forecast real growth of 3.3 per cent for 2003. In January 2003, Kuwait pegged the country's currency (the Dinar) to the U.S. dollar.

Kuwait is aiming to diversify its economy away from near-complete dependence on oil and state subsidies. Currently, the country relies on oil revenues for around 90 per cent -95 per cent of total export earnings and two-fifths of GDP. Kuwait channels around 10 per cent of its oil revenues into the "Future Generations Fund" for the day when oil income runs out. As of 2000, the Fund was worth around $50 billion.

With approximately 65 per cent of the population under the age of 25, and with around 90 per cent of employees in the private sector currently non-Kuwaiti citizens, creating jobs for young Kuwaiti's is a major objective of the government. Kuwait hopes to attract additional foreign investment, and has started a program to privatize state-owned businesses (outside the oil sector) as a way of reducing subsidies. By 2005, Kuwait aims to begin privatizing such key sectors as utilities, ports, oil stations, and telephone service. Privatization is complicated by the need to protect the jobs of Kuwaiti citizens, who traditionally have been employed mostly (93 per cent) by state-owned enterprises and the government.

In March 2001, Kuwait's national assembly passed the "Foreign Direct Investment Act," which aimed at promoting foreign investment. Among other things, the Act eased restrictions on foreign banks, provided long-term protection to foreign investors against nationalization or confiscation, and eliminated the requirement for foreign companies to have a Kuwaiti sponsor or partner. In the oil sector, the Kuwaiti constitution forbids foreign ownership of Kuwait's mineral resources, but the Kuwaiti government is exploring allowing foreign investment in upstream oil development under terms which provide for per-barrel fees to the foreign firms rather than traditional production sharing agreements (PSA's). The idea is highly controversial in Kuwait, and thus far opposition in the Kuwaiti parliament has stalled the initiative.

Kuwait's foreign policy continues to focus on regional security issues. Since Operation Desert Storm in 1991, Kuwait has maintained close military cooperation with Western countries. In October 1994, the United States, the United Kingdom and other countries came to Kuwait's assistance after Iraq moved 70,000 troops and heavy armor close to the Kuwaiti border. In November 1994, Iraq officially recognized Kuwait's sovereignty, territorial integrity and political independence, as well as its borders as defined by the United Nations. Kuwait welcomed the move but continues to demand the release of all Kuwaiti citizens detained in Iraq and called for international sanctions against Iraq to be maintained until it was in compliance with all pertinent U.N. resolutions. In April 2001, Kuwait's cabinet agreed to a joint Gulf Cooperation Council (GCC) defense pact. The security arrangement requires parliamentary approval.

Oil

Kuwait contains an estimated 96.5 billion barrels of proven oil reserves, more than 9 per cent of the world total. The Neutral Zone area, which Kuwait shares with Saudi Arabia, holds 5 billion barrels of reserves, half of which belong to Kuwait. Most of Kuwait's oil reserves are located in the 70-billion barrel Greater Burgan area, which comprises the Burgan, Magwa and Ahmadi structures. Greater Burgan is widely considered the world's second largest oil field, surpassed only by Saudi Arabia's Ghawar field, and has been producing oil since 1938. Kuwait's Rawdhatain, Sabriya, and Minagish fields have large proven reserves as well, with 6 billion, 3.8 billion, and 2 billion barrels of oil, respectively. All of these fields have been producing since the 1950s. They generally contain medium to light crude oil with gravities in the 30°-36° API range. The South Magwa field, discovered in 1984, is estimated to hold at least 25 billion barrels of light crude oil with a 35°-40° API gravity. In November 2000, Kuwait announced the discovery of significant amounts of light crude oil at Sabiryah.

Another Kuwaiti field–Ratqa–has been the subject of controversy. Once thought to be an independent reservoir, Ratqa is actually a southern extension of Iraq's super-giant Rumaila field. During the weeks preceding Iraq's August 1990 invasion of Kuwait, Iraq had accused Kuwait of stealing billions of dollars worth of Rumaila oil, and had refused to negotiate a sharing or joint development arrangement for Ratqa and southern Rumaila. After the Gulf War of 1991, a United Nations survey team made a demarcation of the border between Iraq and Kuwait, and this demarcation put all 11 of the existing wells at Ratqa within Kuwaiti territory. Despite this, in September 2000, Iraq renewed accusations it has made previously that Kuwait was "stealing" its oil. Iraq claimed that Kuwait was doing this through horizontal drilling on fields straddling the border between the two countries, and that Iraq was losing $3 billion per year worth of oil. Kuwait denied the charges. Kuwait produces around 40,000 bbl/d from Ratqa.

In early 2003, Kuwait began taking precautions to protect its oil installations in the event of any military conflict with Iraq. For instance, in mid-February, the Kuwaiti government turned much of the northern part of the country into a military exclusion zone and began to move oil rigs out of the area. Kuwait also reportedly began to limit output at its northern fields, which normally produce around 600,000 barrels per day (bbl/d). Specifically, Kuwait announced on February 17 that it was halting

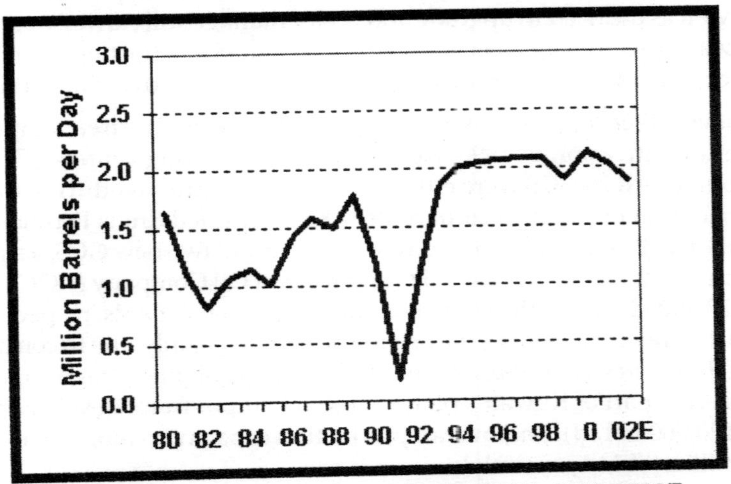

KUWAIT'S CRUDE OIL PRODUCTION, 1980-2002E

production at Ratqa and Abdali, which have a combined production capacity of around 100,000 bbl/d. In early March 2003, Kuwait Petroleum Corporation (KPC) announced that it would halt production completely in the northern fields in the event of war, although other Kuwaiti officials indicated that production would continue normally. Kuwait reportedly has been stockpiling crude oil reserves on tankers outside the Gulf region. Finally, Kuwait reportedly has increased security measures at its oil facilities.

Current Oil Production

The bulk of Kuwait's oil production capacity is located at the southern onshore Greater Burgan field, whose Burgan, Magwa, and Ahmadi structures normally produce around 1.35 million bbl/d. Kuwait's other main producing fields include the northern fields of Rawdhatain and Sabiryah (600,000 bbl/d of production capacity); the southern fields of Minagish and Umm Qudayr (150,000 bbl/d); and Kuwait's share of the Saudi-Kuwaiti Neutral Zone (150,000 bbl/d).

On January 31, 2002, an explosion and fire at an oil-gathering center near Kuwait's northern Rawdhatain oil field killed four workers and injured seventeen, while cutting the field's oil and gas output sharply. The explosion, which reportedly was caused by a leak at a major oil pipeline, knocked out three critical gathering centers (GC-15, GC-23, and GC-25), an electrical substation, and a natural gas booster station (BS-130). Production at Rawdhatain was quickly restored to around 300,000 bbl/d by the end of February 2002, but further repairs are required on GC-15 to restore the rest. Reportedly, U.S.-based Fluor Corp. and South Korea's SK Engineering & Construction are to work on major restoration of GS-15 and BS-130. Total repair costs could run into the $250 million range.

In late February 2002, Oil Minister al-Subaih resigned, taking responsibility for the incident, though reportedly also in protest of cost cutting and political interference in Kuwait's oil industry. Several analysts have blamed a lack of maintenance, insufficient technical expertise, and poorly trained, low-paid foreign workers. In addition, much of Kuwait's infrastructure was damaged during the Iraqi invasion of 1990/1991 and was never completely repaired.

In June 2002, another accident–this time a significant oil spill from a broken heavy fuel oil pipeline–took place near the al-Ahmadi refinery. And in November 2002, a fire at Burgan closed down gathering center (GC) #4, cutting production at the field by 100,000-130,000 bbl/d temporarily. In part motivated by these accidents, as well as the need to upgrade and repair damage still carried over from the 1990/1991 Gulf War, the Kuwaiti government is considering efforts aimed at attracting foreign oil companies into the country on "technical service" contracts, but this move is controversial within Kuwait.

Construction of new gathering centers is a major focus of Kuwait's upstream capacity expansion program. Prior to the Iraqi invasion in 1990, Kuwait had 26 gathering centers (GCs), with a total capacity of 4 million bbl/d. All 26 GCS were either damaged or destroyed during the war. By 1993, operations at 18 GCS had been restored. In January 1996, KPC awarded China Petroleum Engineering Construction Corporation (CPECC) a $390 million contract to build two new GCS, a significant step in Kuwait's efforts to increase its oil production. CPC and Kuwait Oil Company (KOC) constructed the GCS, designated GC-27 and GC-28, at the Umm Gudair and Minagish fields, respectively. Work was completed in early 2001, three years behind the original schedule, mainly due to construction delays. The Umm Gudair GC has capacity of 190,000 bbl/d, and Minagish has design capacity of 210,000 bbl/d. The two facilities are part of Kuwait's plan to raise oil output from West Kuwait to more than 500,000 bbl/d (from 120,000 bbl/d), and overall production capacity to more than 3 million bbl/d (from around 2.1-2.2 million bbl/d currently).

Crude Oil Exports

Kuwait exports the majority of its oil to Asian countries, especially Japan, but also refineries in South and Southeast Asia (India, Singapore, South Korea, Taiwan, Thailand, etc.). Other oil exports go to Europe and to the United States, which averaged 223,000 bbl/d of Kuwaiti oil imports during 2002. This compares to the peak of 353,000 bbl/d (4.1 per cent of U.S. oil imports) reached in 1993. Kuwait's export blend is 31°API (a typical medium Mideast crude), and is considered sour with 2.5 per cent or greater sulfur content.

Kuwait has completed major renovations of Mina al-Ahmadi, the country's main port for the export of crude oil, virtually destroyed during the Gulf War. Kuwait also is planning a $900 million expansion at the port in order to add storage capacity and increase export capacity in conjunction with plans for expanded oil production in coming years. Current plans call for construction of 19 new storage tanks with total capacity of 11.4 milion barrels by 2005.

Besides Mina al-Ahmadi, Kuwait has operational oil export terminals at Mina Abdullah (repairs completed in September 1992), Shuaiba (restored by late 1996) and at Mina Saud. A new terminal is planned for Bubiyan Island, which will handle increased production from northern and western Kuwait under "Project Kuwait".

Production Quotas

Beginning February 1, 2003, Kuwait's crude oil production quota was set at 1.966 million bbl/d. During 2002, Kuwait produced about 2.0 million bbl/d of crude oil, with the next OPEC meeting scheduled for March 11, 2003.

Plans to Expand Oil Production

Kuwait continues to plan for significant expansion in its production capacity. Kuwait hopes to reach capacity of 3.5 million bbl/d by 2005/6 (and 4 million bbl/d by 2020), up from around 2.1-2.2 million bbl/d at present. As part of this plan, known as "Project Kuwait," Kuwait is considering permitting foreign oil companies to invest in upstream production, which would reverse more than two decades of Kuwaiti policy.

"Project Kuwait" is a $7 billion, 25-year plan, first formulated in 1997 by the SPC, to increase the country's oil production (and to help compensate for declines at the mature Burgan field), with the help of international oil companies (IOCs). In particular, Kuwait aims to increase output at five northern oil fields–Abdali, Bahra, Ratqa, Rawdhatain, and Sabriyah–from their current 450,000 bbl/d to around 900,000 bbl/d by 2005. To date, "Project Kuwait" has made little headway, in large part due to political opposition and resistance from parliament to the idea of allowing foreign companies into the country's oil sector. However, there are hopes that this might change in 2003. In February 2003, KPC completed a draft contract and proposed financial terms for Project Kuwait. Shortlisted IOCs reportedly include BP, ChevronTexaco, Conoco, ExxonMobil, Lasmo, Shell, and TotalFinaElf.

The fields which the Kuwaiti government intends to open to foreign investment are all currently operating fields in northern or western Kuwait, including Rawdhatain, Sabriyah, Ratqa, Bahra, Minagish, and Umm Gudair. Kuwait's largest field, Burgan, is to remain off-limits to foreign investment under the new plan. As of December 2002, Kuwait reportedly was planning to invest $6 billion in three areas near the Iraqi border–Abdali, Ratqa, and Rawdhatain. Possible companies involved could be BP, Shell, TotalFinaElf, Sibneft, Lukoil, and Petronas.

Kuwait's current policy, in place since 1975, limits the participation of foreign oil companies to providing technical assistance and construction and maintenance services under contracts, which pay them fixed prices for specific activities. In fact, Kuwait's constitution forbids the award of concessions which give an ownership interest in Kuwait's natural resources to foreign entities. Nevertheless, the government has repeatedly hinted at a desire to find a way to involve foreign oil companies in increasing production without violating the constitution. The Supreme Petroleum Council (SPC), which controls Kuwait's oil and gas sectors on a day-to-day basis, approved foreign cooperation in principle in 1997, but opening upstream activities to deeper involvement by foreign oil companies is highly controversial with opposition members of the Kuwaiti parliament. In February 2000, the Kuwaiti parliament passed a resolution calling on the government not to proceed with the program until legal issues involving foreign interests in the Kuwaiti oil sector were resolved. New legislation dealing with the foreign investment program is currently under consideration in the Kuwaiti parliament. Some in parliament have even accused the government of "selling off" the country.

Unlike PSA's, the structure of the agreements the government is considering, called technical service contracts, allows the Kuwaiti government to retain full ownership of the oil reserves, control over oil production levels, and strategic management of the venture. Foreign firms are to be paid a "per barrel" fee, along with allowances for capital recovery and incentive fees for increasing reserves, in their role as service provider/contractor.

In other oil development news, KOC is planning to award a concession to drill at a major new oil field, called Madinah. The field is located around Kuwait City and Kuwait Bay.

Neutral Zone

The Neutral Zone encompasses a 6,200 square-mile area partitioned equally between Kuwait and Saudi Arabia under a 1992 agreement. The Neutral Zone contains an estimated 5 billion barrels of oil and 1 trillion cubic feet (Tcf) of natural gas. Oil production in the Neutral Zone, which currently is running around 579,000 bbl/d (around half offshore and half onshore), is divided equally between Saudi Arabia and Kuwait. Major Neutral Zone onshore fields include Humma, South Fawaris, South Umm Gudair, and Wafra. Offshore fields include Hout and Khafji. Onshore, U.S.-based ChevronTexaco and KPC produce from Wafra, South Fawaris, and South Umm Gudair. Offshore, the Arabian Oil Company (AOC) of Japan operates Khafji and Hout, both of which are connected to Saudi Arabia's Safaniyah, the world's largest offshore oilfield. AOC has a 40 per cent stake in the Kuwaiti portion, with the remaining 60 per cent held by KPC. In July 2002, Kuwait and Saudi Arabia formed a 50/50 joint venture to maintain and increase oil production at offshore Neutral Zone oilfields.

AOC lost its concession in the Saudi sector when it expired in February 2000, and also surrendered its drilling rights over Hout and Khafji on January 4, 2003. In March 2002, the Kuwait Gulf Oil Company (KGOC) was established as a wholly-owned subsidiary of KPC, with the intention of having it take over operation of offshore Neutral Zone exploration and production when AOC lost its rights. In January 2003, AOC began providing training and technical assistance to KGOC, plus $750 million in financing for Neutral Zone development. In exchange, AOC owns the rights to purchase 100,000 bbl/d of Khafji crude.

Foreign Upstream Operations

Even though Kuwait's overall overseas investments are considerably smaller than before the 1990 invasion by Iraq, Kuwait holds equity interests in oil production in several countries through the Kuwait Foreign Petroleum Exploration Company (KUFPEC). KUFPEC is active in Australia, Indonesia, and Tunisia, among others. Most of the interests are either small fields or minority stakes, though, and KUFPEC's revenues have been under $200 million in recent years, making it a relatively minor part of Kuwait's state oil establishment. According to KUFPEC's 2000 annual report, the company is aiming to increase its production capacity to 100,000 bbl/d oil equivalent by 2010, up from around 35,000 bbl/d oil equivalent currently.

Refining

Kuwait's refining capacity was damaged severely during Iraqi invasion and occupation in 1990-91. After losing most of its pre-war capacity of 820,000 bbl/d, Kuwait had only 200,000 bbl/d of refinery output by early 1992. Kuwait's $400-million downstream reconstruction program was completed in mid-1994. At present, Kuwait's three domestic refineries have a combined capacity of around 773,300 bbl/d, around 100,000 bbl/d less than normal because of damage to Mina al-Ahmadi, the country's largest refinery (with capacity of 442,700 bbl/d). Other large refineries include Mina Abdullah (256,500 bbl/d) and al-Shuaiba (190,000 bbl/d). The bulk of Kuwait's refined products are exported. On August 20, 2001, *Reuters* reported that Kuwait was planning to spend "over $3 billion" to upgrade its three refineries over a five-year period.

On June 25, 2000, the Mina al-Ahmadi refinery, owned and operated by KPC subsidiary Kuwait National Petroleum Company (KNPC), experienced an explosion and fire, causing substantial damage to two of three crude distillation units (CDUs), killing six workers, and forcing the facility to shut down. In December 2000, Kuwait awarded a contract (to Fluor Corp. and Sunkyong) to rebuild Mina al-Ahmadi. As of July 2002, Mina al-Ahmadi was operating at around 300,000 bbl/d, with full capacity not expected to be restored for another year or so, at a cost of around $320 million. In addition, KNPC is working to upgrade atmospheric residue desulphurisation at the refinery.

Foreign Downstream Operations

KPC currently has around 250,000 bbl/d of refining capacity in Europe, including half of Agip's 300,000 bbl/d Milazzo refinery. KPC also owns a 75,500-bbl/d unit in Rotterdam. These two refineries enable KPC to supply a large share of its European retail outlets directly. In September 1998, KPC announced the purchase of 157 service stations in Belgium from BP. The move gives KPC an 8 per cent of the retail market share in Belgium. KPC's subsidiary, Kuwait Petroleum International (KPI), operates approximately 5,500 service stations under the "Q8" banner in 10 countries in Western Europe and about 200 sites in Thailand.

Petrochemicals

Kuwaiti officials have expressed interest in accelerating development of the country's relatively small petrochemical sector. This would accomplish several goals: boosting the value of Kuwait's crude oil reserves; helping to protect Kuwait's revenues during periods of low crude prices; and boosting Kuwaiti revenues while adhering to OPEC crude oil quota limitations. Historically, Kuwait's Petrochemical Industries Company (KPIC) has mainly manufactured low-value products such as urea, ammonia, and fertilizer for export. PIC is now beginning to move upmarket to production of higher-value products.

According to the Kuwait News Agency, PIC may increase production at its polypropylene plant by 20 per cent to 120,000 tons per year if the market price of polypropylene continues to rise. PIC's primary markets are Jordan, Syria, the United Arab Emirates, Morocco, China and Hong Kong, followed by India, Pakistan and countries in eastern Africa.

The EQUATE joint venture, involving PIC and Union Carbide, is the country's largest petrochemical project. The $2 billion industrial complex at Shuaiba, which came online in 1997, includes a 650,000 metric-ton-per-year ethylene cracker, two polyethylene units with a capacity of 450,000 metric tons per year, and a 350,000 metric-ton-per-year ethylene glycol plant, all of which are currently operating. The complex primarily serves the Asian and European markets. PIC and Union Carbide each have a 45 per cent share in the project, with the remainder owned by Boubyan Petrochemical Company. The EQUATE plant was temporarily shut down by the loss of its ethane feedstock from the Mina al-Ahmadi refinery in June 2000, but has since resumed operation. In April 2001, KPIC approved a $1.5 billion plan to construct "Equate II," which would be Kuwait's second petrochemical complex, and would produce olefin (ethylene, polyethylene, ethylene glycol).

Natural Gas

Kuwait produces a relatively modest volume of natural gas (around 335 billion cubic feet–Bcf–in 2001), the vast majority of which is "associated gas" (*i.e.*, found and produced in conjunction with oil). Kuwait hopes to increase its use of natural gas (both domestic and imported) significantly, especially in electricity generation, water desalination, and petrochemcials. A switch to natural gas (from diesel oil) would free up a substantial amount of oil for export, possibly 100,000 bbl/d by 2006. Kuwait also hopes to reduce flaring of associated gas by tying together gathering centers. Finally, Kuwait continues to seek both associated and non-associated gas supplies.

In July 2000, Kuwait and Qatar signed a memorandum of understanding (MOU) for possible export of Qatari gas from its offshore North Field–the largest non-associated natural gas field in the world–to Kuwait. In January 2002, the two countries signed a protocol for "gas sale and purchase agreement" on a $500 million pipeline from Qatar's port of Ras Laffan to Al-Zour South in southern Kuwait. Natural gas could begin flowing through the line by 2006, at a rate as high as 500 Bcf per year. Qatar Petroleum and ExxonMobil (operator of Qatar's North Field) also have signed an agreement on supplying the gas. A territorial dispute between Qatar and Bahrain had held up the pipeline, but this was resolved in 2001. Howver, as of February 2003, Saudi Arabia reportedly was expressing opposition to the pipeline, which would pass through Saudi territorial waters.

Besides Qatar, Kuwait also has signed an MOU with Iran for import of natural gas via pipeline, most likely from its huge South Pars gas field. As of February 2003, it appeared that the two countries were in the final stages of negotiations on a 25-year gas supply deal. Approximately 300 million cubid feet per day (cf/d) of natural gas is to be delivered, beginning in 2005. The gas will be used for power generation at two new power plants (with combined generating capacity of 3,500 megawatts) and water desalination. It remains unclear whether Kuwaiti gas demand could support both Iranian and Qatari export projects simultaneously.

Prior to the 1990/1991 Gulf War, Kuwait received significant volumes of natural gas from Iraq. The gas came from Iraq's southern Rumaila field through a 40-inch, 100-mile, 200 mmcf/d pipeline. The gas was used in Kuwaiti electric power stations and liquefied petroleum gas (LPG) plants.

Aside from imports, Kuwait hopes to increase its domestic natural gas production, both through reduced flaring of associated gas and through new drilling. Exploratory drilling is currently being undertaken at the Rawdhatain oilfield, reaching geological formations much deeper than the oil deposits, which are believed to be gas-rich. (Note: the January 2002 explosion at Raudhatain–see above–significantly disrupted natural gas production at the field).

In another natural gas-related development, Saudi Arabia and Kuwait concluded an agreement in July 2000 on the offshore Dorra natural gas field, which had been claimed by Saudi Arabia, Kuwait, and Iran, and which may contain up to 11 trillion cubic feet (Tcf) of recoverable natural gas reserves. The agreement calls for an equal sharing of the gas resources between Saudi Arabia and Kuwait. Negotiations with Iran over its claims to the Dorra gas field are continuing, while AOC is talking to Kuwait about possibly developing the field.

Electricity

Kuwait has five power stations (Doha East, Doha West, al-Subiya, Shuaiba South, and al-Zour South) and a total electrical generation capacity of about 9.3 gigawatts (GW). Kuwait's electricity demand has been growing rapidly in recent years, and is expected to continue increasing at a 7 per cent -9 per cent rate in coming years, necessitating construction of new generating capacity. A 2,400-megawatt (MW), $2.2 billion thermal plant at al-Subiya came online in 2000, which relieved pressure on the system in the short-term. Over the next 10 years, Kuwait reportedly will need to invest $3.6 billion in its power sector in order to increase generating capacity by another 3,000 MW.

In September 2001, Kuwait's Ministry of Electricity and Water (MEW), which is in charge of Kuwait's power sector, approved construction of three new power plants at a total cost of $2 billion: 1) a 2,500-MW steam turbine station at Al-Zour North; 2) a 600-MW gas-fired plant in Shuaiba; and 3) a 400-MW gas-fired facility in Subiya. The Al-Zour North plant is scheduled to come online in late 2006, and to cost $1 billion.

To reduce excessive power demand and waste, Kuwait is considering trimming some of its power subsidies. Currently, Kuwaitis pay among the lowest prices for power in the world, and the MEW has urged them to use power more judiciously. Meanwhile, Kuwait continues to expand its national power grid, and has accepted a proposal to link up with the grids of other GCC countries. This grid linkup should provide each GCC country with additional spare capacity to handle peak demand periods. Kuwait also has discussed broader linkages and cooperation with other Arab countries and its non-Arab neighbor, Iran.

Country Overview

Independence: June 19, 1961 (from United Kingdom)

Population (7/02E): 2.1 million (includes 1.2 million non-citizens)

Location/Size: Middle East, northwestern corner of the Persian Gulf between Iraq and Saudi Arabia/ 6,900 square miles, slightly smaller than New Jersey

Major Cities: Kuwait City (capital), Salmiya, Hawalli

Languages: Arabic

Ethnic Groups: Kuwaiti (45 per cent), other Arab (35 per cent), South Asian (9 per cent), Iranian (4 per cent), other (7 per cent)

Religion: Sunni Muslim (45 per cent), Shi'a Muslim (40 per cent), Christian, Hindu, other (15 per cent)

Defense (8/98): Army (11,000), Air Force (2,500), Navy (1,800), National Guard (5,000)

Economic Overview

Currency: Kuwaiti Dinar (KD)

Market Exchange Rate (3/04/03): US$1 = 0.3 KD

Nominal Gross Domestic Product (GDP; using market exchange rates) (2002E): $36.2 billion

Nominal Gross Domestic Product (GDP; using Purchasing Power Parity exchange rates) (2002E): $26.8 billion

Real GDP Growth Rate (2002E): 1.8 per cent *(2003F):* 3.3 per cent

Inflation Rate (Consumer Price Index) (2002E): 1.9 per cent *(2003F):* 1.4 per cent

Current Account Balance (2001E): $10.2 billion *(2002E):*$8.8 billion

Major Trading Partners: United States, Japan, Europe, Singapore

Goods and Non-Factor Service Exports (2002E): $9.1 billion

Goods and Non-Factor Service Imports (2002E): $8.4 billion

Goods and Non-Factor Service Trade Balance (2002E): $0.7 billion

Major Export Products: Petroleum (95 per cent of total export revenues)

Major Import Products: Industrial goods, consumer goods, machinery, transport equipment, food

Total External Debt (2001E): $9.6 billion

International Reserves excl. gold (2Q02E): $11.0 billion

Energy Overview

Proven Oil Reserves (1/1/03E): 96.5 billion barrels (including 2.5 billion in the Neutral Zone)

OPEC Crude Oil Production Quota (effective February 1, 2003): 1.966 million bbl/d

Crude Oil Production Capacity (2003E): 2.1-2.2 million bbl/d (includes half of Neutral Zone)

Oil Production (2002E): 2.0 million bbl/d (includes half of Neutral Zone); of which 1.9 million bbl/d was crude oil

Oil Consumption (2002E): 293,000 bbl/d

Net Oil Exports (2002E): 1.7 million bbl/d

Crude Oil Refining Capacity (1/1/03E): 889,200 bbl/d

Major Crude Oil Customers: Asia (around 60 per cent); United States; Europe

Natural Gas Reserves (1/1/03E): 52.2 trillion cubic feet (Tcf) (includes 0.5 Tcf in Neutral Zone)

Natural Gas Production (2001E): 335 billion cubic feet (Bcf)

Natural Gas Consumption (2001E): 335 Bcf

Electric Generation Capacity (2001E): 8.5 gigawatts

Net Electricity Generation (2001E): 31.5 billion kilowatt hours (all thermal)

Environmental Overview

Total Energy Consumption (2001E): 0.92 quadrillion Btu* (0.2 per cent of world total energy consumption)

Energy-Related Carbon Emissions (2001E): 16.4 million metric tons of carbon (0.2 per cent of world total carbon emissions)

Per Capita Energy Consumption (2000E): 453.7 million Btu (vs U.S. value of 351.0 million Btu)

Per Capita Carbon Emissions (2000E): 8.1 metric tons of carbon (vs U.S. value of 5.6 metric tons of carbon)

Energy Intensity (1999E): 29,483 Btu/ $1990 (vs U.S. value of 10,638 Btu/ $1990)**

Carbon Intensity (1999E): 0.53 metric tons of carbon/thousand $1990 (vs U.S. value of 0.19 metric tons/thousand $1990)

Fuel Share of Energy Consumption (2000E): Oil (65 per cent), Natural Gas (35 per cent)

Fuel Share of Carbon Emissions (2000E): Oil (70 per cent), Natural Gas (30 per cent)

Renewable Energy Consumption (1998E): 0.17 trillion Btu* (17 per cent decrease from 1997)

Status in Climate Change Negotiations: Non-Annex I country under the United Nations Framework Convention on Climate Change (ratified December 28th, 1994). Not a signatory to the Kyoto Protocol.

Major Environmental Issues: limited natural fresh water resources; some of world's largest and most sophisticated desalination facilities provide much of the water; air and water pollution; desertification.

Major International Environmental Agreements: A party to Conventions on Climate Change, Desertification, Environmental Modification, Hazardous Wastes, Law of the Sea, Nuclear Test Ban, Ozone Layer Protection. Has signed, but not ratified, Biodiversity, Endangered Species and Marine Dumping.

The total energy consumption statistic includes petroleum, dry natural gas, coal, net hydro, nuclear, geothermal, solar, wind, wood and waste electric power. The renewable energy consumption statistic is based on International Energy Agency (IEA) data and includes hydropower, solar, wind, tide, geothermal, solid biomass and animal products, biomass gas and liquids, industrial and municipal wastes.

Oil and Gas Industries

Organization: The *Supreme Petroleum Council* governs the nationalized oil industry, which is run by *Kuwait Petroleum Corporation* (KPC). KPC subsidiaries include: *Kuwait Oil Company* (KOC)–exploration and production of oil and gas; *Kuwait National Petroleum Company* (KNPC)–refining and shipping; *Kuwait Petroleum International* (KPI)–refining and product marketing; *Petrochemical Industries Company* (PIC)–production and marketing of chemical products; *Kuwait Foreign Petroleum Exploration Company* (KUFPEC)–foreign exploration; and *Kuwait Oil Tanker Corporation* (KOTC)–tanker operations

Major Refineries (capacity–bbl/d, 2003E): Mina Al-Ahmadi (442,700 bbl/d), Mina Abdullah (256,500), Shuaiba (190,000)

Major Oil Fields (reserves–billion barrels): Greater Burgan–Burgan, Magwa, and Ahmadi (50); Raudhatain (6); Sabriya (3.8); Minagish (2); Abdali; Rugei; Bahra; *Neutral Zone:* Al-Hout and Khafji (6.3); Wafra (2); South Fawaris; Umm Gudair

Major Pipelines: Raudhatain-Ahmadi; Minagish-Ahmadi; Umm Gudair-Shuaiba; Wafra-Mina Abdullah; Burgan-Ahmadi

Major Oil Terminals: Mina Al-Ahmadi, Mina Abdullah, Shuaiba, Mina Saud

51

Libya

Libya is a major oil exporter, particularly to Europe. With the suspension of U. N. sanctions against Libya following its extradition of two men suspected in the 1988 bombing of Pan Am flight 103 over Lockerbie, Scotland, oil companies are eager to resume and/or expand operations in Libya.

General Background

Oil export revenues, which account for about 95 per cent of Libya's hard currency earnings (and 75 per cent of government receipts), were hurt severely by the dramatic decline in oil prices during 1998, as well as by reduced oil exports and production–in part as a result of U.S. and U.N. sanctions. With higher oil prices since 1999, however, Libyan oil export revenues have increased sharply, to $10.8 billion in 2002 and a forecast $12.9 billion in 2003, up from only $6.0 billion in 1998. As a result of strong oil export revenues, Libya's fiscal situation is now significantly in surplus (around 16 per cent of GDP in 2002), but remains highly vulnerable to fluctuations in oil prices.

Libya has experienced strong economic growth over the past three years. Real gross domestic product (GDP) grew by around 6.5 per cent in 2000, 3.1 per cent -4.3 per cent in 2001 and 1.2 per cent in 2002. For 2003, real GDP growth of 2.1 per cent is expected. Despite this strong economic growth, Libya's unemployment rate remains high, and inflation appears under control (2.5 per cent growth in prices is expected for 2003). Libya's relatively poor infrastructure (*i.e.* roads and logistics), unclear legal structure, often-arbitrary government decision making process, a bloated public sector (as much as 60 per cent of government spending goes towards paying public sector employees' salaries), huge public works programs (*i.e.*, the "Great Man Made River" project), and various structural rigidities all have been impediments to foreign investment and economic growth.

There are signs that Libya now is moving towards a variety of economic reforms and a reduction in the state's direct role in the economy. In June 2003, President Qadhafi said that the country's public sector had failed and should be abolished, and called for privatization of the country's oil sector, in

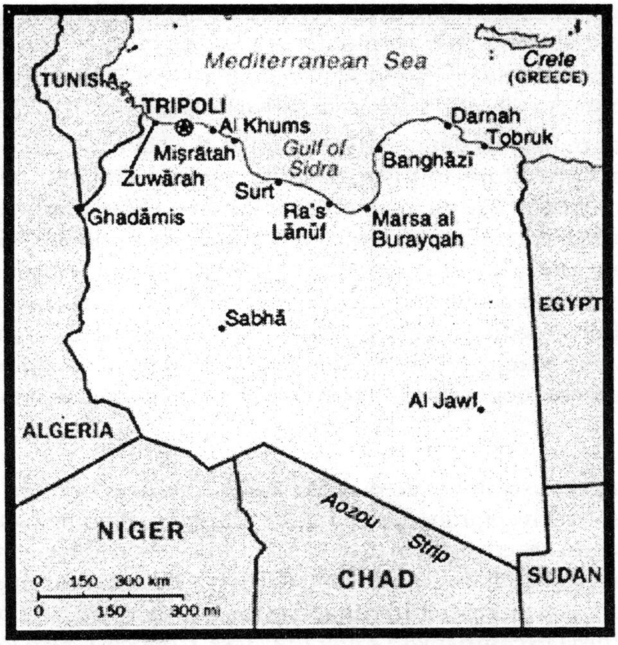

addition to other areas of the economy. Also in June 2003, Libya's Parliament selected former Trade and Economy Minister Shukri Muhammad Ghanim, a proponent of privatization, as Prime Minister. In January 2002, Libya announced its intention to open up its economy and to attract foreign capital to the country, while devaluing the official exchange rate on its currency, the dinar, by 51 per cent as part of a move towards unification of the country's multi-tier (official, commercial, black-market) foreign exchange system. Among other goals, the devaluation aimed to increase the competitiveness of Libyan firms and to help attract foreign investment into the country. Also in January 2002, Libya cut its customs duty rate by 50 per cent on most imports in part to help offset the effects of its currency devaluation.

On April 5, 1999, more than 10 years after the 1988 bombing of Pan Am flight 103 over Lockerbie, Scotland that killed 270 people, Libya extradited two men suspected in the attack. In response, the United Nations suspended economic and other sanctions against Libya which had been in place since April 1992. US sanctions, including the Iran-Libya Sanctions Act (ILSA) of 1996 (which covers foreign companies that make new investments of $40 million or more over a 12-month period in Libya's oil or gas sectors) remain in effect. On July 27, 2001, the US Congress voted to extend ILSA for five more years. UN sanctions since 1992 reportedly have cost Libya billions of dollars in lost income, and have made it more difficult for Libya to develop its energy sector. A full lifting of sanctions can occur 90 days after the UN certifies that Libya has met all requirements, including renunciation of support for terrorist acts. On July 9, 1999, the UN Security Council issued a statement saying that while it "welcomed the significant progress" which Libya had made in complying with UN demands, that at the same time Libya would need to do more (*i.e.*, cooperate with court proceedings, pay compensation to families if the suspects are convicted) before sanctions were lifted permanently. In late May 2002, Libya was reported to be offering $2.7 billion in compensation to families of Pan Am flight 103, with money to be released as UN and US sanctions are lifted. However, a Libyan government spokesman denied the offer, according to the state news agency (JANA). In late April 2003, Libya's foreign minister stated that Libya "has accepted civil responsibility for the actions of its officials in the Lockerbie affair," but the U.S. State Department responded that "What is important is whether Libya meets the U.N. requirements and.not what their officials might say to the press."

Libya is hoping to reduce its dependency on oil as the country's sole source of income, and to increase investment in agriculture, tourism, fisheries, mining, and natural gas. Libya's agricultural sector is a top governmental priority. Hopes are that the Great Man Made River (GMR), a five-phase, $30-billion project to bring water from underground aquifers beneath the Sahara to the Mediterranean coast, will reduce the country's water shortage and its dependence on food imports. Libya also is attempting to position itself as a key economic intermediary between Europe and Africa, has become more involved in the Euro-Mediterranean process, and has pushed for a new African Union. In April 2001, members of the Arab Maghreb Union (Algeria, Libya, Mauritania, Morocco, and Tunisia) agreed to encourage intra-regional cooperation on trade, customs, banking, and investment issues.

Oil

Libya's oil industry is run by the state-owned National Oil Corporation (NOC), along with smaller subsidiary companies, which combined account for around half of the country's oil output. Several international oil companies are engaged in exploration/production agreements with NOC. The leading foreign oil producer in Libya is Italy's Agip-Eni, which has been operating in the country since 1959. Two US oil companies (Exxon and Mobil) withdrew from Libya in 1982, following a US trade embargo begun in 1981. Five other US companies (Amerada Hess, Conoco, Grace Petroleum, Marathon, and Occidental) remained active in Libya until 1986, when President Reagan ordered them all to cease activities there. Conoco, Amerada Hess and Occidental made up the "Oasis Group," which was producing around 850,000 bbl/d in 1986.

In December 1999, US oil company executives from Oasis plus Marathon traveled to Libya, with US government approval, to visit their old oil facilities in the country. The former head of NOC, Abdullah al-Badri, has stated that if US companies return to Libya, they will return to the fields they used to operate in the country. However, in the first part of 2001, Libya contacted the US companies and indicated that, given its desire to develop their fields, Libya was considering transferring them to European companies. In September 2001, Libya stated that the US companies must either make use of their concessions within a year or risk losing them. In March 2002, the US State Department said that it would permit Marathon Oil to hold discussions with Libyan officials while sanctions remain fully in place.

Overall, Libya would like foreign company help to increase the country's oil production capacity from 1.4 million bbl/d at present to 2 million bbl/d over the next five years, at a cost of perhaps $6 billion. This would restore Libya's oil production capacity to the level of the early 1970s. During the 1970s, the country's revolutionary government imposed tough terms on producing companies, leading to a slide in oilfield investments and oil production. In May 2000, Libya invited around 50 foreign oil and gas companies to a meeting to discuss exploration and production sharing agreements. In order to achieve its oil sector goals, Libya will require as much as $10 billion in foreign investment through 2010. Around $6 billion of this is to go towards exploration and production, with the rest going towards refining and petrochemicals. In addition, NOC has earmarked $1.5 billion for oil infrastructure investment. In January 2002, NOC appointed Abdel-Hafez Zleitni as its new chairman, with the specific mission to work on attracting foreign investment into the country's oil sector. Combined with the selection of reform-minded Prime Minister Shukri Ghanim, some privatization of the country's oil sector, particularly the downstream sector, now appears more likely than it has in the past.

Currently, Libya has 12 oil fields with reserves of 1 billion barrels or more each, and two others with reserves of 500 million-1 billion barrels. Libya's onshore oil (where most production currently takes place) is found mainly in three geological trends of the Sirte Basin: 1) the western fairway, which includes several large oil fields (Samah, Beida, Raguba, Dahra-Hofra, and Bahi); 2) the north-center of the country, which contains the giant Defa-Waha and Nasser fields, as well as the large Hateiba gas field; and 3) an easterly trend, which has such giant fields as Sarir, Messla, Gialo, Bu Attifel, Intisar, Nafoora-Augila, and Amal. Despite years of oil production, Libya retains a large untapped oil and gas potential, with only around 25 per cent of Libya's area covered by agreements with oil companies. This potential is due largely to lack of investment mainly as a result of stringent fiscal terms imposed by Libya on foreign oil companies. NOC priorities for exploration include new areas in the Sirte (*i.e.*, Blocks 25 and 36), Ghadames (*i.e.*, Block 20), and Murzuq basins (*i.e.*, Field A in Block NC-186, approved for development in August 2002), plus unexplored areas such as Kufra and Cyrenaica. NOC also hopes to apply modern Enhanced Oil Recovery (EOR) techniques to existing oil fields.

Libya has a relatively narrow continental shelf and slope in the Mediterranean and Gulf of Sirte, which widens in the west in the Gulf of Gabes. The northern part of the Gulf of Gabes, also known as the November Seventh concession, lies on the Libyan-Tunisian border and is rich in oil and gas. As part of a 1988 settlement to a long-standing territorial dispute, the area (which contains an estimated 3.7 billion barrels of oil and nearly 12 trillion cubic feet–Tcf–of natural gas) is set to be exploited by the Libyan-Tunisian Joint Oil Company (JOC), a 50-50 venture of Libya's NOC and Tunisia's ETAP. The Libyan side of the zone contains the Omar structure, which is estimated to contain more than 65 per cent of the zone's total oil and gas reserves. On February 1, 1997, JOC awarded the entire block to a consortium consisting of Saudi Arabia's Nimr Petroleum (55 per cent) and Malaysia's Petronas (45 per cent). The companies have a $30 million, 5-year commitment to explore the block. Full development of the concession could cost more than $1 billion.

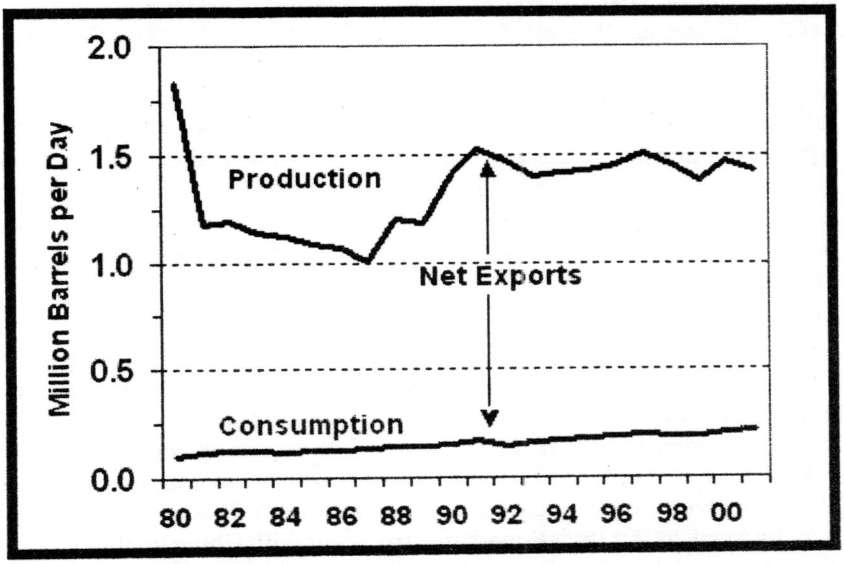

LIBYA'S OIL PRODUCTION AND CONSUMPTION, 1980-2000

Note: Production includes crude oil, natural gas liquids and refinery gain.

Production, Exports, and Reserves

Libya produces high-quality, low-sulphur ("sweet") crude oil at very low cost (as low as $1 per barrel at some fields). Libyan oil is priced off of Dated Brent, and main export grades include Es Sider (36-37° API), El Sharara (44° API), Zueitina (42° API), Bu Attifel (41° API), Brega (40° API), Sirtica (40° API), Sarir (38° API), Amna (36° API), and El Bouri (26° API). Most Libyan oil is sold on a term basis, including to the country's Oilinvest marketing network in Europe; to companies like Agip, OMV, Repsol YPF, Tupras, CEPSA, and Total; and small volumes to Asian and South African companies.

During the first half of 2003, Libyan oil production was estimated at nearly 1.5 million bbl/d, an increase from 2002 levels but still only about two-fifths of the 3.3 million bbl/d produced in 1970. Libya aims to boost oil output capacity by 175,000 bbl/d in 2004 with the help of European companies. The suspension of UN sanctions, along with possible changes to Libya's 1955 hydrocarbons legislation, could be helpful in this regard. Sanctions had caused delays in a number of field development and EOR projects, and had deterred foreign capital investment to an extent. Suspension of sanctions means that Libya now can resume purchases of oil industry equipment.

With reserve replacement slipping since the 1970s, Libya's challenge is maintaining production at mature fields (Brega, Sarir, Sirtica, Waha, Zuetina) while at the same time bringing new fields like Murzuq/El Sharara (online in December 1996; reserves of 2 billion barrels; main operator Repsol-YPF, along with Austria's OMV and Total) and Mabruk (Total) online. Libya currently exports about 1.2 million bbl/d of oil. Nearly all (about 90 per cent) of this is sold to European countries like Italy (485,000 bbl/d in 2002), Germany (188,000 bbl/d in 2002), France (47,000 bbl/d in 2002), Spain and Greece.

With state-operated oil fields undergoing a 7 per cent -8 per cent natural decline rate, Libya depends heavily on foreign companies and workers. Major foreign companies include Spain's Repsol-YPF (mainly at the El Sharara field, plus exploration at blocks NC-186, NC-187, and North-A), Italy's Agip (mainly at Bu Attifel, plus exploration on block NC-174 and in the el-Bouri offshore Mediterranean field), Austria's OMV, PetroCanada's subsidiary Veba Oil Libya (mainly at the Amal field in Block NC-12), Wintershal, and multinational Total. Production from Block NC-115 of the Murzuq basin, being developed by Repsol-YPF, Total, and OMV (with 75 per cent of output going to Libya's NOC), increased to around 75,000 bbl/d in early 1998, and 160,000 bbl/d as of October 2001. In January 1999, Repsol (now Repsol-YPF) said that it had found an "important" petroleum deposit of light, sweet (low sulfur) oil in the Block. In December 2001, Repsol-YPF (along with OMV, Total and Saga Petroleum) announced that it had discovered a significant new oil deposit in Block NC-186 of Murzuq. Also, in April 2002, the same Repsol-YPF consortium announced its first discovery in in the NC-190 block of Murzuq, in the Hawaz formation.

Libya is actively courting foreign oil companies, and is considered a highly attractive oil province due to its low cost of oil recovery, its proximity to European markets, and its well-developed infrastructure. European companies reportedly have been frustrated over the slow pace of progress in awarding Libyan oil concessions, including the 130 exploration blocks offered since UN sanctions were lifted in 1999. However, in May and June 2003, Libya signed oil and gas exploration deals with a Spain's Repsol, Austria's OMV, and Germany's RWE-Dea. Also, in August 2002, India's ONGC and the Turkish Petroleum Overseas Co. agreed to acquire a 49 per cent stake in two blocks in the Sirte and Ghadames basin. In April 2002, Libya signed an agreement with China to offer Chinese companies a wider role in the Libyan oil sector. In August 2002, the China National Petroleum Company (CNPC) won a $230 million contract to build twin oil and natural gas pipelines linking the Wafa field in the south to Melitah, near Tripoli. The pipelines could become operational in 2004.

Of NOC's subsidiaries, the largest oil producer is the Waha Oil Company (WOC), created in 1986 to take over operations from Oasis Oil Co., a joint venture of NOC, Conoco, Marathon, and Amarada Hess. WOC has been among the companies most adversely affected by the US embargo. This is due to the fact that its oilfields are equipped mainly with old US equipment, for which WOC cannot now acquire needed spare parts. As a result, production at WOC's giant Waha field has fallen sharply despite an emergency maintenance program begun in 1992.

After Waha, the next largest NOC subsidiary is the Arabian Gulf Oil Company (Agoco), with production coming mainly from the Sarir, Nafoora/Augila, and Messla fields.

Two other large NOC subsidiaries are the Zueitina Oil Company (ZOC), which operates the five Intisar fields in Block 103 of the Sirte Basin, and the Sirte Oil Company (SOC), originally created in 1981 to take over Exxon's holdings in Libya. In 1986, SOC took over the assets of Grace Petroleum, one of the five US companies forced by the US government to leave Libya in 1996. SOC operates the Raguba field in the central part of the Sirte Basin. The field is connected by pipeline to the main line between the Nasser field, one of the largest in Libya, and Marsa el-Brega. Besides Nasser, SOC is in charge of two other gas fields–Attahadi and Assumud–plus the Marsa el-Brega liquefied natural gas (LNG) plant.

Libya's oilfields are connected to Mediterranean terminals by an extensive network of pipelines. Libya's main crude oil pipelines, all owned by NOC, are: Sarir-Marsa el Hariga (Tobruk); Messla-Ras Lanuf; Waha-Es Sider; Hammada El Hamra-Az Zawiya; Amal-Ras Lanuf; Intisar-Zueitina; Nasser (Zelten)-Marsa El Brega. NOC also has six oil terminals and storage facilities (Marsa El Hariga, Zueitina, Marsa el-Brega, Ras Lanuf, Es Sider, Zawiya), and is considering bids for a $150 million-$300 million expansion of the oil terminal and refinery facility at Az Zawiya.

Exploration and Development

Oil exploration in Libya began in 1955, the key national Petroleum Law No. 25 was enacted in April 1955. Libya's first oil fields were discovered in 1959 (at Amal and Zelten–now known as Nasser), and oil exports began in 1961. After years of little activity due in part to sanctions, Libya now is attempting to attract foreign companies with improved incentives and production terms. Libya has legislation pending which would grant foreign firms better terms, including access to exploration acreage, small field developments, large field incremental production opportunities, and adoption of international competitive bidding practices. Under the current exploration and production sharing agreement (EPSA 3), only around 25 per cent of the country's oil fields have been granted to foreign operators (although Libya does plan to open up some 40 blocks in the Sirte basin and other areas to foreign investment). In July 2000, NOC said that it would open up around 70 per cent of its land to exploration, and that it would bundle exploration blocks into three packages, with the first package to include blocks in the oil-rich Murzuq basin. In May 2003, Libya reportedly awarded a six-block package of onshore and offshore blocks in the Murzuk, Sirte, and Kufra basins, to Repsol-YPF and OMV.

The major component of Libya's expansion plans is development of the el-Bouri oilfield off Libya's western coast, the largest producing oilfield in the Mediterranean Sea (at around 60,000 bbl/d). Italy's Agip-Eni is the developer of the field, discovered in 1976 at a depth of 8,700 feet and estimated to contain 2 billion barrels in proven recoverable crude oil reserves. The first phase of field development, costing $2 billion, was completed in 1990, with el-Bouri producing about 150,000 bbl/d in 1995, followed by a sharp decline thereafter. This decline was due largely to an inability to import EOR equipment under UN sanctions, and possibly could be reversed with an infusion of investment. Besides oil, el-Bouri also contains large amounts (2.5 Tcf) of associated gas.

Since the discovery of the giant, 2-billion barrel el-Bouri field, Agip-Eni has reported a series of oil finds in its various blocks, as have other oil companies in the country. The most significant of these is in the Murzuq basin, in the Sahara south of Tripoli. El Bouri was purchased by Repsol in 1993 for $65 million. Repsol-YPF currently is leading a European consortium, which also consists of OMV and Total. Original expectations were that Murzuq/El Sharara's output of light (44° API), sweet (less than 0.6 per cent sulphur content) crude production would reach 200,000 bbl/d by the end of 1998, but various problems, including difficulties with the pipeline to the port of Az Zawiya, delayed achievement of this target. Currently, oil from Murzuq/El Sharara is being processed by the Az Zawiya refinery.

In October 1997, an international consortium led by British company Lasmo, along with Agip-Eni and a group of five South Korean companies (led by Korea National Oil Corp., replacing Pedco, and including Hyundai), announced that it had discovered large recoverable crude reserves (around 700 million barrels) at the NC-174 Block, 465 miles south of Tripoli, in the remote Murzuq basin. Lasmo, which was purchased by Agip-Eni in 2001 after a deal with Amerada Hess fell through, estimated that production from the field, called Elephant, would cost around $1 per barrel (Repsol-YPF's Murzuq/El Sharara field, with its 30-inch pipeline to the coast, is located only 40 miles to the north). Elephant originally was due to begin production late in 2000 at around 50,000 bbl/d, and to utilize an existing 30-inch pipeline located 42 miles to the north. Production was delayed due to bureaucratic obstacles, with the field now seen reaching full capacity of 200,000 bbl/d in late 2004.

Other foreign companies active in Libya include: Lundin Oil, a Swedish independent, along with its affiliate Red Sea Oil of Canada, has discovered an estimated 84 million barrels of oil at the En Naga North and West fields on block NC-177 in the Sirte basin (in December 1999, Red Sea announced that testing on the block had been suspended); Total, whose Mabruk field is producing around 18,000 bbl/d; and Canadian Occidental, which controls but has not yet developed a potential 200-million-barrel field in Block NC-101 in the Murzuq basin. In 2001, Petro-Canada purchased Lundin's interest in the En Naga block, Concession NC177, with production in 2003 expected at about 6,800 bbl/d. Including En Naga, Petro-Canada also has interests in over 20 producing fields in Libya, with output of more than 40,000 bbl/d.

Refining/Marketing

Libya has three domestic refineries, with a combined nameplate capacity of approximately 343,400 bbl/d, nearly twice the volume of domestic oil consumption (182,000 bbl/d; the rest is exported). Libya's refineries include: 1) the Ras Lanuf export refinery, completed in 1984 and located on the Gulf of Sirte, with a crude oil refining capacity of 220,000 bbl/d; 2) the Az Zawiya refinery, completed in 1974 and located in northwestern Libya, with crude processing capacity of 120,000 bbl/d; and 3) Brega, the oldest refinery in Libya, located near Tobruk with crude capacity of 8,400 bbl/d. In February 2001, bids were submitted by engineering and construction firms on a $400 million project to upgrade Az Zawiya (including construction of a new 120,000-bbl/d refinery). In May 2002, Libya signed a $280 million contract with South Korea's LG Petrochemicals to upgrade the refinery. Ras Lanuf also is slated for upgrading, although that project appears to have been delayed. In March 2002, Ras Lanuf was shut down for several days after a fire broke out at an ethylene storage tank on March 19.

In addition to its domestic refineries, Libya also has operations in Europe. Libya is a direct producer and distributor of refined products in Italy, Germany, Switzerland, and (since early 1998) Egypt. In Italy, Tamoil Italia, based in Milan, controls about 5 per cent of the country's retail market for oil products and lubricants, which are distributed through nearly 2,100 Tamoil service stations. Sanctions have constrained Libya's ability to increase the supply of oil products to European markets, however, as Libya's refineries are badly in need of upgrading, especially in order to meet stricter EU environmental standards in place since 1996. In Egypt, Libya is planning to build gasoline stations on the coastal road linking the two countries as well as in other areas of Egypt. The stations are to be run by Libya's foreign oil investment arm Oilinvest, which maintains 300,000 bbl/d of refining capacity in Europe.

Recently, Libya has been active in various African countries. In June 2003, for instance, Tamoil reportedly secured interests in Zimbabwe's oil pipeline and storage facilities, and reportedly was set to begin pumping oil into Zimbabwe's holding tanks at the Mozambique port of Beira. Zimbabwe has been experiencing an economic and fuel supply crisis in recent months. In September 2002, the Central African Republic (CAR) signed a 99-year deal with Libya to search for oil, uranium and other minerals in the CAR. In August 2002, Libya reached a deal with Ghana to supply that country with an additional 10,000 bbl/d of crude oil to its 45,000-bbl/d Tema refinery.

Libya's refining sector reportedly was hard hit by UN sanctions, specifically UN Resolution 883 of November 11, 1993, which banned Libya from importing refinery equipment. Libya is seeking a comprehensive upgrade to its entire refining system, with a particular aim of increasing output of gasoline and other light products (*i.e.* jet fuel). Possible projects include a new 20,000-bbl/d refinery in Sebha (for which Libya is seeking foreign investment), which would process crude from the nearby Murzuq field, and a 200,000-bbl/d export refinery in Misurata.

Libya has a significant petrochemical industry, centered on Marsa el-Brega. Ethanol, ammonia, and urea are the main petrochemicals produced at present, with Ras Lanouf being developed to produce benzene, butadiene, and MTBE.

Natural Gas

Continued expansion of natural gas production remains a high priority for Libya for two main reasons. First, Libya has aimed (with limited success) to use natural gas instead of oil domestically, freeing up more oil for export. Second, Libya has vast natural gas reserves and is looking to increase its gas exports, particularly to Europe. Libya's proven natural gas reserves in 2003 are estimated at 46.4 Tcf, but the country's actual gas reserves are largely unexploited (and unexplored), and thought by Libyan experts to be considerably larger, possibly 50-70 Tcf. Major producing fields include Attahadi, Defa-Waha, Hatiba, Zelten, Sahl, and Assumud. To expand its gas production, marketing, and distribution, Libya is looking to foreign participation and investment. In recent years large new discoveries have been made in the Ghadames and el-Bouri fields, as well as in the Sirte basin. Libya also produces a small amount of liquefied petroleum gas (LPG), most of which is consumed by domestic refineries.

Libyan natural gas development projects currently underway include as-Sarah and Nahoora, Faregh, Wafa, offshore block NC-41, abu-Attifel, Intisar, and block NC-98. In May 2000, NOC reportedly came out with a framework for gas exploration in the country, under which NOC would have first priority to the foreign company's gas share at an agreed discount. In December 2000, NOC announced that it had discovered a 472-Bcf gas field in the Sirte basin, northwest of Assumud.

Potential exists for a large increase in Libyan gas exports to Europe, although at present the only customer for Libyan gas is Spain's Enagas. A joint venture between Eni and NOC on the Western Libyan Gas Project (WLGP), an approximately $5 billion plan aimed at developing and exporting large volumes of natural gas to Italy, is moving ahead. In June 2002, for instance, Eni affiliate Saipem was awarded a $500-$550 million contract to build and install an offshore natural gas platform northwest of Tripoli. In February 2002, $1 billion worth of engineering, procurement and construction contracts were awarded to a consortium led by Japan's JGC and including France's Sofregaz and Italy's Technimont. The consortium is working on oil and natural gas infrastructure in the Wafa Desert and near Melitah on the Mediterranean coast, with possible first flows starting in 2004 and 2005.

Overall, the WLGP calls for Libya to export 8 billion cubic meters (280 Bcf) per year of natural gas from a processing facility at Melitah to Italy and France over 24 years, beginning in 2006, via a 370-mile underwater pipeline (called "Green Stream") under the Mediterranean to southeastern Sicily and the Italian mainland. To date, Italy's Edison Gas has committed to taking around half (140 Bcf) of this gas, and to use it mainly for power generation in Italy. Besides Edison, Italy's Energia Gas and Gaz de France have each committed to taking around 70 Bcf of Libyan gas. As part of the overall WLGP, Agip-ENI is set to develop huge Libyan gas reserves in offshore Block NC-41 in the Gulf of Gabes, as well as in the Wafa onshore gas (and oil) field on the Algerian border. Feasibility studies have been completed on Wafa and NC-41, and gas is expected to begin flowing by mid-2004. The project also is expected to produce condensates estimated at around 70,000 bbl/d oil equivalent.

Agip-ENI also has promoted linking the reserves of both Egypt and Libya to Italy by pipeline. An agreement in principle to link Egypt and Libya's natural gas grids was reached in June 1997, following a visit to Libya by Egyptian President Hosni Mubarak. In November 2002, an agreement reportedly was reached on construction of a $10 billion project for one pipeline to carry Egyptian natural gas to Libya (for power generation, water desalination, and possible export) and for another to carry Libyan oil to Alexandria, Egypt for refining and consumption there).

Yet another proposal is to build a nearly 900-mile pipeline from North Africa to southern Europe. Such a pipeline could transport natural gas from Egypt, Libya, Tunisia and Algeria, via Morocco and into Spain (a pipeline between Morocco and Spain already exists). Also, Tunisia and Libya agreed in May 1997 to set up a joint venture which will build a natural gas pipeline from the Mellita area in Libya to the southern Tunisian city and industrial zone of Gabes. In late 1998, Tunisia and Libya signed an agreement for around 70 Bcf of gas per year to be delivered from Libyan gas fields to Cap Bon, Tunisia, but no work on a pipeline had begun as of late 2002.

In 1971, Libya became the second country in the world (after Algeria in 1964) to export liquefied natural gas (LNG). Since then, Libya's LNG exports have generally languished, largely due to technical limitations which do not allow Libya to extract LPG from the LNG, thereby forcing the buyer to do so. Libya's LNG plant, at Marsa El Brega, was built in the late 1960s by Esso and has a capacity of about 100 billion cubic feet per year, but due to technical limitations only about one-third of this is available for export, mainly to Enagas of Spain. Work to refurbish and upgrade the El Brega LNG plant in order to deal with the LPG separation problem has been delayed since 1992. If completed, Libyan LNG exports could triple, with likely customers including Spain, Turkey and Italy. On February 1, 2002, Libya joined the Gas Exporting Countries Forum (GECF), formed in 2001 to promote cooperation in the world natural gas industry. Members of the GECF account for around three-fourths of world natural gas reserves and three-fifths of exports.

Electricity

Libya currently has electric power production capacity of about 4.6 gigawatts. Power demand is growing rapidly (around 6 per cent annually), and Libya has plans to more than double installed capacity by 2010 at a cost of over $3.5 billion. Most of Libya's existing power stations are oil-fired, though several have been converted to natural gas.

Plans to utilize natural gas include the 660-megawatt (MW) Western Mountain Power Project (Italy's Enelpower has been announced as the preferred bidder), an 800-MW power plant in Zuwara on the west coast, a 1,400-MW power plant to be located on the coast between Benghazi and Tripoli (Enelpower is bidding on this plant as well), and the 1,200-MW "Gulf Stream" combined power and desalination complex in Sirte (France's Alstom appears to be the lead bidder). In February 2002, Russia's Tekhnopromexport signed a $600 million deal with Libya to build the 650-MW Western Tripoli power plant.

Libya's state-owned General Electricity Company (GECOL) has hinted at the possibility of allowing private investment in the country's power generation and distribution. The country's power sector requires substantial investment, and officials are looking at alternatives to public financing, but despite this, it remains unlikely that Libya will undertake any large-scale power privatization or allow independent power projects (IPPs) anytime soon. Meanwhile, the Export-Import Bank of South Korea reportedly has guaranteed $99 million of the $299 million cost of an expansion and upgrading project at the 450-MW Benghazi North power plant. The project would double the plant's capacity and convert it to combined cycle. GECOL's biggest current project is to expand Libya's network of power substations, which are concentrated mainly in Benghazi, Sebha, and Tripoli. There are also plans to link Libya's power network with those of Egypt, Tunisia, and possibly other countries in Africa.

Country Overview

Independence: December 24, 1951 (from Italy)

Population (2002E): 5.5 million

Location/Size: North Africa/1,775,500 sq km (685,524 sq mi), slightly larger than Alaska

Major Cities: Tripoli (capital), Benghazi, Misurata

Languages: Arabic; Italian and English widely understood in major cities

Ethnic Groups: Arab (97 per cent)

Religions: Sunni Muslim (97 per cent)

Defense (1998E): Army (35,000), Air Force (22,000), Navy (8,000)

Economic Overview

Currency: Libyan Dinar (LD)

Exchange Rate (7/2/03): US$1=1.22 LD

Gross Domestic Product (GDP) (2002E; Purchasing Power Parity exchange rate): $7.0 billion

Real GDP Growth Rate (2002E): 1.2 per cent *(2003F):* 2.1 per cent

Inflation Rate (consumer prices, 2003F): 2.5 per cent

Main Destinations of Exports (2000E): Italy (42 per cent), Germany (19 per cent), Spain (13 per cent), France (6 per cent)

Main Origins of Imports (2000E): Italy (25 per cent), Germany (10 per cent), Tunisia (8 per cent), UK (7 per cent)

Merchandise Exports (2002E): $3.3 billion *(2003F):* $3.4 billion

Merchandise Imports (2002E): $2.3 billion *(2003F):* $2.4 billion

Merchandise Trade Balance (2002E): $1.0 billion *(2003F):* $1.0 billion

Major Export Products: Crude oil, refined petroleum products, natural gas

Major Import Products: Manufactured goods, food and primary products

Total External Debt (non-military) (2001E): $4.4 billion

International Reserves (12/01E): $14.8 billion (17 months worth of import cover)

Energy Overview

Proven Oil Reserves (1/1/03E; Oil and Gas Journal): 29.5 billion barrels

OPEC Crude Oil Production Quota (effective 6/1/03): 1.36 million bbl/d

Oil Production Capacity (2003E): 1.43 million bbl/d

Oil Production (2002E): 1.38 million barrels per day (bbl/d), of which 1.32 million bbl/d was crude oil, and 65,000 bbl/d was natural gas liquids

Oil Consumption (2002E): 218,000 bbl/d

Net Oil Exports (2002E): 1.16 million bbl/d

Major Oil Customers (2002E): Italy, Germany, Spain, and France combined account for around three-quarters of Libya's oil exports; other customers include Austria, Greece, Britain, and Switzerland

Net Oil Export Revenues (2002E): $10.8 billion *(2003F):* $12.9 billion

Net Oil Export Revenues/Total Export Revenues (2002E): 98 per cent

Crude Oil Refining Capacity (1/1/03E): 343,400 bbl/d

Natural Gas Reserves (1/1/03E; Oil and Gas Journal): 46.4 trillion cubic feet (Tcf)

Natural Gas Production (2001E): 218 Billion cubic feet (Bcf)

Natural Gas Consumption (2001E): 191 Bcf

Electric Generation Capacity (2001E): 4.6 gigawatts (100 per cent thermal)

Net Electricity Generation (2001E): 20.2 terawatthours

Environmental Overview

Total Energy Consumption (2001E): 0.65 quadrillion Btu* (0.16 per cent of world total energy consumption)

Energy-Related Carbon Emissions (2001E): 12.5 million metric tons of carbon (0.2 per cent of world carbon emissions)

Per Capita Energy Consumption (2001E): 120.1 million Btu (vs U.S. value of 341.8 million Btu)

Per Capita Carbon Emissions (2001E): 2.3 metric tons of carbon (vs U.S. value of 5.5 metric tons of carbon)

Energy Intensity (2001E): 19,067 Btu/$1995 (vs U.S. value of 10,736 Btu/$1995)**

Carbon Intensity (2001E): 0.37 metric tons of carbon/thousand $1995 (vs U.S. value of 0.17 metric tons/thousand $1995)

Fuel Share of Energy Consumption (2001E): Oil (69.2 per cent), Natural Gas (30.8 per cent)

Fuel Share of Carbon Emissions (2001E): Oil (71.5 per cent), Natural Gas (28.4 per cent)

Status in Climate Change Negotiations: Non-Annex I country under the United Nations Framework Convention on Climate Change (ratified June 14th, 1999). Not a signatory to the Kyoto Protocol.

Major Environmental Issues: Desertification; very limited natural fresh water resources; the Great Manmade River Project, the largest water development scheme in the world, is being built to bring water from large aquifers under the Sahara to coastal cities.

Major International Environmental Agreements: A party to Conventions on Desertification, Marine Dumping, Nuclear Test Ban and Ozone Layer Protection. Has signed, but not ratified, Biodiversity, Climate Change and Law of the Sea.

The total energy consumption statistic includes petroleum, dry natural gas, coal, net hydro, nuclear, geothermal, solar, wind, wood and waste electric power. The renewable energy consumption statistic is based on International Energy Agency (IEA) data and includes hydropower, solar, wind, tide, geothermal, solid biomass and animal products, biomass gas and liquids, industrial and municipal wastes.

Oil and Gas Industries

State Oil Companies: Libyan National Oil Company (NOC)–Manages the state-owned oil industry and controls more than half of Libya's oil production, *Oilinvest*–Manages all international investments

Foreign Energy Company Involvement: Agip-Eni, Canadian Occidental, Husky Oil, Lundin Oil, Nimr Petroleum (Saudi Arabia), OMV, PanCanadian; Pedco (South Korea), Petrobras, Petro-Canada, Petronas (Malaysia), Red Sea Oil Corp. (Canada), Repsol-YPF, Shell, Total, and Wintershall (Germany)

Major Oil Ports: Es Sider, Marsa el-Brega, Tobruk, Ras Lanuf, Zawiya, Zuetina

Major Oil and Gas Fields: Amal, el-Bouri, Bu Attifel, Defa-Waha, Elephant, Kabir, Mabruk, Murzuq, Nasser, Omar, Sarah, Zueitina

Major Pipelines: Amal-Ras Lanuf; Defa-Nasser; Hammada el Hamra-Az Zawiya; Intisar-Zueitina; Intisar -Hatiba; Messla-Ras Lanuf; Nasser-Hatiba; Nasser (Zelten)-Marsa el Brega; Sarir-Marsa el Hariga; Waha-Es Sider

Major Refineries (crude oil capacity): Ras Lanuf (220,000 bbl/d), Az-Zawiya (115,000 bbl/d), Brega (8,400 bbl/d)

52

Malaysia

Malaysia is important to world energy markets because of its 75.0 trillion cubic feet of natural gas reserves and its net oil exports of over 260,000 barrels per day.

General Background

Malaysia is currently in its fourth year of economic growth following a deep recession caused by the Asian financial crisis of 1997-98. Following a 7.5 per cent decline in real gross domestic product (GDP) in 1998, Malaysia experienced real GDP growth of 6.1 per cent in 1999 and 8.3 per cent in 2000. Growth slowed to 0.4 per cent in 2001 on weakened demand for the country's manufactured exports in the face of a global economic slowdown, but it is projected to recover to 3.4 per cent in 2002. Imports of capital goods have been rising, which has lowered the country's merchandise trade surplus.

Malaysia's banking system has been stabilized, after being undermined by a high proportion of nonperforming loans during the financial crisis. The country's banking sector went through a major restructuring in 2000, with many weaker financial institutions being taken over by stronger ones. In order to stimulate the economy, the country's government has increased spending in 2001 and 2002, incurring a budget deficit of 6.5 per cent of GDP in 2001.

Malaysia has maintained its policy of a fixed exchange rate between the ringgit and the U.S. dollar, which was imposed by Prime Minister Mahathir in September 1998, as part of capital controls designed to stem the outflow of short-term capital in the wake of the Asian financial crisis. Malaysian currency is considered somewhat undervalued at the present exchange rate of 3.8 ringgits to one U.S. dollar. Some of the capital controls imposed in 1998 have been relaxed in early 2001, such as the taxes on repatriation of short-term stock market profits by foreign portfolio investors.

Prime Minister Mahathir announced in June 2002 that he intends to relinquish his post, which he has held for two decades, in late 2003. The current Deputy Prime Minister, Abdullah Ahmed Badawi, is expected to succeed him.

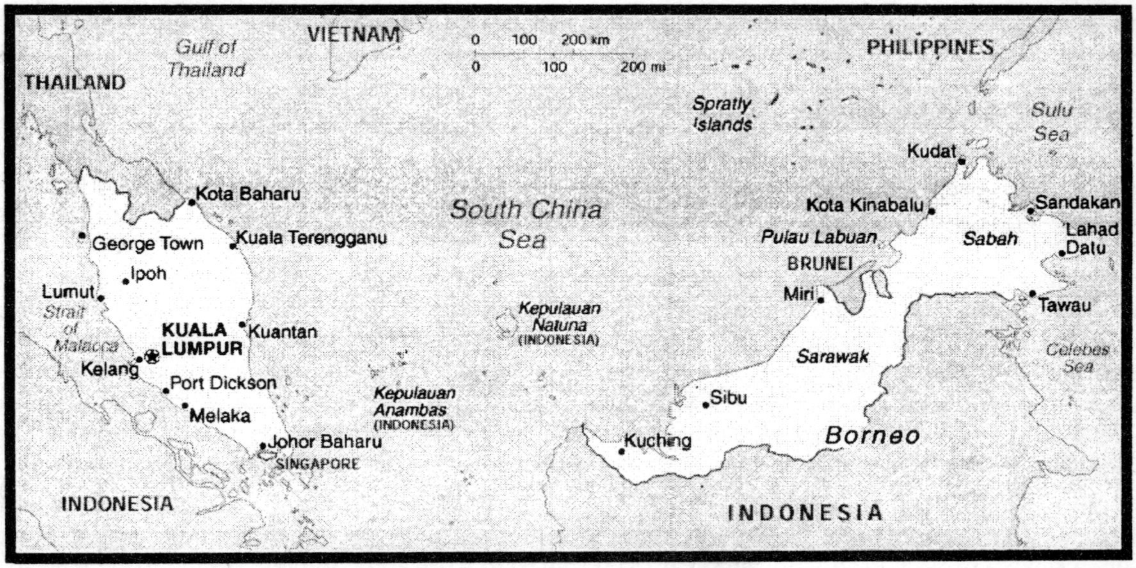

Oil

Malaysia contains proven oil reserves of 3.0 billion barrels, down from 4.3 billion barrels in 1996. Despite this trend toward declining oil reserves (due to a lack of major new discoveries in recent years), Malaysia's crude oil production has been stable in recent years, with monthly production numbers fluctuating between 650,000 barrels per day (bbl/d) and 730,000 bbl/d between 1996 and early 2002. In 2001, crude oil production averaged 659,205 bbl/d. After a pause during the Asian financial crisis, Malaysia's domestic petroleum product consumption is growing again, and the country is expected to become a net oil importer before the end of the current decade.

As a result of declining oil reserves, Petronas, the state oil and gas company, has embarked on an international exploration and production strategy. Currently, Petronas is invested in oil exploration and production projects in Syria, Turkmenistan, Iran, Pakistan, China, Vietnam, Burma, Algeria, Libya, Tunisia, Sudan, and Angola. Overseas operations now make up nearly one-third of Petronas revenue. In 2001, Malaysia exported the majority of its oil to markets in Japan, Thailand, South Korea, and Singapore.

Malaysia's domestic oil production occurs offshore and primarily near Peninsular Malaysia. Most of the country's oil fields contain low sulfur, high quality crude, with gravities in the 35°-50° API range. More than half of the country's oil production comes from the Tapis field, which contains 44° API oil with a low sulfur content. Esso Production Malaysia Inc. (EPMI), an affiliate of ExxonMobil Corporation, is the largest crude oil producer in Peninsular Malaysia, accounting for nearly half of Malaysia's crude oil production. EPMI operates seven fields near the peninsula, and one-third of its production comes from the Seligi field. The Seligi-F platform, with its 28 wells, is the newest satellite in the Seligi field, located 165 miles off the coast of Terengganu, Peninsular Malaysia. Built at a cost of \$155 million, Seligi-F is the seventh production platform on the Seligi field. The platform came on stream in March 1998 and is expected to produce an annual average of 21,000 bbl/d. EPMI holds a 78 per cent interest in the project with Petronas Carigali holding the remaining 22 per cent. In addition,

EPMI began drilling the nearby Raya-A platform in the second quarter 1998. EPMI has invested $96 million in six wells, and holds an 80 per cent interest with Petronas Carigali holding the remaining 20 per cent. EPMI began production from the offshore Larut field in Block PM5 in early 2002, which is expected to reach peak production of 140,000 bbl/d, offsetting some of the future decline in Malaysia's production from mature fields.

In other developments, Sabah Shell Petroleum Company, a unit of Royal Dutch/Shell Group, raised production at the Kinabalu field to 36,000 bbl/d, as well as 28 million cubic feet per day (Mmcf/d) of gas. Production at Kinabalu, located in the SB-1 block 34 miles off the coast of Labuan, Sabah in east Malaysia, began in December 1997. As operator of the SB-1 block, Shell holds an 80 per cent stake in the block, with Petronas holding a 20 per cent stake. In February 1998, Amerada Hess signed two, five-year production sharing contracts (PSCs) with Petronas for blocks PM304 and SK306. The PSCs commit Amerada to $24.9 million of exploration activities on the two blocks. A successful well was reported in the PM 304 block in April 2001, but it is still under evaluation and a reserve estimate has not been announced. Under the PSCs, Amerada holds a 70 per cent stake in PM304, offshore Terengganu, and an 80 per cent stake in SK 306, offshore Sarawak, with Petronas holding the remaining interests in both blocks.

In February 2000, Sweden's Lundin Oil announced that it had signed a sales agreement with Petronas and PetroVietnam which will allow it to proceed with development of its long-delayed Bunga Kekwa project. Production stood at 18,000 bbl/d in April 2001, and is expected to increase to a volume of 40,000 bbl/d when development is completed in 2003. Lundin Oil is the operator of the field, and Petronas and Petrovietnam hold equity stakes in the project. Murphy Oil reported a modest-sized new find in Block SK309 in shallow waters offshore from Sarawak in February 2002. The new field is expected to be in production by early 2003.

Refining and Downstream

Malaysia has six refineries, with a total processing capacity of 514,500 bbl/d. The three largest are the 155,000 bbl/d Shell Port Dickson refinery and the Petronas Melaka-I and Melaka-II refineries, which each have a capacity of 95,000 bbl/d.

The second phase of the $1.4-billion, 200,000-bbl/d Melaka refinery complex, located about 90 miles south of Kuala Lumpur, commenced operation in August 1998. The 100,000-bbl/d Melaka-II second phase is a joint venture between Petronas (45 per cent), Conoco (40 per cent), and Statoil (15 per cent). This second refinery contains a 62,000-bbl/d vacuum distillation unit, 26,000-bbl/d catalytic cracker, 28,500-bbl/d hydrocracker, 35,000-bbl/d desulfurization unit, and 21,000-bbl/d coker. One of the main purposes of this refinery is to supply gasoline to Conoco's service stations in Thailand and a new line of stations planned for Malaysia. The first phase of the Melaka refinery was finished in mid-1994 and consisted of a 100,000-bbl/d sweet crude distillation unit, which is wholly-owned by Petronas and processes Tapis crude oil.

Natural Gas

Malaysia contains 75 trillion cubic feet (Tcf) of proven natural gas reserves. Natural gas production has been rising steadily in recent years, reaching 1.50 Tcf in 2000, up from 1.42 Tcf in 1999. Natural gas consumption in 2000 was estimated at 0.72 Tcf, with LNG exports of 0.74 Tcf (mostly to Japan, South Korea, and Taiwan). Exports had dipped slightly in 1998 as a result of the Asian financial crisis, but began to climb again in 1999 and 2000.

One of the most active areas in Malaysia for gas exploration and development is the Malaysia-Thailand Joint Development Area (JDA), located in the lower part of the Gulf of Thailand and governed by the Malaysia-Thailand Joint Authority (MTJA). The MTJA was established by the two governments for joint exploration of the once-disputed JDA. The JDA covers blocks A-18 and B-17 to C-19. A 50:50 partnership between Petronas and Amerada Hess is developing block A-18, while the Petroleum Authority of Thailand (PTT) and Petronas also share equal interests in the remaining blocks. PTT and Petronas announced an agreement in November 1999 to proceed with development of a gas pipeline from the JDA to a processing plant in Songkla, Thailand, and a pipeline linking the Thai and Malaysian gas grids. Malaysia and Thailand will eventually each take half of the gas produced, though initial production will go just to Malaysia. The project had been controversial in Thailand, facing opposition from local residents in Songkla along the pipeline route. In May 2002, the Thai government announced a final decision to commence construction on the project later in 2002, through the pipeline route was altered slightly to avoid some populated areas.

Block A-18 is operated by a joint venture between Amerada Hess, BP, and Petronas. In December 1997, the MTJA approved a development plan for the Cakerawala gas field, which will be the first JDA field to come on line. In November 1999, a gas sale agreement was signed with Petronas and PTT, which will allow the firms to proceed with development. Gas production of 390 Mmcf/d will begin in early 2003.

ExxonMobil announced in March 2002 that it would move forward with development of the offshore Binting gas field in the South China Sea. The field contains about 1 Tcf of proven reserves, and is expected to reach a peak output of 335 Mmcf/d. Drilling is set to commence in late 2002.

Malaysia accounted for approximately 15 per cent of total world LNG exports in 2000. After a brief downturn related to the Asian financial crisis, demand for LNG is rising again. After much delay, Malaysia is proceeding with a long-planned expansion of its Bintulu LNG complex in Sarawak. In February 2000, Petronas signed a contract with a consortium headed by Kellogg Brown and Root for construction of the MLNG Tiga facility, with two LNG liquefaction trains and a total capacity of 7.6 million metric tons (370 Bcf) per year. The Bintulu facility as a whole will then be the largest LNG liquefaction center in the world, with a total capacity of 23 million metric tons per year (1.1 Tcf). Financing for the MLNG Tiga facility was completed in April 2001, and it is expected be begin operation in 2003. Most of the production from the new LNG trains will be sold under term contracts to utilities in Japan. Tokyo Electric Power (TEPCO), Tokyo Gas, and Chubu Electric all have signed contracts for LNG from the project.

In addition to LNG, Malaysia exports 150 million cubic feet per day (Mmcf/d) to Singapore via pipeline. Surprisingly, Malaysia may also become an importer of gas from Indonesia. Petronas signed an agreement in April 2001 with Indonesia state oil and gas company Pertamina for the import of gas from Conoco's West Natuna offshore field in Indonesian waters. The move is being seen as part of a Malaysian strategy to become a hub for Southeast Asian natural gas integration. Deliveries are scheduled to begin once pipeline construction is complete in August 2002. The pipeline will connect to an existing pipeline from the shore to Malaysia's offshore Duyong field, which will help minimize construction costs.

Electricity

Malaysia currently has approximately 13 gigawatts (GW) of electric generation capacity, of which 84 per cent is thermal and 16 per cent is hydroelectric. In 2000, Malaysia generated around 63 billion kilowatthours of electricity. The Malaysian government expects that investment of $9.7 billion will be required in the electric utility sector through 2010. Much of that amount will be for coal-fired plants, as the Malaysian government is promoting a shift away from the country's heavy reliance on natural gas for electric power generation.

A contract was signed in December 2000 between a local Malaysian independent power producer (IPP), GB3, and Alstom for an additional 650-MW generation unit to be built at the site of the Lumut Power Plant. Siemens received contracts in December 2000 for two 710-MW power plants, one at Teluk Gong for the IPP Powertek, and one in Sepang for Malaysian Resources Corporation.

Some other projects have been called into question, though, in the short-term, due to the decline in the country's economic growth rate. The 2,100-MW Pulau Bunting project and the 1,400-MW Jimah project, both coal-fired, may be delayed as a result of slowing electricity demand growth.

In 1994, the Malaysian government granted approval for the massive 2.4-GW Bakun hydroelectric project in Sarawak. Scheduled for completion in 2002, the Bakun Dam had been slated to send 70 per cent of its generated power from Sarawak to Kuala Lumpur through the construction of 415 miles of overhead lines in eastern Malaysia, 400 miles of submarine cables, and 285 miles of distribution infrastructure in Peninsular Malaysia. In addition, expansion plans included a high voltage line south to Johor Baharu and north to Perlis, near the western Thai border. A local company, Ekran, was awarded a turnkey contract to manage the project in January 1995. In 1996, the construction contract went to Sweden's Asea Brown Boveri (ABB). However, in early September 1997, the Malaysian government announced that it was delaying the project indefinitely, citing an unexpected rise in the dam's cost due to the country's economic difficulties.

In mid-1999, work resumed on the river diversion tunnels, a major component of the project, which have since been completed. The Malaysian government has taken control of the project and negotiated financial settlements with the firms involved. The subsea transmission line concept has been abandoned, and the Malaysian government is exploring the possibility of sales of electricity to Brunei and Indonesia. While it had appeared likely that the project would be scaled back from its 2,400-MW capacity, the Malaysian government announced in February 2001 that it had decided to complete the project on its original scale. Bids were received in July 2002 for the main construction work for the dam, and a contract is expected to be awarded by fall 2002.

Malaysia is considering reforms to its power sector to make it more competitive and lower costs. Currently, three state-owned utilities dominate power generation and distribution in Malaysia. The market was opened to independent power producers (IPPs) in 1994, and 15 IPPs were licensed, though not all of the projects have been built.

In recent developments, Tenaga Nasional Bhd, the main state-owned utility, began in 1999 to divest some of its power generation units. Eventually, Malaysia expects to achieve a fully competitive power market, with generation, transmission, and distribution decoupled, but reform is still at an early stage and the exact process of the transition to a competitive market has not been decided. The issue is still under study, and many observers have voiced caution in light of the experiences of other deregulated utility systems.

Country Overview

Independence: August 31, 1957 (from United Kingdom)

Population (2001E): 22.2 million

Location/Size: Southeast Asia / 127,320 sq. mi. (slightly larger than New Mexico)

Major Cities. Kuala Lumpur (capital), Ipoh, Melaka, Johor Baharu, Penang, Kota Baharu, Georgetown, Kuching

Languages: Malay (official), English, Chinese dialects, Tamil, tribal dialects

Ethnic Groups: Malay and other indigenous (58 per cent), Chinese (26 per cent), Indian (7 per cent), others (9 per cent)

Religion: Islam, Buddhism, Confucianism, Hínduism, Christianity, various tribal

Defense (8/98): Army (85,000), Navy (12,500), Air Force (12,500)

Economic Overview

Currency: Ringgit

Market Exchange Rate (7/29/02): $1 = 3.8 ringgits

Gross Domestic Product (market exhange rates) (2001E): $91.9 billion; *(2002E):*$97.8 billion

Real GDP Growth Rate (2001E): 0.4 per cent ; *(2002E):* 3.4 per cent

Inflation Rate (consumer prices)(2001E): 1.4 per cent ; *(2002E):* 1.6 per cent

Current Account Balance (2001E): $5.6 billion; *(2002E)* $9.4 billion

Major Trading Partners (2002): Singapore, Japan, United States, European Union

Merchandise Exports (2001E): $90.0 billion; *(2002E):* $93.2 billion

Merchandise Imports (2001E): $75.6 billion; *(2002E):* $75.4 billion

Major Export Products: Petroleum and petroleum products, palm oil, rubber, tin, electronic equipment

Major Import Products: machinery equipment, chemicals and food

External Debt (2002E): $34.3 billion

Energy Overview

Proven Oil Reserves (1/1/02E): 3.0 billion barrels

Oil Production (2001E): 730,205 barrels per day (bbl/d), of which 659,205 bbl/d is crude oil

Oil Consumption (2001E): 460,000 bbl/d

Net Oil Exports (2001E): 270,205 bbl/d

Crude Oil Refining Capacity (1/1/02E): 514,500 bbl/d

Natural Gas Reserves (1/1/02E): 75.0 trillion cubic feet (Tcf)

Natural Gas Production (2000E): 1.50 Tcf

Natural Gas Consumption (2000E): 0.72 Tcf

LNG Exports (2000E): 0.74 Tcf

Recoverable Coal Reserves (12/31/96): 4 million short tons

Coal Production (2000E): 0.2 million short tons

Coal Consumption (2000E): 3.3 million short tons

Net Coal Imports (2000E): 3.1 million short tons

Electricity Generation Capacity (1/1/00): 13.0 gigawatts (84 per cent thermal, 16 per cent hydroelectric)

Electricity Generation (2000E): 63 billion kilowatthours

Environmental Overview

Total Energy Consumption (2000E): 1.9 quadrillion Btu* (0.46 per cent of world total energy consumption)

Energy-Related Carbon Emissions (2000E): 29.9 million metric tons of carbon (0.46 per cent of world total carbon emissions)

Per Capita Energy Consumption (2000E): 79.8 million Btu (vs U.S. value of 351.1 million Btu)

Per Capita Carbon Emissions (2000E): 1.3 metric tons of carbon (vs U.S. value of 5.6 metric tons of carbon)

Energy Intensity (2000E): 16,631 Btu/$1995 (vs U.S. value of 10,919 Btu/$1995)**

Carbon Intensity (2000E): 0.27 metric tons of carbon/thousand $1995 (vs U.S. value of 0.17 metric tons/thousand $1995)

Sectoral Share of Energy Consumption (1998E): Industrial (52.2 per cent), Transportation (25.4 per cent), Residential (10.9 per cent), Commercial (11.5 per cent)

Sectoral Share of Carbon Emissions (1998E): Industrial (51.9 per cent), Transportation (30.2 per cent), Residential (7.3 per cent), Commercial (10.6 per cent)

Fuel Share of Energy Consumption (2000E): Oil (50.0 per cent), Natural Gas (40.9 per cent), Coal (4.8 per cent)

Fuel Share of Carbon Emissions (2000E): Oil (56.1 per cent), Natural Gas (36.6 per cent), Coal (7.3 per cent)

Renewable Energy Consumption (1998E): 110.2 trillion Btu* (17 per cent increase from 1997)

Number of People per Motor Vehicle (1998): 5.8 (vs U.S. value of 1.3)

Status in Climate Change Negotiations: Non-Annex I country under the United Nations Framework Convention on Climate Change (ratified July 13th, 1994). Signatory to the Kyoto Protocol (signed March 12th, 1999–not yet ratified).

Major Environmental Issues: Air pollution from industrial and vehicular emissions; water pollution from raw sewage; deforestation; smoke/haze from Indonesian forest fires.

Major International Environmental Agreements: A party to Conventions on Biodiversity, Climate Change, Desertification, Endangered Species, Hazardous Wastes, Law of the Sea, Marine Life Conservation, Nuclear Test Ban, Ozone Layer Protection, Ship Pollution, Tropical Timber 83, Tropical Timber 94.

The total energy consumption statistic includes petroleum, dry natural gas, coal, net hydro, nuclear, geothermal, solar, wind, wood and waste electric power. The renewable energy consumption statistic is based on International Energy Agency (IEA) data and includes hydropower, solar, wind, tide, geothermal, solid biomass and animal products, biomass gas and liquids, industrial and municipal wastes.

Oil and Gas Industries

Organization: Malaysia's national petroleum corporation, Petroliam Nasional Berhad (Petronas), was formed in 1974. Petronas controls oil production through partnerships with Exxon (Esso Production Malaysia) and Shell (Sabah Shell Petroleum, Sarawak Shell Berhad, and Sarawak Shell/Petronas Carigali)

Major Foreign Oil Company Involvement: BP Amoco, Conoco, Enron, ExxonMobil, Lundin Oil, Murphy Oil, Nippon Mitsubishi Oil, Occidental, Shell, Texaco, Triton

Major Oil Fields: Bekok, Bokor, Erb West, Bunga Kekwa, Guntong, Kepong, Kinabalu Pulai, Samarang, Seligi, Semangkok, Tapis, Temana, Tiong

Major Natural Gas Fields. Bedong, Bintang, Damar, Jerneh, Laho, Lawit, Noring, Pilong, Resak, Telok, Tujoh

Major Oil Refineries (capacity–bbl/d): Port Dickson-Shell (155,000), Melaka I (95,000), Melaka II (95,000), Kerteh-Petronas (40,000), Port Dickson-Esso (84,500), Lutong-Shell (45,000)

Major Oil Pipelines: Malaysia-Singapore pipeline, planned Malaysia–Songkhla (Thailand) product pipeline

Major Oil Terminals: Bintulu, Johor Baharu, Kerteh, Kuching, Melaka, Penang, Port Dickson, Port Kelang

53

Mexico

Mexico is a major non-OPEC oil producer and home to one of the world's largest oil companies, Pemex. Most of Mexico's oil exports go to the United States.

General Background

Mexico's real gross domestic product (GDP) grew by only 0.9 per cent in 2002 as the continued slow down in global demand weighed heavily on many of the country's key export industries. However, after contracting by 0.3 per cent in 2001, 2002's growth rate marks a modest recovery for the Mexican economy. In 2002 the country benefited from both higher than expected oil prices for the year, as well as economic recovery in the United States, Mexico's largest trading partner. Economic growth in 2003 will depend largely upon developments in the U.S. economy, as well as the ability of President Vicente Fox and the Mexican Congress to come to an agreement on legislation designed to revitalize Mexico's economy.

Fellow NAFTA member the United States consumes over 80 per cent of Mexico's exports. Mexico's exports to the U.S. include significant quantities of crude oil, as well as manufactured goods and machinery. Mexico's merchandise exports increased modestly in 2002 over the previous year, but were still below the year 2000 level. In addition, Mexico's Maquiladora sector, which comprises Mexican corporations that work in government sanctioned partnerships with foreign investors to export goods, mainly to the United States, suffered declining production in 2002 after being a major contributor to Mexico's economic growth throughout the 1990s.

Relations between President Fox and the Mexican Congress were strained in 2002 as the Congress did not approve several of the President's economic proposals, including legislation to improve tax collection and revitalize the energy industry. Vicente Fox of the National Action Party (PAN) was inaugurated as president of Mexico on December 1, 2000, ending 71 consecutive years of Institutional Revolutionary Party (PRI) rule. The 2003 legislative session will begin in February and Fox's electricity

and labor reform packages are expected to be high on the administration's agenda. The next scheduled legislative elections will take place in July 2003.

Oil

Mexico has the fourth largest proven crude oil reserves in the Western Hemisphere after Canada, Venezuela, and the United States at 12.6 billion barrels. During the first 10 months of 2002, Mexico produced about 3.6 million barrels per day (bbl/d) of oil, of which approximately 3.2 million bbl/d were crude oil. Mexico consumed approximately 1.9 million bbl/d of oil in 2002, resulting in net exports of roughly 1.7 million bbl/d. About 1.5 million bbl/d of these exports were bound for the United States, making Mexico the third largest foreign supplier of petroleum to the United States. During this period, Mexico ranked as the world's fourth-largest oil producer (including crude, lease condensate, natural gas liquids, and refinery gain), behind the United States, Saudi Arabia, and Russia.

Sector Organization: Pemex

The Mexican oil industry was nationalized in 1938. *Petroleos Mexicanos* (Pemex), the state oil company, is the world's fifth largest oil company, the single most important entity in the Mexican economy, and is considered a symbol of Mexican sovereignty and independence. Unlike many other major global oil companies, Pemex retains exclusive rights to oil exploration and production in Mexico.

The government relies on Pemex for approximately 1/3 of its fiscal revenues, as an estimated 60 per cent of the company's revenues are turned over to federal authorities. This financial responsibility has left the company with little ability to invest in exploration and development in recent years, resulting in a 9 per cent drop in proven crude oil reserves between 1998 and 2002, as the consumption of hydrocarbons increasingly outpaces discovery. After lobbying from President Fox in 2001, the Mexican Congress allocated $23.9 billion to the Pemex budget in 2002, providing $14.7 billion for new investment. The 2002 budget represented Pemex's largest annual investment increase in twenty years, and resulted in a modest increase in oil production according to preliminary statistics. The 2003 Pemex budget is the biggest in the country's history, and President Fox aims to increase Mexican crude oil output by 34 per cent between 2001 and 2006.

But many observers, including President Fox, consider underfunding to be only one of the many problems facing Pemex. Pemex, with its workforce of 139,000, employs three and a half times as many people as does Venezuela's PdVSA, to produce only slightly more oil and gas. Moreover, as the country's existing fields mature, some observers consider the company unequipped to discover and monetize new natural resources. Since early in his presidential campaign, President Fox has made efforts towards modernizing Pemex by proposing that the company open to more foreign involvement. Upon taking office in 2000, President Fox appointed Raul Munoz Leos, a former Dupont executive to head the company. During 2002, Fox's efforts included the drafting of new prototype contractual arrangements as well as opening certain geographic areas to increased foreign involvement. Currently, new investment in Mexico's oil industry, including some upstream exploration, is implemented under the auspices of Pidiregas program, by which private partners finance strategically important state-owned projects. Pidiregas was first implemented in 1995 in an effort to attract new foreign capital to Mexico's energy industry. In 2002, Pemex devoted $1.54 billion to Pidiregas, of which approximately $1 billion went towards upstream development.

Reserves and Production

In September 2002, Pemex revised its crude oil reserve estimate downwards by 26 per cent to 12.6 billion barrels. The revision was made to comply with U.S. Securities and Exchange Commission (SEC) filing guidelines, which require that hydrocarbon reserves qualifying as "proven reserves" be under commitment for exploration in the short term. Pemex's dollar bond sales require the company to report to the SEC. This revised proved crude oil reserve estimate is 14.4 billion barrels below last year's level. Despite dwindling crude oil reserves, EIA expects Mexican petroleum production to increase to 3.8 million bbl/d in 2003, slightly exceeding 2002 levels as production increases at the country's existing fields.

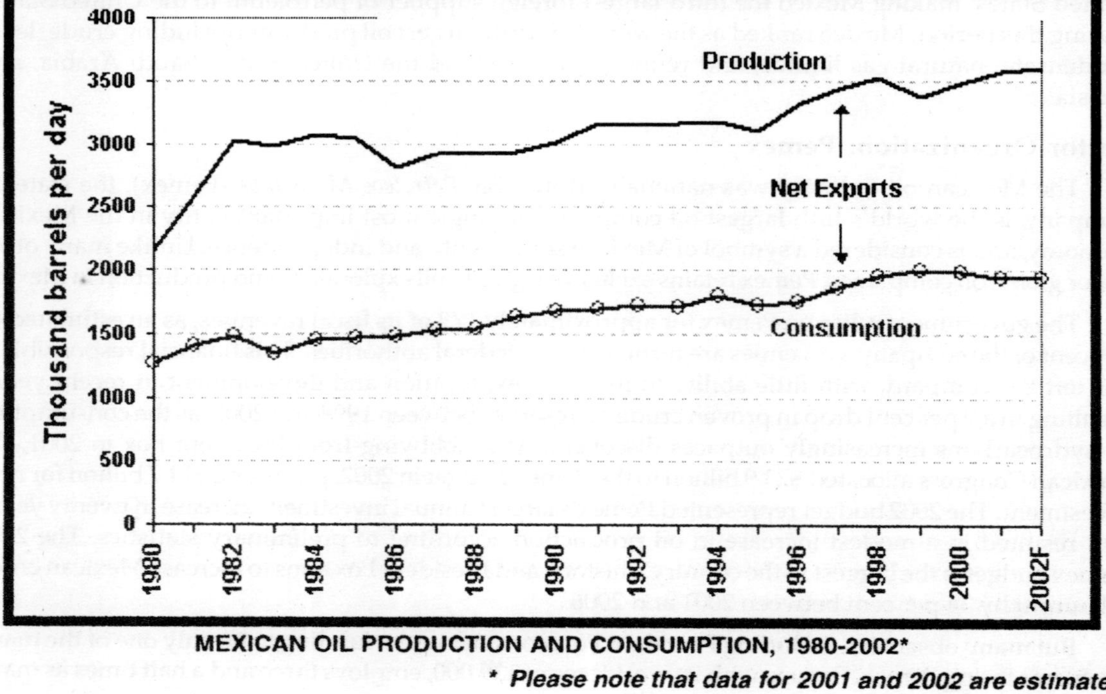

MEXICAN OIL PRODUCTION AND CONSUMPTION, 1980-2002*

Please note that data for 2001 and 2002 are estimates.

Approximately three-quarters of Mexico's crude oil production comes from the Bay of Campeche, located in the Mexican Gulf along the western side of the Yucatan Peninsula. Most of the oil produced in Campeche is a characteristically heavy crude oil (22 API) known as Maya-22. Maya accounts for approximately 70 per cent of the country's total production, and approximately 80 per cent of Mexico's oil exports. The most productive field in the Bay of Campeche, and the mainstay of the Mexican oil industry, is Cantarell, which is located about 80 miles offshore and yields approximately 1.7 million bbl/d of Maya-22 heavy oil. The Cantarell field, (which refers to a complex four major sites: Akal, Nohoch, Chac and Kutz) has lost essential natural pressure since its initial discovery in 1976 and until recently was experiencing declining rates of production. In 1997, Pemex began a nitrogen injection program, designed to maintain the necessary pressure in the field's wells by injecting nitrogen into the reservoir. The project was completed in late November 2001, concomitantly with Mexico's decision to reduce exports in cooperation with OPEC. After the expiration of Mexico's self-imposed export cap in mid-2002, Cantarell began showing significant signs of recovery, and by November 2002, daily production from Cantarell was almost 370,000 bbl/d above the 2001 level. Pemex continues to develop Cantarell, and expects to reach production of 2.23 million barrels per day in 2004. Pemex is also hopeful that the untapped Sihil field, located underneath the Cantarell field, holds as much as 1.4 billion barrels of oil and associated natural gas.

Mexico also produces two other grades of crude oil; light, low-sulfur Isthmus-34, accounting for less than one-third of total production; and extra-light Olmeca-39, which is about one-fifth of total production. Fields yielding these grades are also located in the South, mostly near the Gulf of Mexico. Pemex continues to explore for new fields, and in August 2002 the company announced plans to drill 5 exploratory wells near Coatzacoalcos, located on the continental platform of the Gulf of Mexico, south of Vera Cruz. Pemex also has exploration projects in the wetlands of the Grijalva River Delta, located in Southeast Mexico.

Mexico, Oil Exports, and OPEC

Underscoring its key position in world oil markets, Mexico has had more contact and involvement with the Organization of Petroleum Exporting Countries (OPEC) than any other non-member country. Mexico has worked in conjunction with OPEC since 1998 and Pemex export data from 1998-2001 show that the country's self-imposed targets have been kept. In January 2002, Mexico committed to an export cap of 1.66 million barrels per day for the first half of 2002, a cut of 100,000 bbl/d in order to coordinate with OPEC. Similar cuts were made by Russia, Norway, Oman and Angola. On January 22, 2002, Pemex declared Force Majeure on the entirety of its foregone exports. At the expiration of its self-imposed export cap, Mexican exports began to rise, and in November 2002, Mexico became the top foreign crude supplier to the United States for the month. On January 13, 2003, following OPEC's announcement to raise the OPEC-10 output ceiling by 1.5 million bbl/d, Mexico announced its intentions to increase its own production platform by 120,000 bbl/d to 1.88 million bbl/d. Mexico's Isthmus crude oil is the only non-OPEC crude oil included in the "OPEC basket", the arithmethic average of seven selected crudes that is used to as an indicator of the average price per barrel of OPEC's oil.

Downstream

Mexico has six refineries, with a total throughput capacity of 1.7 million bbl/d as of January 1, 2003. During 2002, Pemex reported that renovation at two of the country's refineries, Ciudad Madero (320,000 bb/d) and Cadereyta (275,000 bbl/d) had increased total capacity by almost 200,000 bbl/d, with the largest increase (125,000) at Ciudad Madero. The Ciudad Madero refinery had been under renovation since 1999 by a consortium led by Sunkyong Engineering and Construction. Two other refineries, Tula Hidalgo (320,000 bbl/d) and Salamanca (245,000 bbl/d) have also reportedly been targeted for upgrades. In 2001, the government began a $3.9 billion long-term upgrading plan designed to augment total refinery capacity by 350,000 bbl/d by 2006. Mexico's downstream oil sector was nationalized in the 1950s, and privatization is not planned, although private companies now are allowed to sell retail refined petroleum products.

Despite having sizable crude oil reserves, Mexico's insufficient refining capacity compels the country to import petroleum products. In 2000, Mexico imported 448,000 bbl/d of petroleum products, accounting for approximately 22 per cent of total consumption. In 2000, petroleum products from the United States averaged 358,000 bbl/d, representing almost 80 per cent of Mexican imports and 17 per cent of the country's total consumption. During the first 10 months of 2002, U.S. petroleum product exports to Mexico totaled 237,000 bbl/d.

Mexico's retail gasoline prices are amongst the highest in the region, causing many inhabitants of border towns to cross into the United States and Belize to purchase motor gasoline. In May 2002, the Mexican government began a pilot program aimed at reducing this cash outflow by reducing prices at test locations along the country's Northern and Southern borders. The pilot program proved successful, as the Northern border town of Ciudad Juarez reportedly experienced a 40 per cent increase in gasoline sales after the government lowered gasoline prices. On December 1, 2002, President Fox lowered retail gasoline prices at 500 of the government's nationally owned stations located along the border with California and Texas.

Natural Gas

Mexico has proven natural gas reserves of 8.8 trillion cubic feet (Tcf), with 2000 production of about 1.33 Tcf and consumption of about 1.38 Tcf. Mexico's natural gas reserves were revised downwards along with its crude oil reserves in September 2002 (see Reserves and Production). Although the country is home to the Western hemisphere's sixth-largest natural gas reserves (after the United States, Venezuela, Canada, Argentina, and Bolivia), Mexico has until recently neglected natural gas exploration and development, and production has increased only modestly since 1980.

The consumption of natural gas in Mexico first began to outpace production in 1988. Later, in 1995, structural changes to the Mexican economy encouraged the construction of numerous new gas-fired power plants. In 1999, tariffs on Mexican imports of U.S. natural gas were eliminated. As a result of this growing gasification of the Mexican economy, Mexico's imports of natural gas have risen significantly in recent years. Most of these imports come from the United States. In 2000, approximately 8 per cent of Mexico's natural gas demand was met through imports from the United States. During the first 10 months of 2002, Mexico imported approximately 0.7 billion cubic feet per day (Bcf/d) from the United States, more than doubling the 2000 levels. Demand for natural gas in Mexico is expected to grow by 8 per cent annually through 2010, resulting in an estimated daily consumption of approximately 9 Bcf/d by the end of the decade-double daily consumption in 2000. The Fox administration has made meeting Mexico's overwhelming natural gas demand an integral part of country's overall energy policy. In 2002, the administration undertook measures designed to both increase domestic production as well as facilitate imports from a more diverse array of foreign suppliers.

Upstream

Mexico's existing natural gas reserves are located primarily in the southwestern states of Tabasco and Chiapas. In March 2002, Pemex announced the discovery of three new natural gas fields in the

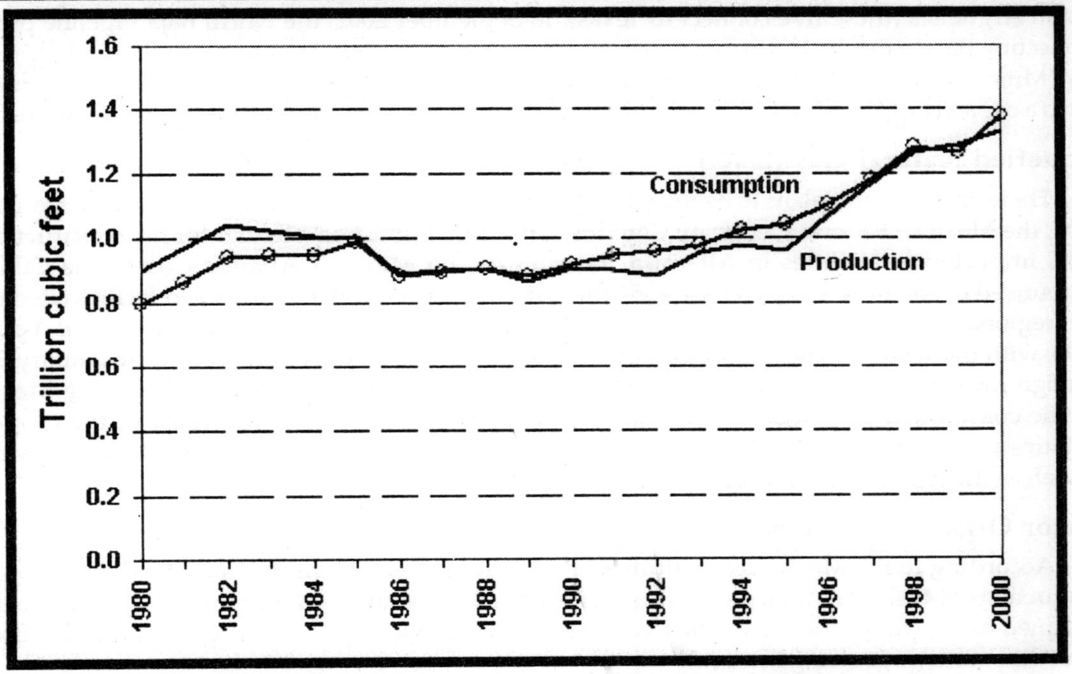

MEXICAN NATURAL GAS PRODUCTION AND CONSUMPTION, 1980-2000

state of Veracruz, located along the Mexican Gulf. Preliminary drilling completed in May 2002 has indicated that one of the fields, Lankahuasa, located offshore, might hold up to 1 Tcf of natural gas. Pemex estimates that when combined with the other two fields, Playuela and Hap, these new discoveries could account for one quarter of total Mexican natural gas reserves. Pemex has estimated that the Lankahuasa field could begin producing in 2003. The Cantarell field also holds significant associated natural gas reserves, most of which currently is flared.

Mexico also has sizable natural gas reserves in the Burgos Basin, located in northeastern Mexico along the border with Texas. The Burgos Basin, which has been producing since 1945, currently produces approximately 1 Bcf/d of dry natural gas, approximately 1/4 of total annual production. The Fox administration has announced its intentions to double production from the Burgos Basin by 2006, by opening the fields to increased foreign involvement. In June 2002, President Fox unveiled the prototype contractual arrangement designed to govern foreign involvement.

Transportation and Distribution

Mexico's natural gas network currently extends about 7,500 kilometers (4,660 miles), with eight compression stations. Although private companies have been allowed to participate in natural gas transport since 1995, and private companies also have had access to the Pemex network since 1997, Pemex's monopoly over production has prevented further private involvement.

Pemex has nine interconnections with the United States' natural gas network, which can either import or export gas between the United States and Mexico, including: Mexicali (connecting to California), Naco (connecting to Arizona), Piedras Negras, Reynosa, Samalayuca, Ciudad Juarez, and two in Arguelles (these five connect to Texas). In September 2002, the North Baja Pipeline project, connecting Baja to the the U.S. natural gas network began initial deliveries of 200 million cubic feet per day (Mmcf/d). Throughput at the North Baja Pipeline is expected to continue in 2003 and ultimately reach a capacity of 500 Mmcf/d. There are several more interconnections under consideration.

Liquefied Natural Gas (LNG)

There are currently plans to develop LNG import facilities on both coasts of Mexico. In December 2002, the Mexican Federal Electricity Commission (CFE), published a tender for the construction of LNG importation facilities in Altamira, Tamaulipas, on Mexico's Atlantic Coast. The CFE has announced its intentions to purchase 425 Mmcf/d for 15 years from the proposed facility which will fuel regional independent power producers. The CFE will collect technical proposals until April 29, 2003, with the winner to be announced in May and the facilities to be constructed by January 2006. Foreign firms have also shown interest in constructing LNG importation facilities along Mexico's Pacific coast, including sites on the Baja peninsula and Lazaro Cardenas. Many of the proposed facilities are close to the U.S.-Mexican border, and are intended to supply markets in Northern Mexico as well as the American Southwest.

Sector Organization: Pemex and the Private Sector

According to the Mexico's constitution, Pemex retains exclusive rights to the exploration and production of Mexican natural gas. However, the Fox administration has proposed significant changes designed to encourage the involvement of foreign firms in Mexico's natural gas sector in an effort to meet the country's growing demand.

President Fox's plan entails the development of a new arrangement between Pemex and private investors known as the Multiple Service Contract (MSC). Currently, private companies are allowed to provide oil industry services to Pemex (*i.e.* drilling), but are strictly excluded from the ownership of Mexico's natural resources in accordance with the country's constitution. In December 2001, discussion began of a new contractual arrangement that could open up a broader range of services to private firms, without attaching the companies' investments to the reserves or their profits (the companies would be paid fees rather than profit shares). In June 2002, President Fox unveiled the prototype MSC. The initial draft contract was criticized by both the Mexican congress and private investors and was revised and resubmitted in September.

The downstream Mexican natural gas market has been open to private investors since the passage of the 1995 Natural Gas Law. This legislation modified the constitution to allow private companies to become involved in natural gas transportation, storage, and distribution in Mexico, although it prohibits a company from ownership in more than one function within the industry. The legislation also liberalized exports and imports and established the regulatory framework for building and expanding transmission and distribution pipelines.

The Mexican Energy Regulatory Commission (CRE) regulates the gas industry. CRE's powers include enforcement of regulations, inspections of facilities, issuance of permits, regulation of prices, overall supervision of the industry, ensuring an adequate supply, security, the promotion of competition, and the elimination of cross-subsidies. Private-sector participation in these areas currently is subject to permits granted by CRE for 30 years, based on competitive bidding.

Coal

Mexico has recoverable coal reserves of about 1.3 billion short tons, just over 70 per cent of which is anthracite and bituminous, and just less than 30 per cent of which is lignite and subbituminous.

The majority of the country's coal reserves are located in Coahuila, in the northeast of the country, bordering the United States. Coal production has remained stable in the past few years and is used mostly for steel production and electricity generation. U.S.-based Mission Energy, which purchased the previously government-owned company *Minera Carbonifera Rio Escondido* (MICARE) when it was privatized, is now Mexico's largest coal producer. A small volume of imports from the United States, Canada, and Colombia augments domestic coal supplies.

Coal-fired plants supply roughly 10 per cent of Mexico's electricity, but this percentage is slated to fall as natural gas-fired power plants are favored to meet rising demand. Mexican coal has very high ash content and therefore is mixed with lower-ash imported coal. Local coal has higher production costs than imported coal.

Electricity

Mexico has installed electric power generating capacity of 38.9 million kilowatts and in 2000 generated 194 billion kilowatthours (bkwh). Oil-fired power plants make up the largest share of Mexico's electricity generation, and thermal (oil, gas, and coal) electricity generation in 2000 accounted for 76 per cent of total generation. Hydropower accounted for 17 per cent, nuclear power 4 per cent, and other renewable sources (wind, solar, biomass) 3 per cent. Mexico's industrial energy policy calls for the conversion of many oil-fired power plants to natural gas by 2005. Most new power plants will be run on natural gas.

Currently, demand for electricity in Mexico is only partially met. Mexico's electricity producers expect programmed power outages for industrial customers in 2003 owing to both the country's insufficient transmission system and two consecutive years of low rainfall. And although about 95 per cent of Mexican households currently are electrified, many thousands of rural towns are still without electricity. Electricity demand is expected to grow by 6.7 per cent annually beginning in 2003,

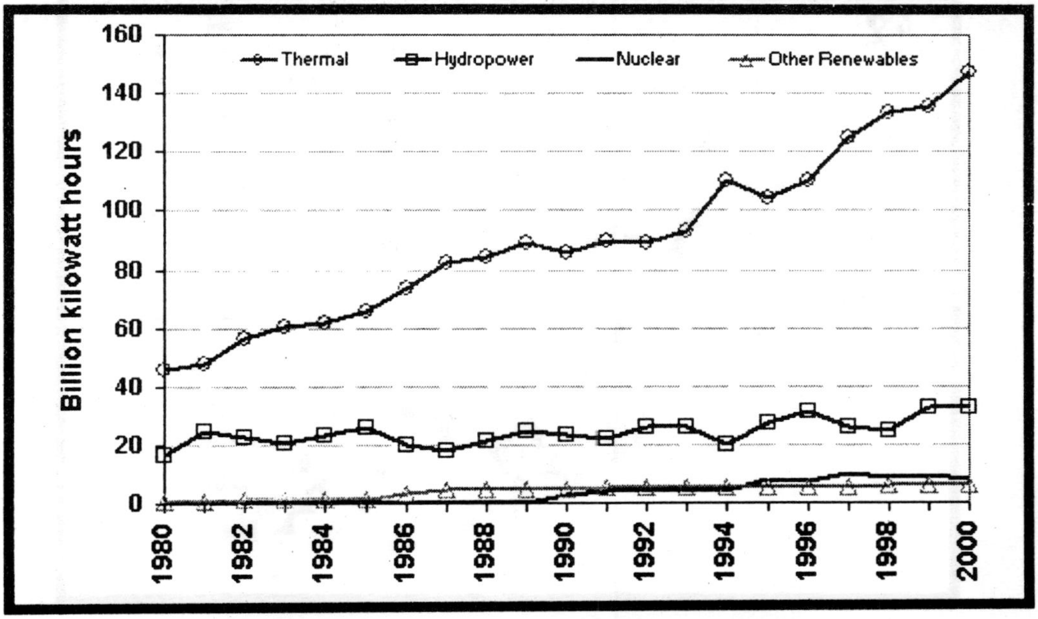

MEXICAN ELECTRICITY GENERATION, 1980-2000

with demand in some high-growth regions such as Baja expected to grow by as much as 8-9 per cent per year. According to government estimates, the country will need $65 billion in investment by 2010 to meet the country's growing electricity demand.

Sector Organization

The Federal Electricity Commission (CFE) and *Luz y Fuerza Centro* (LFC) are Mexico's two state-owned electricity companies. CFE has enjoyed a monopoly in the electric power sector for decades, although reforms instituted in 1992 allow independent power producers (IPPs) and cogenerators limited involvement. CFE generates about 92 per cent of Mexican electricity. LFC contributes about 2 per cent, with most of its customers in Mexico City. Pemex generates 4 per cent, while the remainder is generated by the private sector.

IPPs are allowed to build and own power generation facilities, and the power can be used at related industrial companies or sold under long term contracts to the public utilities. A small amount of excess power can be sold to CFE. IPPs also can gain permits to build larger plants, for electricity sales to "off-takers," which are large groups of consumers or industrial firms. The off-taker market is most active in Baja California.

President Fox made has proposed reform of Mexico's electricity sector as a step towards meeting the country's growing demand. However, stiff resistance from the Mexican Congress has stalled changes thus far. In April 2002, the Mexican Supreme Court nullified an executive order issued by Fox in May 2001 that allowed private generators to sell their excess electricity to CFE and LFC. The court cited articles 27 and 28 of the Mexican Constitution, which stipulate that Mexico's power sector belongs exclusively to the state. Later in August 2002, President Fox submitted an electric power reform package to the Mexican Senate that would change the country's constitution to allow more

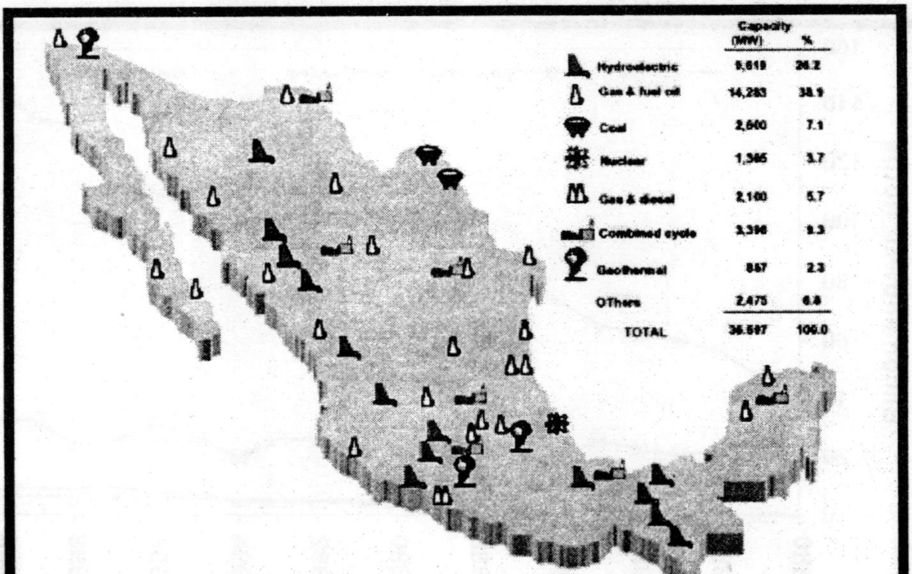

MAJOR ELECTRIC POWER STATIONS, 2000

room for private investment without privatizing the national industries. In December 2002, the Mexican congressional session closed without discussing Fox's reform package. The Mexican Congress will reconvene in February, 2003.

Infrastructure

Mexico has a national interconnected power grid with four regional divisions: Northern, North Baja, South Baja, and Southern (the largest). In the southeast and northeast, the grids are stretched such that new generation cannot be added without bolstering the transmission network. Accordingly, CFE has undertaken projects designed to make needed improvements to the national grid by working with private companies to install hundreds of miles of new high voltage transmission lines over the next few years. Mexico's electricity grid is also connected to to the U.S. in the North, and to Belize in the South. Northern Mexico is connected to the U.S. grid in nine places. Two of the connections are to California, with a third under construction, while the remaining seven are to Texas. There is a smaller grid connection between Mexico's Yucatan peninsula and Belize.

Several new border power connections are in the works. There are multiple projects in various planning stages that could link Mexico and the United States. While potential projects span the length of the Mexico-U.S. border, there are several aiming to connect Baja California (Mexico) to California, as well as northwestern Mexico to Arizona.

Environment

While Mexico contributes only 1.6 per cent of the total world energy-related carbon emissions, it is a major contributor, along with Brazil, Argentina, Venezuela, Colombia and Chile, in Latin America. Mexico's carbon intensity is relatively high in comparison with most other countries of Central and South America. If policies are implemented as envisioned, however, carbon intensity likely will decrease in the future. Per capita energy consumption and carbon emissions are also high in comparison with other countries in the region; however, Mexico's per capita energy consumption and carbon emissions are low when compared to fellow OECD members. Mexico's status as a developing country, as well as a member of OECD, has placed it in a unique position in the current climate change negotiations.

There are many opportunities to exploit the use of renewable energy technologies in Mexico. In March 1998, the U.S. Initiative on Joint Implementation selected a project aimed at building a hybrid system to utilize solar, wind, and diesel capacity in San Juanico, Baja California Sur. The project involves Arizona Public Service Co., Niagara Mohawk Corp., and CFE. In April 1998, Switzerland's Asea Brown Boveri (ABB) announced that it would supply an advanced, high-efficiency, low-emissions power plant to CFE for $250 million.

Country Overview

Independence: September 16, 1810 (from Spain)

Population (December 2002E): 102.5 million

Location/Size: Southern N. America / 762,000 square miles (nearly three times the size of Texas)

Major Cities: Mexico City (capital), Guadalajara, Monterrey, Púebla

Languages: Spanish, various Mayan, Nahuatl, and other regional indigenous languages

Ethnic Groups: Mestizo (Indian-Spanish), 60 per cent ; Amerindian, 30 per cent ; Caucasian, 9 per cent ; Other, 1 per cent

Religions: Roman Catholic, 89 per cent ; Protestant, 6 per cent, Other, 5 per cent

Defense (8/98): Army: 130,000, Navy: 37,000, Air Force: 8,000, Rural Defense Militia: 14,000

Economic Overview

Currency: 1 Peso = 100 centavos

Market Exchange Rate (01/31/03): US$1 = 10.97 pesos

Nominal Gross Domestic Product (GDP, 2002E): $643.9 billion

Real GDP Growth Rate (2002E): 0.9 per cent *(2003F):* 3.1 per cent

Inflation Rate (consumer prices, 2002E): 5.0 per cent *(2003F):* 4.5 per cent

Major Trading Partners: United States, Canada, Japan, Germany, and United Kingdom

Merchandise Trade Balance (2002E): -$8.5 billion

Exports: $160.0 billion

Imports: $168.5 billion

Major Export Products: products from maquiladoras, oil

Major Import Products: products for maquiladoras, consumer goods, capital goods, raw materials and intermediate goods

Total External Debt (2001E): $191 billion

Energy Overview

Proven Oil Reserves (1/1/03E): 12.6 billion barrels (see Reserves and Production)

Oil Production (2002E): 3.6 million barrels per day (bbl/d), of which 3.18 million bbl/d was crude

Oil Consumption (2002E): 1.93 million bbl/d

Net Oil Exports (2002E): 1.68 million bbl/d

Crude Oil Refining Capacity (1/1/03E): 1.7 million bbl/d

Natural Gas Reserves (1/1/03E): 8.8 trillion cubic feet (Tcf) (see Reserves and Production)

Natural Gas Production (2000E): 1.33 Tcf

Natural Gas Consumption (2000E): 1.38 Tcf

Recoverable Coal Reserves (2000E): 1.3 billion short tons

Coal Production (2000E): 10.86 million short tons

Coal Consumption (2000E): 13.41 million short tons

Net Coal Imports (2000E): 2.55 million short tons

Electric Generation Capacity (2000E): 38.9 million kilowatts

Net Electricity Generation (2000E): 194.37 billion kilowatthours (bkwh); 74 per cent thermal, 18 per cent hydro, 5 per cent nuclear, 3 per cent other

Net Electricity Consumption (2000E): 182.8 bkwh

Net Electricity Imports (2000E): 2.07 bkwh

Environmental Overview

Total Energy Consumption (2000E): 6.18 quadrillion Btu* (1.6 per cent of world total energy consumption)

Energy-Related Carbon Emissions (2000E): 103.2 million metric tons of carbon (1.6 per cent of world total carbon emissions)

Per Capita Energy Consumption (2000E): 62.5 million Btu (vs U.S. value of 351.0 million Btu)

Per Capita Carbon Emissions (2000E): 1.0 metric tons of carbon (vs U.S. value of 5.6 metric tons of carbon)

Energy Intensity (2000E): 16,509 Btu/$1995 (vs U.S. value of 10,918 Btu/$1995)**

Carbon Intensity (2000E): 0.28 metric tons of carbon/thousand $1995 (vs U.S. value of 0.18 metric tons/thousand $1995)

Sectoral Share of Energy Consumption (1998E): Industrial (54.7 per cent), Transportation (24.8 per cent), Residential (15.9 per cent), Commercial (4.6 per cent)

Sectoral Share of Carbon Emissions (1998E): Industrial (50.9 per cent), Transportation (31.1 per cent), Residential (13.2 per cent), Commercial (4.8 per cent)

Fuel Share of Energy Consumption (2000E): Oil (63.2 per cent), Natural Gas (23.7 per cent), Coal (4.0 per cent)

Fuel Share of Carbon Emissions (2000E): Oil (73.5 per cent), Natural Gas (20.4 per cent), Coal (6.2 per cent)

Renewable Energy Consumption (1998E): 713.7 trillion Btu* (1 per cent decrease from 1997)

Number of People per Motor Vehicle (1998): 6.9 (vs U.S. value of 1.3)

Status in Climate Change Negotiations: Non-Annex I country under the United Nations Framework Convention on Climate Change (ratified March 11th, 1993). Ratified the Kyoto Protocol on September 7th, 2000.

Major Environmental Issues: Natural fresh water resources scarce and polluted in north, inaccessible and poor quality in center and extreme southeast; raw sewage and industrial effluents polluting rivers in urban areas; deforestation; widespread erosion; desertification; serious air pollution in the national capital and urban centers along US-Mexico border.

Major International Environmental Agreements: A party to Conventions on Biodiversity, Climate Change, Desertification, Endangered Species, Hazardous Wastes, Law of the Sea, Marine Dumping, Marine Life Conservation, Nuclear Test Ban, Ozone Layer Protection, Ship Pollution, Wetlands and Whaling.

The total energy consumption statistic includes petroleum, dry natural gas, coal, net hydro, nuclear, geothermal, solar, wind, wood and waste electric power. The renewable energy consumption statistic is based on International Energy Agency (IEA) data and includes hydropower, solar, wind, tide, geothermal, solid biomass and animal products, biomass gas and liquids, industrial and municipal wastes.

Energy Industries

Organization: Oil and natural gas–Petroleos de Mexicanos (Pemex), four operating subsidiaries (Exploration and Production, Refining, Gas and Basic Petrochemicals, Secondary Petrochemicals),

Petroleos Mexicanos Internacional (PMI); *Electric power and distribution*–CFE and LFC; *Natural gas and electric power regulation*–Comission Reguladora de Energia (CRE)

Major Ports: Gulf Coast–Cayo Arcos, Dos Bocas, and Pajaritos (handle most of Pemex's oil exports), Tuxpan, Ciudad Madero; Pacific Coast–Salina Cruz, Rosarito

Major Oil-Producing Fields: Cantarell, Abkatun, Ku, Caan, Pol, Chuc

Major Refineries (Crude Capacity) : Salina Cruz (330,000 bbl/d), Tula Hidalgo (320,000 bbl/d), Salamanca (245,000 bbl/d), Cadereyta (275,000 bbl/d), Minatitlan (194,000 bbl/d), Ciudad Madero (320,000 bbl/d)

54

Nigeria

An OPEC member, Nigeria is one of the world's largest oil exporters. Nigeria is a major oil supplier to Western Europe and was the 5th largest supplier of crude oil to the United States in 2002.

General Background

The election and inauguration of President Olusegun Obasanjo's administration in 1999 returned Nigeria to civilian rule. On January 6, 2003, it was announced that President Obasanjo won the nomination for president in the ruling People's Democratic Party (PDP) primaries in Abuja. Nigeria's presidential elections are scheduled to be held on April 19. Local and legislative elections are also planned to be conducted around the same time. Nigeria has not held successful elections under a civilian government since independence from the United Kingdom in 1960, with attempts in 1966 and 1983 ending in violence and military coups.

In April 2000, President Obasanjo signed into law a new revenue sharing formula with the nine oil-producing states by which the latter receive 13 per cent of oil revenues versus the previously allotted 3 per cent. In February 2003, a resolution of the dispute between the federal government and the littoral states over the allocation of offshore oil and gas revenues was reached. The Obasanjo government had taken the unprecedented step of asking Nigeria's Supreme Court to intervene in its argument with regional state governments over control of the country's offshore oil and gas resources. The federal government's position was that all natural resources within the territorial waters of Nigeria are derived from the federation and not from any one state. The compromise between President Obasanjo and the governors of the six South-South states: Akwa Ibom, Bayelsa, Cross River, Delta, Ondo, and Rivers removed all references to the continental shelf and the Exclusive Economic Zone and was replaced by a 200-meter water depth isobar limit. The vast majority of Nigeria's current producing fields are located within water depths of 200 meters or less. Deep-water developments will

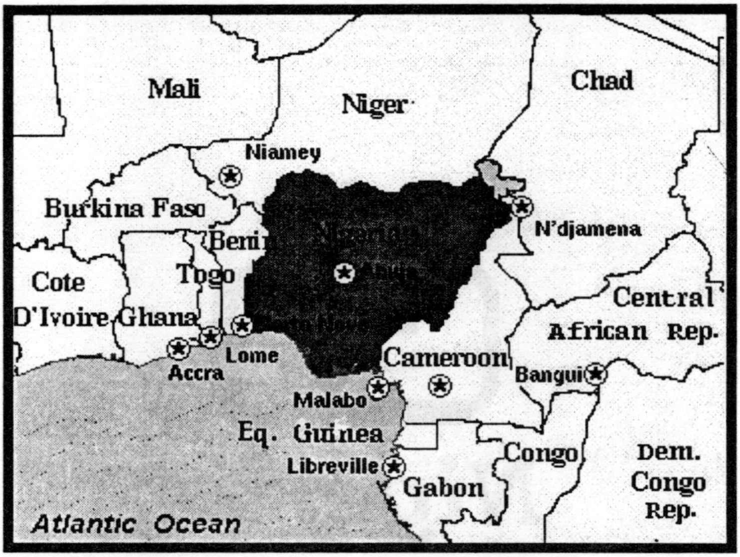

generate revenues that will be subject to direct federal government control. The revised Dichotomy Bill was sent back to the legislature in early March 2003.

Nigeria's economy is heavily dependent on the oil sector, which accounts for 90-95 per cent of export revenues, over 90 per cent of foreign exchange earnings and nearly 80 per cent of government revenues. Real GDP is estimated to have grown by 3.0 per cent in 2001, but International Monetary Fund (IMF) estimates for 2002 see GDP declining by 0.9 per cent. Nigeria's central bank (CBN) estimates that the economy grew by 3.3 per cent in 2002, below the targeted 5.0 per cent. Inflation is estimated to have slowed to 13.4 per cent in 2002, compared with 18.9 per cent in 2001.

The Obasanjo administration is working on a number of economic reforms, including the privatization of parastatals, exchange rate management, and the phasing out of subsidies. There have been several delays in the privatization program. The government is some 23 per cent behind its 2002 target of $520 million in revenues from privatization. The major reason for the shortfall was the failure to sell Nigeria Telecommunications (Nitel). The government had, by August 2001, successfully sold four Global System for Mobile Communications (GSM) licenses and brought Nitel to the point of sale, but no investors came forth. The Bureau for Public Enterprises (BPE) estimates it raised $360-$400 million from privatization proceeds during 2002. Privatization of the National Electric Power Authority (NEPA), a part of the final portion of the privatization program, is likely to be shelved in the near term given the proximity of elections and the failure to privatize Nitel. The BPE has selected two international consulting firms, including US-based group Nexant,Inc, to advise on the proposed privatization of the country's downstream petroleum sector. The Nigerian National Petroleum Corporation's (NNPC) four oil refineries, petrochemicals plants, and the Pipelines and Products Marketing Company (PPMC) are due to be sold. As with NEPA's privatization, the planned sales will not be pursued until after the national elections. Considerable opposition to the proposal measures have been voiced by the National Union of Petroleum and Natural Gas Workers (NUPENG) and the Petroleum and Natural Gas Senior Staff Association of Nigeria (PENGASSAN), who fear job losses and sharply higher product prices resulting from the privatizations.

Oil

Estimates of Nigeria's estimated proven oil reserves range from 24 billion (*Oil and Gas Journal*) to 31.5 billion barrels (OPEC). The majority of these reserves are found in relatively simple geological structures along the country's coastal Niger River Delta, but newer reserves have been discovered in deeper waters offshore Nigeria. The majority of the oil lies in about 250 small (*i.e.*, less than 50 million barrels each) fields. At least 200 other fields are known to exist and contain undisclosed reserves. Nigeria's crude oil reserves have gravities ranging from 21° API to 45° API. Nigeria's main export crude blends are Bonny Light (37° API) and Forcados (31° API). Approximately 65 per cent of Nigerian crude oil production is light (35° API or higher) and sweet (low sulfur content).

Production

Nigerian crude oil production averaged 2.118 million barrels per day (bbl/d) in 2002. On January 12, 2003 OPEC decided to raise the "OPEC 10" (excludes Iraq) production quota from 23 million bbl/d to 24.5 million bbl/d, effective February 1, 2003, in order to ensure adequate supplies of crude in response to the oil supply shortfall in Venezuela. Nigeria's new quota was set at 2.018 million bbl/d. At the OPEC meeting, convened on March 11, 2003, OPEC decided to maintain, for the time being, the current "OPEC 10" production ceiling.

Production from joint ventures (JVs) accounts for nearly all (about 95 per cent) of Nigeria's crude oil production. The largest JV, operated by Shell, produces nearly 50 per cent of Nigeria's crude oil. The NNPC, the state-owned oil firm, has a 55 per cent interest in the Shell JV. The other JVs, in which the NNPC has a 60 per cent stake, are operated by ExxonMobil, ChevronTexaco (formerly as Chevron), ENI/Agip (Agip), ChevronTexaco (formerly as Texaco) and TotalFinaElf.

A major problem facing Nigeria's upstream oil sector has been insufficient government funding of its JV commitments. The Nigerian government has two major funding arrangements for oil producing in the country: JV and production sharing contract (PSC) arrangements. Under the JV arrangements, the Nigerian government and its partners contribute to these projects according to their equity holding. Under a PSC, oil companies fund the operations and the profits are shared according to agreed

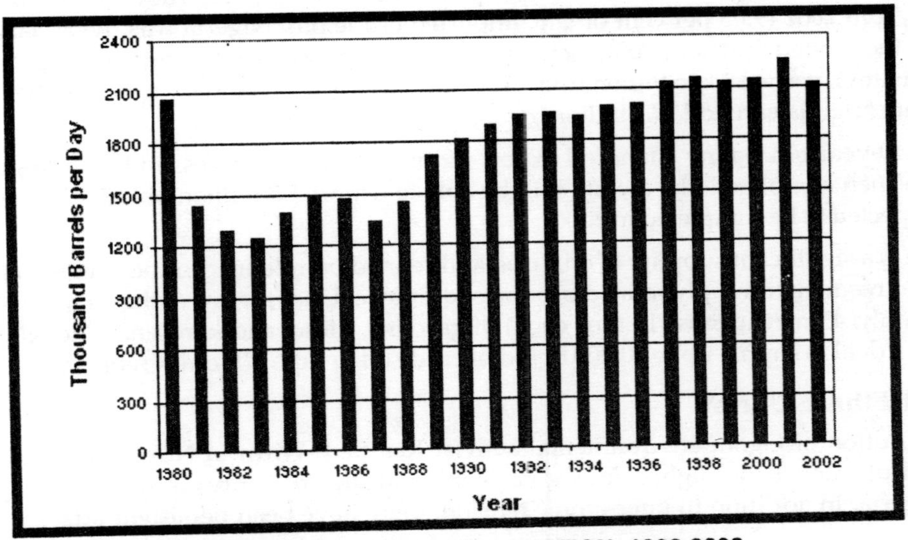

NIGERIAN CRUDE OIL PRODUCTION, 1980-2002

arrangement after the company has recouped its expenditure. The NNPC has proposed a JV cash call budget for 2003 of $4.6 billion in its proposals to the Nigerian legislature. NNPC officials stated its funding request is consistent with the federal government objectives of achieving 40 billion barrels of proven reserves and 4.0 million bbl/d of production capacity by 2010.

Political and ethnic strife in the Niger Delta region, including violence, kidnapping, sabotage and the seizure of oil facilities, often disrupts Nigerian oil production. On March 19, 2003, ChevronTexaco suspended its oil production in the Niger Delta region, declaring force majeure on its exports following violent clashes between the Ijaw and the Itsekiri ethnic groups which has threatened the operations of the oil companies in the area. Shell, the other company with facilities in the region, removed its non-essential staff the previous weekend, and both companies then moved to shutdown their operations and evacuate all personnel on March 19, 2003. ChevronTexaco initially shut-in production of 140,000 bbl/d and Shell closed down its flowstations with a combined capacity of 126,000 bbl/d. The 266,000 bbl/d of lost oil, represented approximately 13 per cent of Nigeria's total average production of 2.1 million bbl/d. ChevronTexaco has stated that it will be unable to meet its March and April 2003 obligations to its clients due to the problems.

Shell evacuated four oil facilities, oil pipeline pumping stations at Ogbotobo, Opukushi, Tumo and Benisede, on March 24, 2003 raising the number of closed Shell facilities to 14. These actions have shut-in 320,000 bbld, or nearly one-third of Shell's Nigerian output. Shell stated that an additional 50,000 bbl/d of Nigerian Bonny Light oil production is being lost in the east Niger Delta. Incidents of illegal bunkering, not related to the ethnic violence in the western Delta led to the Bonny shut-in. Chevron has shut-in oil production totalling 440,000 bbl/d and it was forced to evacuate its workers from offshore platforms and its Escravos export terminal. TotalFinaElf has shut down operations producing 7,500 bbl/d in the area. A total of 817,500 bbld, nearly 40 per cent of Nigeria's production is currently shut-in.

Exports

The majority of Nigeria's crude exports are destined for markets in the United States and Western Europe, with Asia becoming an increasingly important market as well. In 2002, Nigerian crude exports to the United States averaged 567,000 bbl/d (6.27 per cent of U.S. imported crude oil) a decline from the 842,000 bbl/d in 2001 (9.03 per cent of U.S. imported crude oil). Nigeria was the 5th largest crude exporter to the United States in 2002, behind Saudi Arabia, Mexico, Canada and Venezuela. The NNPC estimates it earned $4.15 billion from crude oil exports from January to September 2002 (EIA estimates that Nigeria earned $17.2 billion in oil export revenues in 2002).

Shell's 40-year-old Bonny Island oil export terminal is in the process of being upgraded and expanded. When completed, the facility will be able to export 1.5 million bbl/d. The $600 million project is expected to be completed in 2006.

Nigeria is a significant supplier of crude oil and refined petroleum products regionally. Nigeria reached an agreement to supply refined products to Guinea. Under terms of the deal the NNPC will supply monthly, 45 metric tons (15 tons each) of gasoline, diesel and kerosene. The NNPC is also considering a deal to supply up to 50,000 bbl/d of crude oil to Cote d'Ivoire's refinery.

Crime and Ethnic Unrest

The execution of Ogoni environmental activist Ken Saro-Wiwa in November 1995 attracted international attention to the plight of the Ogoni people and other minority groups in the oil-producing areas of Nigeria. In addition to inter-ethnic tension, there have been persistent attacks against oil companies by youths protesting the environmental degradation of indigenous homelands and their marginalization in terms of federal resource allocation. The attacks have resulted in disruptions of oil production, domestic supply and exports.

Illegal fuel siphoning as a result of a thriving black market for fuel products has increased the number of oil pipeline explosions in recent years. The most serious disaster was the October 1998 Jesse fire in which over 1,000 people died. In July 2000, a pipeline explosion outside the city of Warri caused 250 deaths. An explosion in Lagos in December 2000 killed at least 60. The NNPC reported 800 cases of pipeline vandalization from January through October 2000. In January 2001, The Nigerian navy announced plans to clamp down on arson attacks on oil facilities following the loss of about $4 billion in oil revenues last year due to vandalism. The Federal government also has ordered the navy to sink any ship conveying crude products that cannot be accounted for. The government estimates that as much as 300,000 bbl/d of Nigerian crude is illegally bunkered (freighted) out of the country.

Exploration and Field Development

In March 2000, Nigeria opened competitive bidding on 22 new oil blocks, including 11 in the Niger Delta deep and ultra-deep offshore, in which 46 oil companies participated and 14 blocks received a total of 51 bids. Awards for eight of the offered exploration blocks were announced in December 2000. The licensing round was the Nigeria's first in over a decade.

In July 2002, Shell was awarded Block OPL 245 after the license had been withdrawn from a local firm, Malabu Oil & Gas. The government has said the license was revoked in May 2001 because Malabu, owned by former Nigerian President Sani Abacha and his former oil minister Dan Etete, was improperly given the concession and also that it failed to pay the $20-million signature bonus by the agreed deadline. US-independent Ocean Energy was awarded offshore OPL 256 in November 2002. The block is adjacent to two highly prospective areas, Shell's OPL 245 and OPL 246, operated by TotalFinaElf and South Atlantic Petroleum.

Nigeria's next deepwater licensing round, originally scheduled to begin following the April 2003 elections, will probably be delayed until 2004 according to officials at Nigeria's Department of Petroleum Resources (DPR). The delay is being caused by legal confusion over acreage awarded under the previous regimes of Presidents Abubakar and Abacha.

The Nigerian government held a special bidding round to offer marginal fields to local firms and in June 2002 the government pre-qualified 71 companies out of 150 that submitted bids The round was held to allow more participation by indigenous oil companies in Nigeria's upstream oil exploration and production activities. Local firms currently account for approximately 150,000 bbl/d (roughly 5 per cent -6 per cent) of Nigeria's oil production. The government has identified 116 marginal fields located within the leaseholds of the multi-national oil companies in the Niger Delta. These fields, which according to the government had collective reserves totalling about 1.3 billion barrels, have not been developed and are considered uneconomical to produce by the current leaseholders.

Nigeria has the potential to increase its crude oil production significantly in the next few years as recent discoveries come on stream. ExxonMobil has launched its construction work on Nigeria's deep-water Erha field, which is estimated to contain 500 million barrels of reserves. ExxonMobil announced the discovery in 1999 on Block OPL 209. At maturity Erha should yield 150,000 bbl/d, and first oil is expected in 2005. ExxonMobil holds a 56.25 per cent share and operates Block OPL-209 under a production-sharing contract (PSC) with NNPC. Shell is the other partner on OPL-209. Erha's $1.1-billion field development is expected to include a Floating Production Storage and Offloading vessel (FPSO), 15 producing wells, five water injection wells, and four gas injection wells.

ExxonMobil is also developing the 400-million-barrel Yoho field in the shallow waters of Block OML-104. ExxonMobil holds 40 per cent and NNPC 60 per cent of the $1.2 billion project. Initial production, producing at 90,000 bbl/d, began in February 2003. Yoho utilizes an FPSO to allow oil production almost two years ahead of the full-field production capacity. When full-field output of 150,000 bbl/d is achieved in late 2004, the FPSO will be replaced by a floating storage and offloading (FSO) vessel. Associated gas will be re-injected to maintain field pressures and eliminate flaring.

In February 2003, the ChevronTexaco/NNPC JV announced its plan to spend $4 billion on developing three oil and gas projects in Nigeria including the deepwater Agbami field. In November 2001, ChevronTexaco and its partners signed a 30-year PSC with Nigeria for the exploitation of OPL 250. Tests confirm that Agbami contains nearly 1 billion barrels of recoverable hydrocarbons. Agbami, which covers 45,000 acres, is located about 70 miles from Nigeria's coast. Oil production, utilizing a FPSO, should begin by mid-2006 and peak at 225,000 bbl/d. The majority of the Agbami field lies in Block OPL 216, where ChevronTexaco holds a 32 per cent interest. ChevronTexaco's partners on OPL 216 are NNPC (40 per cent) Petrobras (8 per cent) and the indigenous firm, Famfa Oil (20 per cent). Approximately one-third of Agbami lies in the adjacent Block OPL 217. Statoil (54 per cent) is the operator on the block and ChevronTexaco (46 per cent) is its partner.

Work is continuing on Shell's massive Bonga deep-water development, which has the potential to produce 225,000 bbl/d. Bonga is estimated to contain 700 million barrels of oil and it is scheduled to come on stream in early 2004. Shell is the operator (55 per cent), partnered with ExxonMobil (20 per cent), Agip (12.5 per cent), and TotalFinaElf (12.5 per cent). The field will be serviced by a 2.0-million-barrel FPSO. In May 2001, Shell announced a second major discovery on Block OPL 212 (OML 118), which contains the Bonga Field. Bonga South West, is estimated to contain reserves of nearly 600 million barrels of oil. Drilling on the Bonga South West development will be completed in 2003 with the first oil produced, also by FPSO, expected in 2005.

Limited production from Shell's EA field (Block OML 79) began in December 2002. With estimated reserves of 350 million barrels of oil, the EA field is expected to produce at peak more than 125,000 bbl/d of oil and up to 100 million cubic feet per day (Mmcf/d) of natural gas.

Agip is developing the Abo discovery on Block OPL 316. Production, utilizing an FPSO, was expected to begin by the end of 2002, but it is now scheduled to begin in March 2003. Agip (40 per cent and operator), BP (35 per cent) and ExxonMobil (25 per cent) hope to produce 30,000 bbl/d from the Abo field.

In May 2000, TotalFinaElf announced the Akpo-1 discovery on Block OPL 246. Initial tests on the field yielded output of 9,000 bbl/d of very light oil. Akpo was originally estimated to contain reserves of 200 million barrels of crude and condensate, but additional appraisal drilling has upped the estimate to nearly 1 billion barrels of oil and 4 trillion cubic feet (Tcf) of natural gas. TotalFinaElf is the operator with a 24 per cent share, and its partners are South Atlantic Petroleum (60 per cent) and Petrobras (16 per cent). The Akpo field is very near the boundary of the Joint Development Zone (JDZ) recently established between Nigeria and Sao Tome and Principe, and it may extend into the JDZ.

TotalFinaElf (operator) is also developing the Amenam/Kpono field, which straddles Blocks OML 99 and OML 70. Development drilling began in early 2003, and the field is scheduled to began production by July 2004. Production is expected to at 125,000 bbl/d. The Amenam/Kpono project is said to be the single largest scheme in the world in the traditional shallow water environment. Total reserves on Blocks OML 99 and OML 70 are over one billion barrels. In 2002, TotalFinaElf took delivery of the FSO for the project. The new vessel has a storage capacity of 2.2 million barrels of oil, is designed to handle up to 230,000 bbl/d of liquid production and has the capacity to export crude oil at up to 50,000 barrels per hour to loading tankers. TotalFinaElf's (31.2 per cent interest) partners on Amenam/Kpono are ExxonMobil (8.8 per cent), and NNPC (60 per cent).

TotalFinaElf plans to drill additional appraisal wells on Block OPL 222, the location of the Ukot-1 discovery. Ukot was discovered in 1999 in 1,800 feet of water and flowed 15,000 bbl/d on test. A second discovery, Usan was made in May 2002, and it tested at 5,000 bbl/d in 2,450 feet of water. The Ukot/Usan discoveries could be on stream within four years if further delineation and exploration work prove successful. TotalFinaElf, the operator on OPL 222 with a 20 per cent interest, is partnered with ChevronTexaco (30 per cent), ExxonMobil (30 per cent) and Nexen (20 per cent).

Border Disputes

In October 2002, the International Court of Justice (ICJ) awarded the Bakassi peninsula to Cameroon. Both Cameroon and Nigeria had claimed the Bakassi peninsula, a 1,000-square-kilometer (400-square-mile) area located in the Gulf of Guinea that is believed to contain significant reserves of oil. In February 1994, Cameroon submitted the dispute to the ICJ for settlement, and Nigeria later followed with its own suit. In March 1998, the ICJ began formal hearings on the case, and in June 1999, Equatorial Guinea applied for permission to enter the proceedings. The ICJ authorized Equatorial Guinea to intervene in the case in October 1999 on the basis of the country's arguments to protect its legal rights in the Gulf of Guinea as the maritime boundary between Nigeria and Cameroon is determined.

The Nigerian government had questioned Equatorial Guinea's sole ownership of the Zafiro field (Block B). At issue was whether Zafiro was a separate field, or part of an oil structure that straddled the territorial waters of both countries. In late September 2000, the two countries' presidents signed a pact delineating their maritime boundaries. One of the agreement's key provisions was Nigerian recognition of Equatorial Guinea's sovereignty over the Zafiro field.

The governments of Sao Tome and Principe (STP) and Nigeria have agreed to exploit petroleum reserves jointly in a once-disputed offshore region. The JDZ will be overseen by a joint commission. On February 21, 2001, the two countries signed an agreement in which Nigeria will receive 60 per cent of JDZ revenues and STP gets the remainder. STP broke off talks with Nigeria in November 2002 due to differences over the implementation of the agreement. STP had been seeking to renegotiate the treaty, which it claimed was unfair due to a 100 square kilometer area of the zone over which Nigeria had exclusive rights. In February 2003, both parties agreed that this area should be returned to the JDZ, with future oil revenues from it split on a 60-40 basis. On March 17, 2003, Environmental Remediation Holding Corporation(ERHC), announced the signing of a Memorandum of Understanding (MOU) with STP settling all outstanding disputes. Under terms of the MOU, ERHC will increase its rights to participate in the JDZ from a total of 30 per cent paid working interest in two blocks to a total of 125 per cent working interest spread over six exploration blocks. Additionally, ERHC will not be required to pay signature bonuses on four of the blocks. In exchange, ERHC will relinquish its rights to an overriding royalty interest, share of signature bonus and share of profit oil in the JDZ.

Refining and Downstream

Nigeria's four refineries (Port Harcourt I and II, Warri, and Kaduna) have a combined nameplate capacity of 438,750 bbl/d, but problems including sabotage, fire, poor management and lack of turnaround maintenance have sharply decrease actual output. In March 2003, NNPC officials stated that maintenance work at its Port Harcourt refinery has been completed and that the plant is now operating at an improved production capacity of approximately 90,000 bbl/d. However, repairs are ongoing at the Kaduna and Warri refineries. The NNPC signed a Memorandum of Understanding (MOU) with Venezuela's state oil firm PDVSA in December 2000 to help maintain Nigeria's refineries. In March 2003, ChevronTexaco signed an agreement with NNPC to take over the management of the Warri and Kaduna refineries and its crude oil tanks in Delta State. The NNPC will retain ownership of the refineries; only the management of the facilities has been signed over to ChevronTexaco.

While Nigeria's state-held refineries are slated for privatization, plans for several small, independently-owned refineries are being developed. Nigeria has awarded 18 private refinery licences after opening up the country's downstream sector to private investment. President Obasanjo laid the foundation stone of the $1.5 billion Tonwei Refinery (in October 2002) in a ceremony marking the start of construction of Nigeria's first private refinery. The Tonwei Refinery will have an initial capacity of 100,000 bbl/d and it can be expanded to 200,000 bbl/d.

The government of Lagos State has announced that it is studying the possibility of establishing a refinery. Lagos is estimated to consume more than 50 per cent of Nigeria's petroleum products. The refinery, if built, will serve not only Lagos but also Nigeria's other southwestern states.

The Akwa Ibom state government announced that it had concluded plans to build a 12,000-bbl/d refinery. U.S.-based Ventech announced that design and construction of the refinery, to be built in prefabricated modules in the United States and then shipped to Nigeria for assembly, had begun. The facility will be located in Eket, adjacent to the Qua Iboe crude terminal.

The Edo State government has obtained approval from the federal government to build an oil refinery. Capacity is expected to be 50,000 bbl/d, and a consortium of Nigeria's independent, local petroleum marketers stated that the government had approved their plan for the construction of a refinery in Nigeria's Federal Capital Territory.

The lack of refinery capacity in the country has been a key factor in Nigeria's latest fuel crisis, which began in February 2003 and coincided with a strike by oil workers. Panic-buying from the strike increased demand at a time when the Port Harcourt refineries were offline, and the country was unable to import fuel in order to cover the shortfall.

Natural Gas

Nigeria has an estimated 124 trillion cubic feet (Tcf) of proven natural gas reserves (9th largest in the world). Due to a lack of utilization infrastructure, Nigeria flares 75 per cent of the natural gas it produces and re-injects 12 per cent to enhance oil recovery. President Obasanjo announced in January 2003 that the official date to end natural gas flaring has been reset for 2004. This date represents a four-year revision of the previous target of 2008. World Bank estimates that Nigeria accounts for 12.5 per cent of the world's total gas flare.

Nigeria's most ambitious natural gas project, the $3.8 billion LNG (liquefied natural gas) facility on Bonny Island, was completed in September 1999. The facility is expected to process 252.4 billion cubic feet (Bcf) of LNG annually. The consortium developing the project, Nigeria Liquefied Natural Gas Corporation (NLNG), is comprised of the NNPC (49 per cent), Shell (25.6 per cent), TotalFinaElf (15 per cent), and Agip (10.4 per cent). Initially, the facility is to be supplied from dedicated natural gas fields, but within a few years it is anticipated that half of the input gas will consist of associated (currently flared) natural gas.

The third LNG production train, with an annual capacity of 130.6 Bcf, began operations in November 2002. The third LNG train will increase NLNG's overall LNG processing capacity to 383

Bcf per year. NLNG has begun construction of two additional liquefaction trains and they are expected to be in operation by the end of 2005. NLNG has secured a $1.06 billion loan towards the construction of trains 4 and 5, which it says is the largest private sector loan ever in sub-Saharan Africa. The total cost of the fourth and fifth trains is $2.1 billion, excluding ship acquisition costs. NLNG officials stated in January 2003 that it hoped to have a sixth train operational in 2006. The two new trains will have a combined capacity of 363.5 Bcf per year. In March 2003, NLNG obtained $460 million in loans to expand its fleet of vessels. An additional eight vessels are needed to meet the shipping requirements for the fourth and fifth trains.

Several customers have signed long-term purchase agreements with NLNG. Four sales agreements, covering over half of trains four and five's planned output, have been signed. NLNG has delivered 21 spot LNG cargoes to the Lake Charles, Louisiana LNG receiving terminal since 2000, and an additional 17 cargoes are expected to be delivered in 2003.

Plans for additional LNG facilities are being developed. In February 2001, it was announced that Nigeria and the U.S. oil firms of ChevronTexaco, Conoco, and ExxonMobil had signed an MOU to conduct feasibility studies for a second LNG facility, West Niger Delta LNG. The plant is expected to be on stream by 2005. An MOU for a third LNG plant in Nigeria was signed in September 2001. Phillips and Agip signed the agreement with the Nigerian government for the establishment of the Brass River LNG plant. The facility, which is expected to be operational in 2007, will be the world's first offshore LNG plant and have a capacity of 850 Mmcf/d. Shell and Norway's Statoil are set to sign a MOU in the fourth quarter of 2003 with the NNPC to construct a new LNG plant. The plant will be a floating offshore LNG processing facility. The feedstock gas will come from Shell and Statoil's Nnwa/Doro (OPL 218) gas fields. Gas reserves on OPL 218 are estimated to be nearly 10 Tcf. If approved and constructed, the Nnwa LNG facility could begin operations in 2007.

The Escravos gas project (EGP), in which the NNPC holds a 60 per cent share and ChevronTexaco a 40 per cent share, is another project that has expanded Nigeria's natural gas industry. The first phase of the EGP (EGP-1) processes 165 Mmcf/d of associated natural gas, which is supplied to the domestic market by pipeline. Phase two of the EGP (EGP-2), which processes an additional 135 Mmcf/d of natural gas, began operations in late 2000. The gas is currently used domestically, but gas from EGP-2 also will be exported to Benin, Togo and Ghana through the West African Gas Pipeline (WAGP). The third phase of the EGP (EGP-3) will increase gas processing to 400 Mmcf/d of 7natural gas from Chevron's northern offshore fields. Gas from EGP-3 will serve as feedstock for the $1.3 billion, Escravos gas-to-liquid (EGTL) plant, scheduled to come online in 2005. The EGTL will utilize technologies developed by ChevronTexaco and South Africa's Sasol, and will produce nearly 35,000 bbl/d of synthetic fuels (diesel, kerosene, jetfuel & naphtha) which are sulfur and particulate-free. EGTL's capacity can be expanded to 120,000 bbl/d. Other American and Canadian firms also are considering establishing GTL plants in Nigeria.

Several distribution schemes are planned to help promote Nigerian consumption of natural gas. The proposed $580 million Ajaokuta-Abuja-Kaduna pipeline will supply natural gas to central and northern Nigeria, while the proposed Aba-Enugu-Gboko pipeline will deliver natural gas to portions of eastern Nigeria. The Lagos State government and Gaslink Nigeria Limited (Gaslink), a local gas distribution company, are developing a pilot program to deliver natural gas to nine residential neighborhoods in the state. Gaslink, which supplies natural gas to nearly 30 industrial customers in Lagos' Ikeja industrial district, plans to expand operations to include 150 industrial customers, 250,000 residential/commercial customers and 25 independent power plants. In January 2003, Shell and its partner, Nigerian Gas Company (NGC), connected 30 firms in the Agbara/Ota industrial areas of Ogun State to gas supplies.

Nigeria and Algeria continue to discuss the possibility of constructing a "Trans-Saharan Gas Pipeline". The 2,500-mile (4,000-kilometer) pipeline would carry gas from oil fields in Nigeria's Delta region via Niger to Algeria's Beni Saf export terminal on the Mediterranean. It is estimated that construction of the $7 billion project would take six years.

Electricity

Nigeria has approximately 5,900 megawatts (MW) of installed electric generating capacity, in the form of three hydro-based stations and five thermal stations. Nigeria faces a serious energy crisis due to declining electricity generation from domestic power plants. Power outages are frequent and the power sector operates well below its estimated capacity.

The Nigerian government is hoping to increase foreign participation in the electric power sector and is looking for independent power producers to generate and sell electricity to NEPA. In October 2000, NEPA signed a partnership agreement with South Africa's Eskom to help improve electricity supply. Eskom will help develop NEPA's repair capabilities, execute transmission line projects, and participate in rehabilitating, operate and transfer (ROT) schemes for the running of Nigeria's power stations. Shell announced in December 2001 that it had been awarded a 15-year ROT contract for units 1-4 of the Afam power plant, and a lease operate and transfer contract for Afam's fifth unit. Shell will refurbish the Afam power plant at a cost of about $500 million, and with capacity expanding from 400 MW to 900 MW. Eskom is expected to provide management of maintenance and operations at Afam.

Although the privatization of NEPA has been pushed back to some time after the round of national elections, but the determination of the new companies to be created from NEPA has been determined. NEPA's transmission network will remain a single entity, the Nigeria Transmission Company, but the generation sector will be split into six independent companies. 11 companies will be created from NEPA's current distribution operations. No time table on the creation of the new companies has been announced.

Currently, only 10 per cent of rural households and approximately 40 per cent of Nigeria's total population have access to electricity. NEPA plans to boost this share to 85 per cent by 2010. NEPA's plan would call for an additional 15,000 kilometers (9,000 miles) of transmission lines, 16 new power plants, and new distribution and marketing facilities. The government awarded contracts for three 335-MW gas-fired plants, valued at $1.1 billion, in November 2002. China's CMEC will build the facility at Okitipupa in Ondo State. A second plant, to be built Chinese firm, SEPCO, will be located at Papalanto in Ogun State. Seimens will build the plant at Ajaokuta in Kogi state. The government also is expanding its rural electrification program. The government has begun 1,400 rural electrification projects, and has awarded 410 new contracts.

Environment

Nigeria's main environmental challenges result from oil spills, gas flaring and deforestation. Oil extraction in the Niger Delta region, especially, has caused severe environmental degradation, owing to the legacy of oil spills, lax environmental regulations, and government complicity during previous military regimes that governed the country. Although the situation is improving with more stringent environmental regulations for the oil industry, marine pollution is still a serious problem, and air pollution from gas flaring, exhaust emissions from the explosion in car ownership, and electricity generators continues to leave Lagos shrouded in smog.

Nigeria's total primary energy consumption has more than doubled since 1980 as the country's population has multiplied. Nevertheless, both its energy consumption and its energy intensity are still far below other African countries such as Egypt and Algeria, and Nigeria's carbon emissions are still below the OPEC average. Nigeria's plans to end the process of gas flaring by 2008 should cut the country's level of carbon emissions, further reducing the country's carbon intensity.

Although use of solid biomass, such as fuelwood, is prevalent and constitutes a major energy source for rural Nigerians, the production and consumption of renewable energy in Nigeria remains quite limited. With Nigeria's population continuing to increase, the pressure on the country's environment appears likely to increase, even with the added focus on cleaning up the Niger Delta and tightening environmental laws and regulations. The outlook for Nigeria's environment will depend on the country's ability to balance its desire to boost oil production in order to generate additional revenue with its needs to provide a clean and healthy living climate for its growing populace.

Country Overview

Independence: October 1, 1960 (from United Kingdom)

Population (2002E): 129.9 million

Location/Size: West Africa, bordering the Atlantic Ocean (on the south and west), Cameroon (on the south), Chad (on the east), Benin (on the west) and Niger (on the north)/923,770 square kilometers (356,700 square miles), slightly more than twice the size of California

Major Cities: Abuja (capital), Lagos, Ibadan, Kano, Kaduna, Port Harcourt

Languages: English (official), Hausa, Yoruba, Ibo (Igbo), Fulani

Ethnic Groups: Hausa, Fulani, Yoruba, Ibo, and over 250 others

Religion (2000E): Islam (50 per cent), Christianity (40 per cent), traditional beliefs (10 per cent)

Defense (9/02): Army (62,000), Navy (15,000), Air Force (15,000)

Economic Overview

Currency: Naira

Market Exchange Rate (3/21/03): US$1 = 128.4 Naira

Gross Domestic Product (2001E): $40.9 billion *(2002E):* $41.1 billion *(2003F):* $41.9 billion

Real GDP Growth Rate (2001E): 2.9 per cent *(2002E):* -0.9 per cent *(2003F):* 3.5 per cent

Inflation Rate (2001E): 18.9 per cent *(2002E):* 13.4 per cent *(2003F):* 12.2 per cent

Current Account Balance (2001E): $1.2 billion *(2002E):* -$3.7 billion *(2003F):* -$2.5 billion

Major Trading Partners: United States, France, India, United Kingdom, Spain, Germany, Brazil

Merchandise Trade Balance (2001E): $5.6 billion *(2002E):* $1.6 billion

Merchandise Exports (2001E): $17.9 billion *(2002E):* $14.9 billion

Merchandise Imports (2001E): $12.3 billion *(2002E):* $13.3 billion

Major Export Products: Crude oil, natural gas, cocoa, rubber, timber, manufactured goods

Major Import Products: Petroleum products, food, machinery and equipment, manufactured goods

Oil Export Revenues (2002E): $17.2 billion

Oil Export Revenues/Total Export Revenues (2002E): 90 per cent

Total External Debt (2002E): $30.2 billion

Energy Overview

Proven Oil Reserves (1/1/03E): 24.0 billion barrels (Oil & Gas Journal) 31.5 billion barrels (OPEC Statistical Bulletin)

Oil Production (2002E): 2.12 million barrels per day (bbl/d), of which 2.1 million bbl/d is crude oil

OPEC Crude Production Quota (beginning February 1, 2003): 2.018 million bbl/d

Oil Consumption (2002E): 257,000 bbl/d

Net Oil Exports (2002E): 1.9 million bbl/d

Crude Refining Capacity (1/1/03E): 438,750 bbl/d

Major Crude Oil Customers (2001): United States, EU, India

Natural Gas Reserves (1/1/03E): 124 trillion cubic feet (Tcf)

Natural Gas Production (2001E): 0.55 Tcf

Natural Gas Consumption (2001E): 277 billion cubic feet (Bcf)

Recoverable Coal Reserves: 209 million short tons (Mmst)

Coal Production (2001E): 0.07 Mmst

Coal Consumption (2001E): 0.07 Mmst

Electric Generation Capacity (1/1/01E): 5.9 gigawatts

Electricity Generation (2001E): 15.67 billion kilowatthours

Environmental Overview

Total Energy Consumption (2001E): 0.92 quadrillion Btu* (0.2 per cent of world total energy consumption)

Energy-Related Carbon Emissions (2001E): 23.5 million metric tons of carbon (0.4 per cent of world carbon emissions)

Per Capita Energy Consumption (2001E): 7.8 million Btu (vs U.S. value of 341.8 million Btu)

Per Capita Carbon Emissions (2001E): 0.2 metric tons of carbon (vs U.S. value of 5.5 metric tons of carbon)

Energy Intensity (2001E): 8,315 Btu/ $1995 (vs U.S. value of 10,736 Btu/ $1995)**

Carbon Intensity (2001E): 0.21 metric tons of carbon/thousand $1995 (vs U.S. value of 0.17 metric tons/thousand $1995)

Fuel Share of Energy Consumption (2001E): Oil (61.4 per cent), Natural Gas (31.7 per cent), Coal (0.2 per cent)

Fuel Share of Carbon Emissions (2001E): Natural Gas (53.2 per cent), Oil (46.6 per cent), Coal (0.2 per cent)

Status in Climate Change Negotiations: Non-Annex I country under the United Nations Framework Convention on Climate Change (ratified August 29th, 1994). Not a signatory to the Kyoto Protocol

Major Environmental Issues: Soil degradation; rapid deforestation; desertification; recent droughts in north severely affecting marginal agricultural activities

Major International Environmental Agreements: A party to Conventions on Biodiversity, Climate Change, Desertification, Endangered Species, Hazardous Wastes, Law of the Sea, Marine Dumping, Marine Life Conservation, Nuclear Test Ban, Ozone Layer Protection and Whaling

The total energy consumption statistic includes petroleum, dry natural gas, coal, net hydro, nuclear, geothermal, solar and wind electric power. The renewable energy consumption statistic is based on International Energy Agency (IEA) data and includes hydropower, solar, wind, tide, geothermal, solid biomass and animal products, biomass gas and liquids, industrial and municipal wastes. Sectoral shares of energy consumption and carbon emissions are also based on IEA data.

Oil and Gas Industries

Organizations: The Nigerian National Petroleum Corporation (NNPC) manages the state-owned oil industry. NNPC controls majority interests (between 55 per cent -60 per cent) in all joint ventures with foreign oil companies. The NNPC holds 49 per cent in the Nigeria Liquefied Natural Gas (NLNG) Company.

Major Foreign Oil Company Involvement: British Gas, BP, ChevronTexaco, Conoco, Deminex, ENI/Agip, ExxonMobil, Petrobras, Shell, Statoil, Sun Oil, Tenneco, TotalFinaElf

Major Oil Fields: Cawthorn Channel, Edop, Ekulama, Escravos Beach, Forcados Yorki, Jones Creek, Meren, Nembe, Okan, Oso, Ubit

Refineries (nameplate capacity bbl/d) (1/1/03): Port Harcourt-Rivers State (150,000), Warri (118,750), Kaduna (110,000), Port Harcourt-Alesa Eleme (60,000),

Major Terminals: Bonny Island, Brass River, Escravos, Forcados, Odudu, Pennington, Qua (Kwa) Iboe

55

North Central Europe

North Central Europe is important to world energy markets because it is a key transit point for Russian oil and natural gas pipelines.

General Background

Poland, the Czech Republic, the Slovak Republic (commonly referred to as Slovakia), and Hungary are members of the Visegrad Group, created in February 1991 at the northern Hungarian town of Visegrad. After World War II until 1989-1990, these countries were Communist states, as well as members of the Warsaw Pact. (On January 1, 1993, the Czech and Slovak Republics, previously Czechoslovakia, split to form two separate states).

During the last decade, the Visegrad group has made the transition to democracy and to market-based economies. They also have reached the final stage of attaining membership in the European Union (EU). The EU Athens Summit on April 16, 2003 ended with the signing of an accession treaty. Poland, the Czech Republic, the Slovak Republic and Hungary can now attend EU meetings with observer status and will become full members on May 1, 2004, assuming the treaty is ratified in each country. In 1999, Hungary, Poland, and the Czech Republic became the first former-Warsaw Pact countries to join the North Atlantic Treaty Organization (NATO). Slovakia is a member of NATO's Euro-Atlantic Partnership Council. The Czech Republic became a member of the Organization for Economic Co-operation and Development (OECD) in 1995, Hungary and Poland joined in 1996, and Slovakia in 2001. As members of the Visegrad Group, the four countries also belong to Central European Free Trade Agreement (CEFTA). Slovenia, Romania, and Bulgaria are members too.

The Visegrad countries are dependent on trade with the EU, in particular with Germany. These four countries also continue to face economic restructuring challenges, including: modernizing large, and to a certain extent, antiquated agricultural sectors (especially in Poland); implementing more energy efficient processes for industry in order to decrease energy consumption (although energy

intensity is decreasing); absorbing the costs from cleaning up heavily-polluting industries; and adapting industries and services to EU standards.

Table 55.1: Economic and Demographic Indicators for North Central Europe

Country	*Gross Domestic Product (GDP), 2002E (Billions of U.S. $)*	*Real GDP Growth Rate, 2002 Estimate*	*Per Capita GDP, 2002 Estimate (U.S. $)*	*Population 2002E (Millions)*
Poland	190.2	1.3%	4,966	38.6
Czech Republic	68	2.5%	6,655	10.3
Slovak Republic	23	4.0%	4,346	5.4
Hungary	61.4	3.3%	6,051	10.2
Total/Weighted Average	**342.6**	**2.1%**	**5,242**	**64.5**

Regional Energy Issues

The Visegrad countries are neither large producers nor consumers of energy. Coal is the single abundant fossil fuel in the region, with only Poland and the Czech Republic having significant quantities. The Visegrad countries therefore import most of their crude oil and natural gas requirements, mainly from Russia. The exceptions are Poland and Hungary, which met roughly 39 per cent and 24 per cent of their natural gas consumption demand with domestic sources in 2001. This dependence on Russian gas and oil imports is also a point of contention for these countries, particularly as they privatize their energy markets in preparation for EU accession. Some government officials have argued that giving up stakes in state energy companies will compromise national energy security, as the state will no longer have control over production. During the last decade, the Visegrad countries have tried to diversify their energy supplies to reduce their dependence on Russia. The strategic importance of the region, however, lies largely in the crude oil and natural gas pipelines which traverse the Visegrad countries on their way to Western Europe.

Table 55.2: Energy Consumption and Carbon Dioxide Emissions in North Central Europe, 2001

Country	*Total Energy Consumption (Quadrillion Btu, 2001)*	*Oil (Thousand Barrels per day, 2001)*	*Natural Gas (Billion Cubic Feet)*	*Coal (Million Short Tons)*	*Electricity (Billion Kilowatt Hours)*	*Energy-Related CO_2 Emissions (million metric tons of carbon, 2001)*
Poland	3.54	424	489	150.7	118.8	78.6
Czech Republic	1.53	176	349	67.8	55.6	29.0
Slovak Republic	0.83	82	280	10.24	24.4	10.8
Hungary	1.09	149	472	17.54	35.1	15.2
Total	**6.99**	**831**	**1,590**	**246.3**	**234**	**133.6**

Table 55.3: Energy Supply Indicators in North Central Europe

Country	*Crude Oil Reserves, 1/1/03 (Million Barrels)*	*Natural Gas Reserves, 1/1/03 (Trillion Cubic Feet)*	*Coal Reserves (Million Short Tons), 2001*	*Petroleum Production, 2002E (Thousand Barrels Per Day)*	*Natural Gas Production, 2001 (Billion Cubic Feet)*	*Coal Production, All Types, 2001 (Million Short Tons)*	*Electric Generating Capacity, 2001 (Kilowatt Hours)*	*Crude Oil Refining Capacity, 1/1/03 (Thousand Barrels Per Day)*
Poland	96.4	5.83	24,427	16.8	193.2	178.9	135	350
Czech Republic	15	0.14	6,259	6.4	5.7	72.9	70	198
Slovak Republic	9	0.53	190	1	10.3	3.77	30.3	115
Hungary	102.5	1.2	1,209	31.2	114.1	16	34.4	161
Total	**222.9**	**7.7**	**32085**	**55.4**	**323.3**	**271.3**	**269.7**	**824**

Oil Transit

Druzhba (Friendship) pipeline transports Russian crude oil to the Visegrad countries and to Western Europe. The pipeline splits in Belarus into northern and southern branches. The 1-million-barrel-per-day capacity northern branch brings oil to Poland and Germany. The 1.2-million-barrel-per-day-capacity southern branch splits in the Ukraine, with one section going through Slovakia and the Czech Republic and the other section going to Hungary, where it connects to the Adria pipeline.

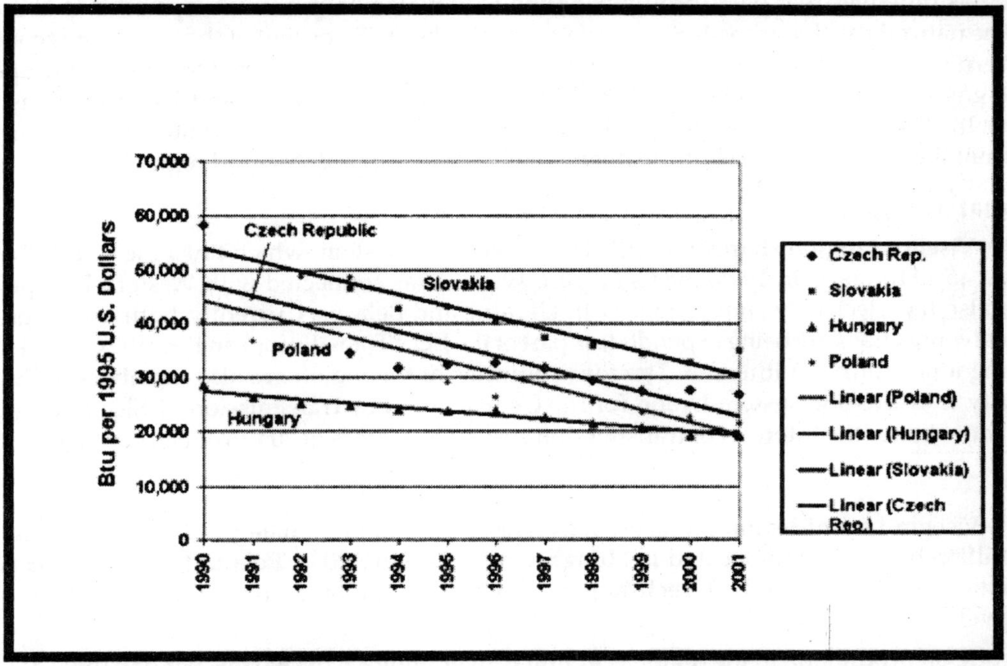

ENERGY CONSUMPTION PER DOLLAR OF GDP, LINEAR REGRESSION TREND AND DATA POINTS, 1990-2001

The Adria pipeline in turn transfers oil to Serbia and to Croatia. In December 2002, the governments of Russia, Belarus, Ukraine, Slovakia, Hungary and Croatia signed an agreement to integrate and expand the capacity of the Druzhba and Adria pipeline systems in order to facilitate the transportation of Russian crude oil to the Croatian deepwater port of Omisalj. This would allow direct shipments of Russian oil to North America, which Russia's oil company Yukos, one of the main supporters of the project, envisions. However, the 110-mile segment of the Adria mainline between Omisalj and Sisak, Croatia, can only accomodate imports.

The Odessa-Brody pipeline allows Caspian Sea region oil that is piped to Black Sea ports to be shipped across the Black Sea to a terminal near Odessa. The oil is then transported to Brody, where it connects with the southern Druzhba pipeline for shipment to Slovakia, Hungary, and onward. There is discussion of extending the Odessa-Brody pipeline to Gdansk, Poland, allowing Caspian crude oil to reach Poland, Germany, and other Baltic states.

Natural Gas Transit

The Visegrad region is a key transit center for Russian natural gas exports to Western Europe. The Yamal-Europe pipeline, which is routed through Belarus and Poland to Germany, is Russia's only natural gas export pipeline to Europe that does not cross Ukrainian territoty. The pipeline has an expected capacity 1.1 Tcf per year. Most of this natural gas is destined for German markets. A second natural gas pipeline, the Yamal II, is planned, but the pipeline has not been formally approved. If built, the combined annual capacity of the two pipelines would be 2.3 Tcf. The Brotherhood and Soyuz natural gas pipelines that pass through Ukraine to Slovakia have annual capacities of about 1 Tcf each. The natural gas that transits Slovakia represents about 25 per cent of the natural gas consumed in Western Europe and about 70 per cent of the Russian natural gas exported to Western Europe. With natural gas demand increasing in Europe, Russia is attempting to expand its natural gas export capacity. In addition, Russia is seeking to build new pipelines to diversify its natural gas export routes away from the Ukraine, increasing the potential importance of the Visegrad region.

Regional Integration

The Visegrad region shares the CENTREL electricity system, which links the Czech Republic, Slovakia and Hungary. In 1995, the CENTREL system was connected with Western Europe's grid. Poland also has electricity connections with Ukraine and Belarus. Currently, both north-south and east-west connections are being expanded, as part of the EU's Trans-European Energy Network project, including a new link to Lithuania. The four countries of the region are also members of European electricity transmission system Union for the Coordination of Transmission of Electricity (UCTE). UCTE coordinates the interests of transmission system operators in 20 European countries.

Oil

The Visegrad countries have total proven oil reserves of approximately 222.9 million barrels, with 102.5 million barrels of that located in Hungary, as of January 2003. Poland has proven reserves of 96.4 million barrels, while the Czech Republic and Slovakia have only 15 and 9 million barrels, respectively.

Crude oil production in the region is minimal, averaging 47,158 barrels per day (bbl/d) in 2002. Hungary is the largest producer of crude oil out the Visegrad Group, with about 24,083 bbl/d, followed by Poland with 16,800 bbl/d, Czech Republic with 5,275 bbl/d and Slovakia with 1,000 bbl/d. Despite almost negligible oil reserves from a global point of view, firms continue to explore the region for oil deposits. For example, in the Czech Republic, exploration has been taking place in the Western Carpathians, an area bordering Austria and Slovakia (See: figure 2). Australia's Carpathian Resources is leading this exploration.

According to its three-year strategic plan, the Hungarian Oil and Gas Company (MOL) aims to double its oil exploration and extraction. The company plans to spend $40-$50 million on exploration in Hungary. The company estimates that only 60 per cent of the country has been thoroughly explored.

In 2002, the Visegrad countries met 5.7 per cent of their total oil demand of 828,000 bbl/d from domestic production, making them heavily dependent on imports. Most of the imports came from Russia via the Friendship pipeline. Poland also receives oil from the "Naftoport" terminal at Gdansk and from the Odessa-Brody pipeline. The Czech Republic imports oil from Russia, as well as from Germany, via the Mero pipeline, which allows the land-locked country to import crude oil from the Italian port of Trieste via the Trans-alpine pipeline network. The Mero pipeline also enables the Czech Republic to reduce its reliance on Russian oil.

Downstream

Poland has 350,000-bbl/d in refining capacity, the largest in the region. Ceská rafinérská is the Czech Republic's largest crude oil refinery, owning and operating two refineries: Litvinov and Kralupy. The two refineries have a combined capacity of 178,000 bbl/d. Hungary has one crude oil refinery in operation, the 161,000-bbl/d Szazhalombatta refinery. Slovnaft is Slovakia's only refinery, with a

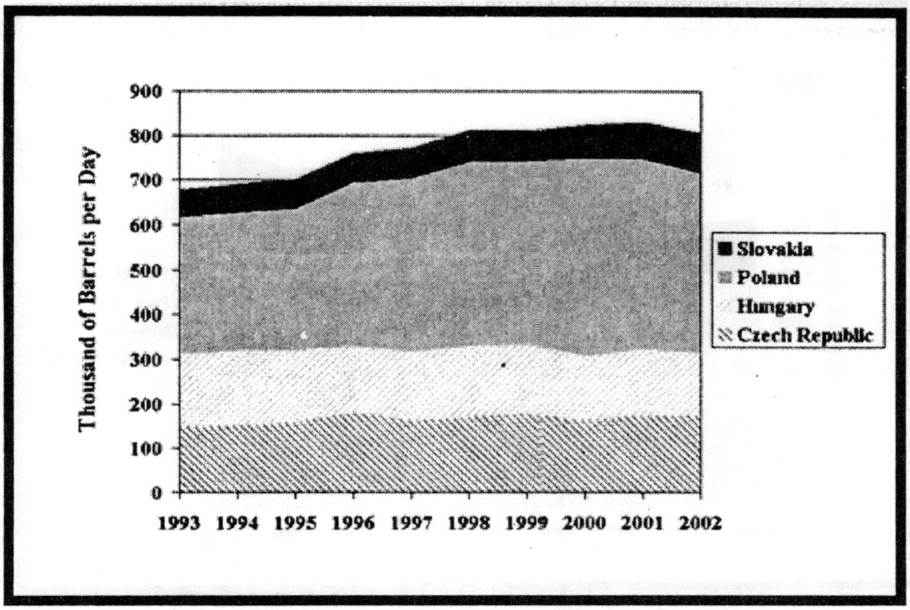

TOTAL OIL CONSUMPTION, 1993-2002

capacity of 115,000 bbl/d. Hungarian oil and gas concern MOL owns 70 per cent of Slovnaft after increasing its stake in March 2003.

As a prerequisite for admission to the EU, each of the Visegrad countries must have 90 days of oil storage capacity. In November 2002, Hungary and Slovenia signed an agreement on an oil storage partnership, which will allow both countries to meet the EU rules. Poland expects its strategic oil reserve to be completed in 2008.

Natural Gas

Proven reserves of natural gas are also minimal in the Visegrad countries, with a combined total of 7.7 trillion cubic feet (Tcf), as of January 2003. Poland, with roughly 75 per cent of the Group's total, has an estimated 5.8 Tcf of natural gas reserves, with Hungary at 1.2 Tcf. Slovakia and Czech Republic contain 530 billion cubic feet (Bcf) and 140 Bcf, respectively. In 2001, Poland produced 193 Bcf, which met 39 per cent of its domestic natural gas demand. Hungary produced 114 Bcf, accounting for a quarter of its demand. Slovakia produced only 10 Bcf while the Czech Republic supplied a mere 5.1 Bcf.

As the Visegrad countries strive to meet EU membership criteria, natural gas is becoming increasingly important to the region's energy mix. Increased consumption of natural gas, as an alternative to coal, is considered to be a key component of the region's plan to meet the stricter EU regulations.

Slovakia's per capita natural gas consumption is the highest among the Visegrad Group countries. About 80 per cent of Slovak households are connected to the natural gas network. In the Czech Republic, natural gas consumption has increased by 35 per cent between 1993 and 2001. In 2001,

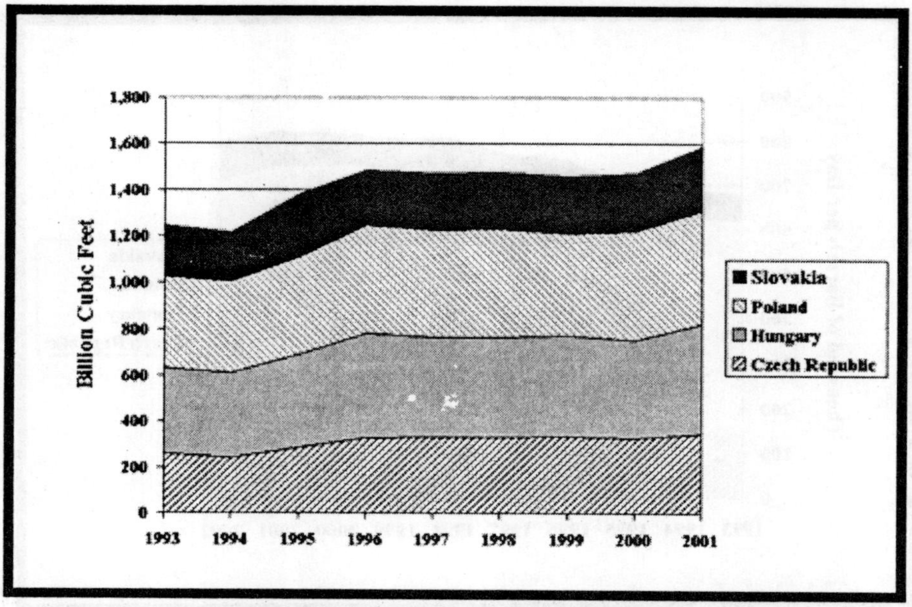

NATURAL GAS CONSUMPTION, 1993-2001

natural gas represented about 43 per cent of total energy consumption in Hungary and 13 per cent in Poland.

Imports

Similarly to crude oil, Russia supplies most of the Visegrad group's natural gas requirements via the Yamal and Brotherhood pipelines. Poland and the Czech Republic import small amounts of natural gas from Germany and Norway. About 80 per cent of Hungary's natural gas imports come from Russia through part of the Brotherhood pipeline. Hungary also imports natural gas via the Gyor-Baumgarten pipeline, which is connected to Western Europe's natural gas grid.

Poland Imports: Russia

Given increased domestic natural gas production and flat demand, Poland has had difficulty in maintaining its Russian, Danish, and Norwegian contracts in their present state. The Polish government has already amended or even deferred some of these contracts. On February 12, 2003, PGNiG, Poland's state oil and natural gas firm, and Gazprom renegotiated the original Yamal pipeline contract, reducing Poland's annual imports from Russia by a third in the years 2003-2022.

The Yamal pipeline, which began operations in September 1999, transports natural gas from the Yamal (West Siberia) field in Russia to Poland, where it is further distributed to Germany and to other Western European countries. EuroPol Gaz operates the Polish section, in which both PGNiG and Gazprom have a 48 per cent share. A consortium of Polish firms called Gas Trading owns the remaining 4 per cent. The pipeline is not yet transporting natural gas at capacity, as it requires the construction of three compressor plants. Plans to build a second pipeline (Yamal II) have also been postponed. Under the original agreement, the second pipeline was to become operational in 2002.

Poland Imports: Denmark and Norway

In July 2001, PGNiG reached an agreement with Dansk Olie og Naturgas (DONG) to import 565 Bcf of natural gas over eight years, starting in 2003. This would be done through a planned $330 million, 186-mile Baltic pipeline. The project, however, was deferred because Poland's natural gas demand was less than expected. The September 2001 contract, in which PGNiG and Norway's Statoil agreed to the delivery 2.6 Tcf over 16 years, has also been postponed. Norwegian exports to Poland would require construction of the $1.1 billion, 683-mile Austerled pipeline.

In February 2003, PGNiG along with DONG and Lithuania's Lietuvos Dujos announced the "Amber" project. The project is a joint Polish-Danish-Lithuanian venture for the construction of a gas pipeline to supply natural gas from Poland to Lithuania. Its main goal is to connect the Polish and the Lithuanian gas system with the EU network in order to ensure gas supplies security to Poland and to Lithuania, as well as possibly to Latvia, Estonia and Finland.

Sector Liberalization

The EU accession treaty requires the Visegrad countries to liberalize their natural gas markets according to the EU 1998 natural gas directive. Along with divesting and unbundling state owned natural gas companies, governments are required to open the natural gas market to outside competition, thus allowing customers to choose their own supplier. The Slovak government, for instance, has been incrementally opening up the country's natural gas market. The process began in July 2002, which allowed customers with an annual consumption of more than 882 million cubic feet to choose their supplier. The next stage took place in January 2003 (more 530 million cubic feet). The third stage will take place in 2008 for customers with an annual consumption of more than 177 million cubic feet.

In March 2003, Hungarian parliament's budget and economic committees approved a draft Gas Act, which will set in place the regulatory framework for a liberalized natural gas market in Hungary. The Act is set to move to the parliament for general debate. If the legislation is passed, it would partially liberalize the gas market as of 2004, allowing power plants and those consuming more than 17,500 cubic feet per hour to choose their supplier. The Gas Act included a clause for compensating household consumers for the increase in natural gas prices. The clause would shield household consumers from a part of the 15 per cent increase in gas prices that will come into effect as of May 15, 2003.

The Czech Republic market should open up for all consumers except for households in mid 2004. Households should enter the market in mid 2007. In Poland, PGNiG is supposed to relinquish its monopolist position on the natural gas market, enabling industrial companies to being choosing their supplier before joining the EU on May 1, 2004.

Coal

Coal is the most prevalent energy source in the Visegrad countries, although its role as a fuel and an industry is declining. Poland is the exception, where coal accounted for 93 per cent of the country's primary energy production in 2001, and remains one of the country's most important employers. Coal also remains significant in the Czech Republic, where it constituted 47.7 per cent of the primary energy consumption.

The region holds 32,090 million short tons (Mmst) of proven recoverable coal reserves, of which Poland has 24,400 Mmst. The Czech Republic contains 6,300 Mmst; Hungary 1,200 Mmst; and Slovakia 190 Mmst. In 2001, the region produced 270.1 Mmst, of which Poland was responsible for 66 per cent.

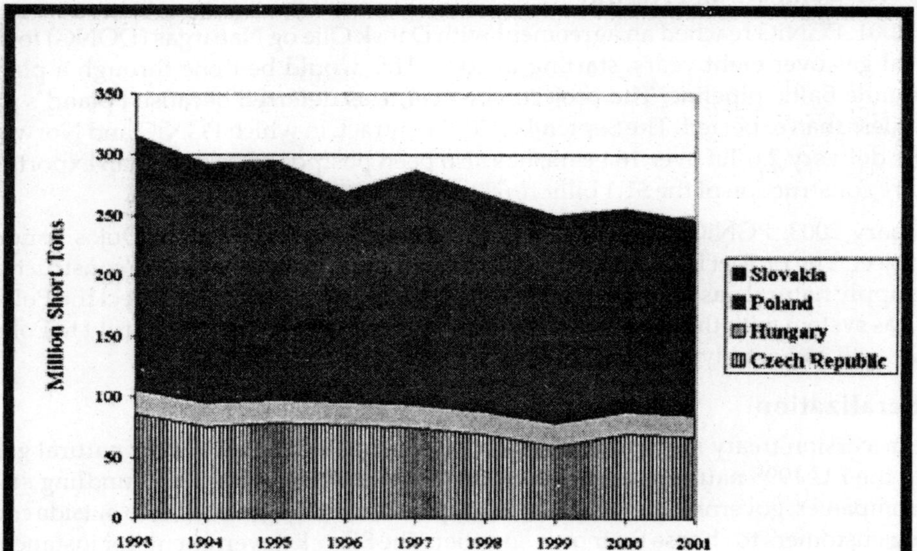

COAL CONSUMPTION, 1993-2001

Slovakia was the smallest producer (2.5 Mmst), preceded by the Czech Republic (72.8 Mmst) and Hungary (15.9 Mmst).

Consumption has decreased sharply in the region in recent years. Between 1993-2001, coal consumption fell by 26 per cent in Slovakia, 21 per cent in the Czech Republic, 22 per cent in Poland and 12 per cent in Hungary. Overall consumption for the region in 2001 was approximately 246.3 Mmst.

Restructuring

Over the last decade, the Visegrad countries have continually restructured and downsized their coal industries. Governments have reduced the number of inefficient mines in operation, cut the labor force associated with coal mining, and increased awareness of environmental issues related to the industry in order to bring their countries in line with EU standards. The Hungarian Economic and Transport Ministry recently reported that, during the last decade, the number of operating coal mines has fallen from 41 to 9, and the number of jobs from 50,000 to 8,300. CEZ, the largest Czech power company, operates 10 coal power plants, which will be decommissioned between 2008 and 2010.

On February 1, 2003, the Polish government created Kompania Weglowa (KW), Europe's largest coal company. The original plan called for the closing of the country's 39 mines; the consolidation of mines from five failing coal firms (Bytomska, Rudzka, Gliwicka, Nadwislanska and Rybnicka); and the reduction of 35,000 jobs by 2006. After protests from coal unions in December 2002, the government toned down initial restructuring plans. The government changed the total employment reduction to 27,200; gave workers job placement guarantees in surviving mines, if their own unit goes under; and required KW to be responsible for 24 mines held by the country's five worst mining firms. Kompania

Weglowa should be relieved of its five worst mines, provided that an expert group demonstrates that they are no longer economically viable.

Electricity

The Polish power generation sector is the largest in Central and Eastern Europe. In 2001, Poland's installed electric capacity was about 30.6 million kilowatts. Electric generation reached 135 billion kilowatt hours (Bkwh), while consumption was 119 Bkwh in 2001. Coal-fired power plants meet most of Poland's annual electricity demand. Poland is also a net exporter of electricity.

Most of Hungary's capacity and generation is thermal (oil, gas, and coal), although the country's sole nuclear plant at Paks produces slightly less than 40 per cent of total electricity generated. Hydropower generates less than 1 per cent of Hungary's electricity. In 2001, Hungary generated about 34.4 Bkwh and consumed about 35.1 Bkwh of electricity. Consumption peaked at 37 Bkwh in 1989, but declined in the early 1990s as Hungary's post-Communist economy grew less energy-intensive. Hungary is a net importer of electricity, mostly from Slovakia.

Both electricity generation and consumption have been rising in the Czech Republic in recent years. Between 1993 and 2001, electricity production in the country rose 26 per cent, from 55.6 Bkwh to 70.0 Bkwh, while electricity consumption increased 12 per cent, from 49.6 Bkwh to 55.6 Bkwh. In 2001, the country exported an estimated 9.45 Bkwh, primarily to Germany, Austria and Slovakia. Electricity exports are becoming increasingly important for the Czech Republic, particularly with the commissioning of the Temelín nuclear power plant in 2001. The Czech government also aims to increase the contribution of renewable sources to the total consumption of primary energy sources to about 3 per cent -6 per cent as of the year 2010 and about 4 per cent -8 per cent as of the year 2020.

In 2001, Slovakia's installed electric generating capacity was about 7.5 million kilowatts. During the same year, the country consumed 24.4 Bkwh while producing 30.3 Bkwh. Since two nuclear reactors came on line in 1998 and 2000, Slovakia has become more reliant on nuclear generation and less reliant on fossil fuels for electricity generation. In 2001, nuclear power plants produced 54 per cent the country's electricity generation while thermal plants provided 30 per cent and hydro 16 per cent. The addition of the nuclear power plants has also allowed Slovakia to become a net exporter of electricity, beginning in 1999.

Sector Liberalization

In preparation for EU accession, the Visegrad countries are required to liberalize their electricity sectors. All of the countries have introduced legislation to fulfill this requirement. Most of the policies include establishment of a legal framework to define the rights and duties of producers, distributors, and users of energy; foundation of an independent regulatory entity to ensure competition within the energy sector; and guarantee of Third Party Access (TPA) of enterprises to energy distribution grids. The policies also include scheduled opening of the electricity markets and the privatization of the large state-owned electricity power companies.

In November 2000, the Czech government adopted an Energy Act, which began incrementally opening up the country's electricity market to competition in January 2002. On January 1, 2003, the government lowered the consumption threshold from 40 gigawatthours (Gwh) to 9 Gwh, allowing an additional 500 Czech companies to choose freely their supplier. In 2006, all customers will be free to choose their supplier. Additionally, the Czech Energy Regulatory Office eliminated subsidies for household electricity prices in 2002.

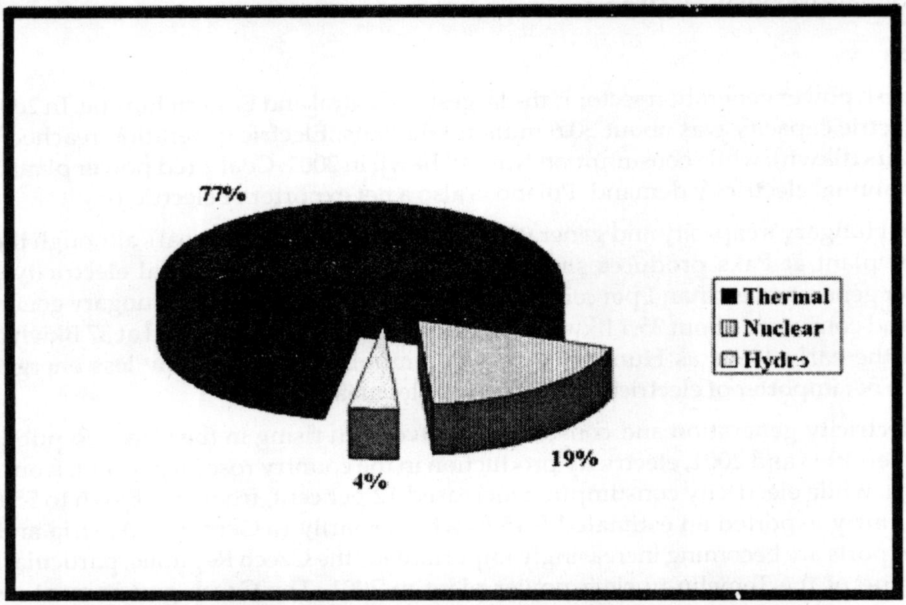

REGIONAL ELECTRICITY GENERATION BY SOURCE, 2001

Slovakia began liberalizing the internal electricity market in January 2002, allowing the largest consumers -more than 100 Gwh–to choose their supplier. In October 2002, the government began the external liberalization of the electricity market, allowing larger consumers to import a twelfth of electricity demand. The market will incrementally open up until 2005, when all consumers will be able to choose their supplier.

Hungary's electric power liberalization legislation went into effect on January 1, 2003. Large consumers, with annual consumption above 6.5 Gwh, are allowed to choose their power suppliers; however, only 50 per cent of their power may come from imports. Residential users remain in the government-controlled system, which maintains a cap on retail electricity prices. As of March 2003, fourteen large energy consumers have diversified suppliers, including Hungary's largest electricity consumer, Borsodchem.

Poland began liberalizing its electricity sector in 1998. As of now, companies consuming over 10 Gwh annually can choose their suppliers. Next year the threshold will fall to 1 Gwh and by 2006, the market will be completely open. In April 1997, the Polish government passed a new Energy Act, which required the Government Economic Committee to pass "Guidelines on Poland's Energy Policy Through 2020." The document spells out long-term energy forecasts and action plans for the Polish government. The key objectives include: increased security of energy supplies, (including diversification of sources); increased competitiveness for Polish energy sources in domestic and international markets; environmental protection; improving energy efficiency; and reducing energy-related carbon emissions.

Nuclear

The Czech Republic has two nuclear power plants, Dukovany and Temelín. After years of delay, on October 9, 2000, the Czech Nuclear Safety Authority cleared Temelín nuclear for operation, located only 37 miles from the Austrian border. The first reactor was connected to the national grid in December 2000. However, the reactor has been shut down a number of times due to technical problems. The second unit of the Czech Temelín nuclear power plant was put into trial operation on April 18, 2003. The two units are expected to reach 100 per cent of their operational capacity at the end of April 2003. When the plant is fully operative, it will provide over 20 per cent of the Czech Republic's power needs.

Temelín has been controversial since construction first began in 1986. Opponents have argued that the plant is unnecessary, noting that the Czech Republic already produces more electricity than it consumes, and that additional electricity can be generated by improving the existing distribution network rather than installing new generating capacity. Although Temelín meets and even exceeds EU safety standards for nuclear power plants, Czech and Austrian environmentalists claim that it is not safe because it combines Soviet design and western fuel and safety technology.

Slovakia has two nuclear power plants, which generated an estimated 54 per cent of Slovakia's electricity in 2001. The Jaslovske Bohunice plant at Trnava has four, 408-MW reactors that are functioning, and one decommissioned reactor. The plant's two older reactors are due to be decommissioned in 2006 and 2008 as part of the energy chapter of Slovakia's accession agreement with the EU. The Mochovce plant has two 412-MW reactors in operation and two uncompleted reactors. Construction of these reactors has been halted, as government financial support for them has ended.

The Paks nuclear power plant in Hungary consists of four Soviet-design, second generation VVER-440/213 reactor units. There are plans not only to expand generation capacity of the reactors by 8 per cent but also to extend the life-cycle of the reactor units by 20 years. The normal lifespan of the four units ends between 2012 and 2017. In order to ensure continuous operation of the plant, the necessary modernization improvements would have to begin in 2007. On April 10, 2003, some fuel rods in unit 2 were damaged, causing a radioactive gas leak. According to reports, the leak constitutes no threat to the environment, but the incident could result in major financial losses to the company.

56

North Korea

The Democratic Peoples Republic of Korea (North Korea) occupies a strategic location bordering China, South Korea, and Russia. In 1994, North Korea signed an agreement (the "Agreed Framework") sponsored by the United States, Japan, and South Korea to halt its graphite technology nuclear program in exchange for two light water nuclear reactors. In early October 2002, North Korea disclosed to U.S. officials that it had violated the terms of the Agreed Framework by pursuing a nuclear weapons program, and that North Korea considered the agreement nullified.

General Background

North Korea's economy, which remains under tight state control (collectivized agriculture and state-owned companies account for about 90 per cent of all economic activity) grew by 3.7 per cent in 2001, after growing 1.3 per cent in 2000, according to estimates by South Korea's central bank. Strong increases in agricultural and mining output, both of which are major sectors of North Korea's economy, contributed to this growth. The GDP growth of the last three years followed a steady economic contraction from 1990 through 1998.

North Korea's communist ideology has been based on the concept of "juche," or self-reliance. Severe economic problems have, however, forced the country to accept international food aid and embark on a series of limited market reforms. Famine in North Korea has reportedly killed hundreds of thousands of people over the last decade. Several governments, including the United States, have provided funding to the United Nations' World Food Program (WFP) for emergency food aid to North Korea.

North Korea has permitted a small amount of foreign investment in recent years. Manufacturing plants of South Korean companies operating joint ventures in North Korea now generate over $100 million in annual revenue.

North Korean-South Korean Relations

From June 13th to 15th, 2000, the leaders of the two Koreas held their first summit meeting in Pyongyang. The summit led to a joint statement by the two leaders which supported, in general terms, the goal of eventual reunification of the two Korean states, reunification of families divided since the Korean War, and economic cooperation. A planned followup visit to South Korea by North Korean leader Kim Jong-Il, however, has been repeatedly delayed. Developments in inter-Korean relations in 2002 have been mixed. A naval clash near the two countries maritime frontier in June 2002 heightened tensions, but progress has been made in some areas, such as the commencement of work in September 2002 on clearing parts of the Demilitarized Zone (DMZ) to facilitate possible road and rail links. It is unclear what effect the disclosure of the North's violations of the Agreed Framework will have on this work.

United States Economic Sanctions

The United States announced on June 19, 2000 that it was easing some of the economic sanctions which have been in place with North Korea under the Trading With the Enemy Act since the start of the Korean War in 1950. Licenses are still required from the Treasury Department's Office of Foreign Assets Control (OFAC) for many transactions, and sales of military and "dual-use" items are still restricted. (See the OFAC Fact Sheet on North Korea Sanctions for more information.) The designation of North Korea as a state supporter of terrorism by the United States also effectively precludes lending by international financial institutions such as the World Bank.

Domestic Energy Resources

North Korea relies on two domestic sources of commercial energy–coal and hydropower–for most of its energy needs. In 2000, coal accounted for about 86 per cent of primary energy consumption.

North Korea's electric generating capacity is split nearly evenly between coal-fired thermal plants and hydroelectric plants. In 2000, hydroelectric power plants generated about 67 per cent of North Korea's electricity and thermal plants about 33 per cent. The country's thermal generating capacity is underutilized due to a lack of fuels. The country's total electricity consumption in 2000 was only 65 per cent of what it had been in 1991, though it showed an increase of nearly 9 per cent over the figure for 1999.

As a result of the electricity shortage, North Korea has resorted to a rationing system. The country often experiences blackouts for extended periods of time, and power losses due to an antiquated transmission grid are high. Rainfall improved in 2001, but some hydroelectric facilities are believed to be out of operation due to flood damage from major flooding in 1996.

The prospect of South Korean help for North Korea's electricity shortage was discussed at the June 2000 summit meeting, and North Korea has since been seeking electricity from South Korea since then. A linkage between the electric grids of the two Koreas is one possibility, but it would be of limited immediate value due to the North's poor transmission infrastructure. At present, South Korea has not agreed to supply the North with electricity from its own transmission grid. North Korea also has reportedly discussed the possibility of electricity aid with Russia.

Problems with the North Korean coal industry are closely related to the problem with electricity supply. Coal-fired power plants have been running well under capacity in recent years, due in part to problems with rail transportation of coal.

Oil

North Korea lacks domestic petroleum reserves, but the West Korea Bay may contain hydrocarbon reserves, as it is considered to be a geological extension of China's Bohai Bay. Sweden's Taurus Petroleum holds an oil exploration concession for two blocks, B and C, off the west coast of North Korea. Britain's Soco International holds the concession for Block A. Initial seismic surveys conducted by Taurus have identified possible oil-bearing geological structures. The two companies held separate discussions with South Korea's Korean National Oil Company (KNOC) and Hyundai about possibly forming a consortium for joint exploration of the areas, but no new activity has been reported in these concessions in the last year. Australia's Beach Petroleum also holds a concession for one block off the east coast of North Korea, which is considered less promising.

In November 2001, the North Korean government awarded its first concession for an onshore block to Sovereign Ventures of Singapore, which covers an area in the Tachon-Rajin area near the Chinese border. In September 2002, the company reported that initial seismic survey results had indicated probable oil and natural gas deposits of modest size.

In the meantime, North Korea must import all of the oil it consumes. Oil accounts for about 6 per cent of total North Korean primary energy consumption, and is mostly limited to non-substitutable uses such as motor gasoline, diesel, and jet fuel. With the exception of the heavy fuel oil which was being provided under the Agreed Framework, most petroleum is imported as crude oil and processed at domestic refineries. Under the 1994 "Agreed Framework," the United States had assumed responsibility for providing 500,000 metric tons (approximately 3.3 million barrels) of heavy fuel oil annually through Korean Peninsula Energy Development Organization (KEDO).

Natural Gas Transportation

South Korea has held discussions with China, Russia, and BP about the possibility of importing natural gas from Russia's huge Kovykta gas field near Irkutsk. While China would be a major importer of gas through the pipeline, the project could be made more economical by adding a link to South Korea, which currently gets the vast majority of its natural gas by tanker as liquefied natural gas (LNG). North Korea is one possible route for the pipeline link to South Korea, and would be less costly than the subsea alternative.

Nuclear Energy

Prior to 1994, North Korea's nuclear program had been a major concern for regional security, since its graphite reactor technology produced fissionable plutonium which can be used in nuclear weapons. North Korea withdrew from the Nuclear Non-Proliferation Treaty in 1993.

Under the Agreed Framework negotiated with the United States in 1994, North Korea had agreed to freeze its nuclear program in exchange for two new pressurized light-water reactors (which are considered less capable of producing weapons-grade plutonium) and 500,000 metric tons per year (about 3.3 million barrels) of heavy fuel oil to meet its energy needs until the first new reactor becomes operational. KEDO, an international consortium led by the U.S. government (with South Korea, Japan, the European Union, and others), was established to implement the agreement. The European Union joined KEDO in September 1997.

Japan signed a contract in May 1999 committing to provide its $1 billion contribution to KEDO to fund the new light-water reactors, an action which had been delayed by North Korea's missile test in August 1998. The project was expected to cost a total of $4.6 billion, with South Korea providing the greatest share of funding at $3.2 billion. The United States and the European Union also had pledged to contribute funds.

Construction of the light water reactors was to be performed under a turnkey contract with KEPCO, which was awarded in December 1999. Initial site preparation work had begun, and the concrete foundation at the site was laid in August 2002. Training of North Korean technicians who were to operate the reactors had begun in June 2002. The project had faced many delays, and the completion date for the first reactor has been pushed back to at least 2008, from an original completion date of 2003. One hurdle the project had faced was the issue of indemnity for potential liabilities created by the plant. General Electric had originally been chosen to supply the generators, but pulled out of the project when the issue was not resolved to its satisfaction. In January 2001, it was announced that a consortium of Japanese firms, led by Hitachi and Toshiba, would supply the generators.

With the North Korean admission in early October 2002, according to press reports, of its violations of the Agreed Framework by pursuing a program to enrich uranium for nuclear weapons, the light water reactor project is currently stalled.

Country Overview

Independence: September 9, 1948

Population (2002E): 22.2 million

Location/Size: Eastern Asia / 120,540 sq. kilometers (46,800 sq. miles), about the size of Pennsylvania

Major Cities: Pyongyang (capital), Hamhung, Chongjin

Language: Korean

Ethnic Groups: Korean

Religions: Buddhism, Christianity, Chundo Kyo

Defense (6/98): Army, (923,000); Air Force, (85,000); Navy, (46,000); security/border troops, (115,000); workers' and peasants' militia (Red Guard), 3.8 million

Economic Overview

Currency: 100 Chon = 1 Won

Official Exchange Rate (10/02): US$1 = 2.20 Won

Gross Domestic Product (2001E): $15.5 billion

Real GDP Growth Rate (2001E): 3.7 per cent (Central Bank of South Korea estimate)

External Debt (1996E): $12 billion (much of this debt currently is in default)

Merchandise Exports (2001E): $700 million

Merchandise Imports (2001E): $1.6 billion

Major Import Products: Petroleum, grain, coking coal, machinery and equipment, and consumer goods

Major Export Products: Minerals, metallurgical products, agricultural and fishery products, manufactured goods (including armaments)

Trading Partners: China, Japan, Russia, Germany, Singapore, South Korea, Hong Kong

Energy Overview

Oil and Natural Gas Production/Proven Reserves: None

Oil Consumption (2000E): 80,000 barrels per day (bbl/d)–all imported

Crude Oil Refining Capacity (1/1/02E): 71,000 bbl/d

Recoverable Coal Reserves (12/31/99): 661 million short tons

Coal Production (2000E): 101.4 million short tons

Coal Consumption (2000E): 103.6 million short tons

Electric Generating Capacity (1/1/00E): 9.5 gigawatts (about 53 per cent hydropower)

Electricity Generation (2000E): 33.4 billion kilowatthours

Major Ports: Chongjin, Hamhung, Nampo

Environmental Overview

Total Energy Consumption (2000E): 2.8 quadrillion Btu* (0.7 per cent of world total energy consumption)

Energy-Related Carbon Emissions (2000E): 66.1 million metric tons of carbon (1.0 per cent of world total carbon emissions)

Per Capita Energy Consumption (2000E): 127.8 million Btu (vs U.S. value of 351.0 million Btu)

Per Capita Carbon Emissions (2000E): 3:0 metric tons of carbon (vs U.S. value of 5.6 metric tons of carbon)

Energy Intensity (1997E): 71,200 Btu/$1990 (vs U.S. value of 13,900 Btu/$1990)

Carbon Intensity (1997E): 1.7 metric tons of carbon/thousand $1990 (vs U.S. value of 0.22 metric tons/thousand $1990)

Sectoral Share of Energy Consumption (1998E): Industrial (70.9 per cent), Transportation (28.7 per cent), Residential (0.4 per cent), Commercial (0.0 per cent)

Sectoral Share of Carbon Emissions (1998E): Industrial (80.3 per cent), Transportation (19.3 per cent), Residential (0.4 per cent), Commercial (0.0 per cent)

Fuel Share of Energy Consumption (2000E): Coal (85.8 per cent), Oil (5.9 per cent), Natural Gas (0.0 per cent)

Fuel Share of Carbon Emissions (2000E): Coal (95.1 per cent), Oil (4.9 per cent), Natural Gas (0.0 per cent)

Renewable Energy Consumption (1998E): 249.5 trillion Btu* (7 per cent decrease from 1997)

Status in Climate Change Negotiations: Non-Annex I country under the United Nations Framework Convention on Climate Change (ratified December 5th, 1994). Not a signatory to the Kyoto Protocol.

Major Environmental Issues: Localized air pollution attributable to inadequate industrial controls; water pollution; inadequate supplies of potable water.

Major International Environmental Agreements: A party to the Antarctic Treaty, Biodiversity, Climate Change, Environmental Modification, Ozone Layer Protection and Ship Pollution. Has signed, but not ratified, the Antarctic-Environmental Protocol and Law of the Sea.

The total energy consumption statistic includes petroleum, dry natural gas, coal, net hydro, nuclear, geothermal, solar, wind, wood and waste electric power. The renewable energy consumption statistic is based on International Energy Agency (IEA) data and includes hydropower, solar, wind, tide, geothermal, solid biomass and animal products, biomass gas and liquids, industrial and municipal wastes.

57

North Sea

The North Sea contains Europe's largest oil and natural gas reserves and is one of the world's key non-OPEC producing regions. Norway and the United Kingdom hold the majority of the North Sea's reserves and production. Denmark, the Netherlands, and Germany have smaller North Sea oil and gas resources.

General Background

North Sea oil and natural gas were first discovered in the 1960s. The North Sea, however, did not emerge immediately as a key non-OPEC oil producing area until the 1980s and 1990s, when major discoveries began coming online. Oil and natural gas extraction in the North Sea's inhospitable climate–cold and windy–and at great depths requires sophisticated offshore technology. Consequently, the region is a relatively high cost producer, but its political stability and proximity to major European consumer markets have allowed it to play a major role in world oil and natural gas markets.

A key feature of North Sea oil is its role as one of the major "benchmark" crude oils, the Brent price marker. Because Brent crude is traded on the International Petroleum Exchange (IPE) in London, fluctuations in the market are reflected in the price of Brent. Therefore, the many other crude oils linked to Brent can be priced according to the latest market conditions.

However, with production in the UK North Sea Brent system declining to only around 350,000-400,000 barrels per day, many felt that such low output was too small for a benchmark against which a large portion of the world's internationally traded crude oil was directly or indirectly priced. In order to address the issue of declining Brent production, the industry price assessor Platts widened the assessment definition of dated Brent by including two similar North Sea crude oils–North Sea Forties (UK) and Oseberg (Norway). According to Platts, the new method of calculating dated Brent would more accurately reflect the North Sea oil market conditions. The change took effect on July 10, 2002, and initially got off to an uneven start, with Royal-Dutch Shell (co-operator of the Brent system)

opposed to the change, and BP (operator of the 750,000 barrels per day Forties system) in favor. The IPE began using the Platt's Brent formulation (BFO, or Brent-Forties-Oseberg) with its October contract on September 13, 2002.

North Sea offshore crude oil production reached new heights in 1999, averaging 5.94 million barrels per day (bbl/d). But production declined in both 2001 and 2002, to about 5.72 million bbl/d and 5.66 million bbl/d, respectively. Total offshore oil production averaged 6.21 million bbl/d in 2002, about a 1 per cent decrease over the previous year. In contrast, natural gas production, including onshore production, has steadily increased, reaching 9.5 trillion cubic feet (Tcf) in 2001. Most of Germany and the Netherlands' natural gas production is located onshore.

In the near term, improved recovery technologies and higher oil prices are expected to boost briefly North Sea oil production, halting the decline which began in 1999. Although the region will continue to be a sizable producer of oil and natural gas for the foreseeable future, oil production in the UK and Norwegian sectors has reached a plateau and is expected to begin a long-term decline.

Vertically integrated oil majors have already begun to de-emphasize the importance of the North Sea and turn their attention to other areas of the world. Small, independent oil companies, however, have become more active in the region, acquiring producing properties and applying for exploration and production licenses, particularly in the UK. A survey that came out in May 2003 ranked the North Sea as one of the best areas for oil exploration, with the UK Continental Shelf named the most-favored area for new ventures.

However, the high cost of decommissioning oil production platforms could hinder the entrance of independents. The cost of oil production, which most likely will increase as independents attempt to extract oil from matured fields, could also make it more difficult for these smaller companies to make acquisitions profitable.

Norway and the UK Joint Development

In August 2002, the UK and Norwegian governments released a report, "Unlocking Value through Closer Relationships," which called for greater cooperation between the countries in further developing oil and natural gas in the North Sea. The UK and Norway share a 37.5-mile-wide corridor which is located on either side of the UK/Norway North Sea meridian. The aim of this initiative is to pool infrastructure resources and technology practices to develop this area, which contains an estimated 13 billion barrels of oil equivalent, of which two-thirds is oil.

Although both countries have agreed to write a treaty to support this initiative, legal issues have delayed the process. One issue includes the use of the UK pipeline infrastructure to import Norwegian natural gas, particularly from Norsk Hydro's Ormen Lange giant natural gas field in the Norwegian Sea.

Oil

Norway and the United Kingdom are the largest producers of North Sea oil, representing 92 per cent of total production in 2002. The UK's production fields are more mature and numerous in comparison to Norway, which has fewer but larger fields.

Norway

Norway has been the world's third-largest oil exporter for several years. With a population of 4.5 million, the country produced about 4.3 per cent of the world's oil and 7 per cent of non-OPEC (Organization of Petroleum Exporting Countries) oil in 2002. In recent years, the country has been in the top ten world oil producers and exporters, while its consumption is ranked about 50th in the world. In 2002, Norway's total oil production was an estimated 3.3 million bbl/d, a 3.3 per cent decrease in comparison to 2001.

Underlying its importance in world markets, Norway may adjust its output in coordination with OPEC production increases or decreases. For instance, Norway reduced its oil output by 150,000 bbl/d from January 1 to June 30, 2002, cutting the government's production from 3.17 million bbl/d to 3.02 million bbl/d, as part of a plan by OPEC and major non-OPEC producers to bolster oil prices in the face of weak world oil demand.

Licensing

In order to stimulate more efficient exploration in mature parts of the Norwegian continental shelf, the government has changed the licensing system. Beginning in 2004, companies will be able to choose from preselected areas from January 1 to October 1 each year. It is expected that this new process will bring more predictability to a previous system which allowed companies to chose a block for development and then wait to hear for approval. The Norwegian government hopes that the new licensing scheme will ensure continued development in mature areas.

United Kingdom

The United Kingdom is the European Union's (EU) only significant energy exporter. Unlike non-EU member Norway, the UK also has some onshore production and is one of the world's largest oil consumers, ranking in the top 15. UK net oil exports were a little less than one quarter of Norway's exports in 2002, at about 850,000 bbl/d. Waters in the central North Sea off the east coast of Scotland contain nearly half of the UK's remaining oil reserves, with about a quarter of reserves located in the northern North Sea near the Shetland Islands.

Exploration and Production

In 2002, the UK's total oil production declined by 1.4 per cent in comparison to the previous year. Since peaking at nearly 3.0 million bbl/d in 1999, production has fallen 14 per cent. Nonetheless, the UK's North Sea production is expected to increase slightly in coming years. New fields recently brought on stream, in addition to improvement in output from other UK offshore fields, have resulted in a steady increase in crude oil production in the first quarter of 2003.

Independent Producers

Because of the UK oil sector's maturity, exploration in recent years has focused primarily on smaller fields and on incremental development of existing fields. Larger companies have begun selling off maturing assets to smaller operators. Most notably, independent producer Apache purchased the North Sea Forties oilfield network from BP in April 2003. Other small, independent companies, such as Talisman Energy and Paladin Resources, have recently acquired production blocks from oil majors. In many cases, these fields are more economically viable for smaller independent operators.

Government Promotional Programs

Government programs have also been key in attracting new oil independents to the UK sector. In January 2000, the government created the PILOT program to help secure long-term production of oil and natural gas from the UK Continental Shelf. The Pilot program created the *Promote License* and the *Fallow Initiative* programs. The *Promote License* program allows companies to assess the value of a field before committing to it. The *Fallow Initiative* seeks to rejuvenate activity in dormant acreage and assets by inviting third parties to propose technical ideas that would stimulate new exploration and development. The License Information For Trading (LIFT) lists 77 blocks (40 fallow blocks and 37 fallow discoveries) which are to be made available under the *Fallow Initiative.* BP's Enhanced Oil Recovery project for the Magnus field is an example of the technological initiatives taking place in North Sea to increase production. The project, which involves reinjecting gas into the field, seeks to recover an estimated 60 million barrels of oil.

The government's investment in these programs has increased interest in the UK North Sea. After the closing of the 21st licensing round on May 8th, 2003, companies submitted offshore production licenses applications for 140 different blocks in the North Sea, up from under 40 blocks from the previous year.

Other North Sea Oil Producers–*Denmark, The Netherlands and Germany*

Denmark is the only net oil exporter out this group, with an estimated 175,000 bbl/d in 2002. In 2002, Denmark's total oil production reached the country's highest annual output to date, with 374,750 bbl/d, an 8 per cent increase over 2001. Since 1998, crude oil production has increased by 56 per cent. New fields–Halfdan, Siri, and Syd Anre–which came on stream in 1999, have helped bolster Denmark's oil production. In 2003, two new fields, Nina and Cecilie, are expected to come on stream. The Danish Energy Authority predicts that oil production will breach 400,000 bbl/d in 2004.

The Netherlands also experienced an increase in oil production in 2002. Offshore production more than doubled over the previous year, reaching 41,000 bbl/d. However, The Netherlands's offshore production has been in decline since 1986, when it peaked at about 77,000 bbl/d. The Hanze field (F2a) in the Dutch North Sea that came online in August 2001 contributed to the jump in Dutch production output for 2002, producing about 31,000 bbl/d.

Germany operates one offshore oil field, Mittelplate, which is located of the coast of the German state, Schleswig-Holstein. Most of Germany's oil production and reserves are located onshore.

Table 57.1: Major North Sea Oil Production Fields*

Country	*Field*	*Est. 2001 Production*	*Operator*
Norway	Troll (I & II)	347,101	Norsk Hydro
Norway	Ekofisk	285,447	Phillips
Norway, 85.5%; UK, 14.5%	Statfjord	263,851	Statoil
Norway	Draugen†	205,632	Shell
Norway	Snorre	199,967	Statoil
Norway	Norne†	195,501	Statoil
Norway	Oseberg	194,902	Norsk Hydro
Norway	Gullfaks	181,986	Statoil
Norway	Heidrun†	176,853	Statoil
Norway	Åsgard†	157,456	Statoil
UK	Schiehallion	140,000	BP
Denmark	Dan	118,454	Maersk Oil
UK	Foinaven	106,000	BP
Norway	Jotun	94,430	ExxonMobil

* These two fields are located in the Norwegian Sea.

Note: Some "fields" actually are systems including multiple adjacent fields.

Oil Infrastructure

Ekofisk was the first North Sea oil field to be discovered and developed, in the late 1960s, with production beginning in 1971. The *Norpipe pipeline*, which connects Ekofisk to Teesside, England, began operation in 1975. Since then, oil companies have built an extensive network of oil pipelines to transfer oil to the mainland, as well as to connect individual fields. Additional pipeline connections to the UK include a major pipeline from the Nelson/Forties field area to Cruden Bay, north of Aberdeen. The *Piper Flotta* pipeline connects northern North Sea production to Scotland's Orkney Islands and two pipelines–*Brent* and *Ninian*–to the Shetland Islands.

Norway has two major oil pipeline connections, *Troll I* and *II*, which link the Troll fields to Mongstad crude oil terminal. The *Oseberg Transport System* connects the Oseberg area to Kollsnes. Two new pipelines will connect Grane and Kvitebjørn fields to the Norwegian mainland.

Denmark's oil pipeline system, which is controlled by the Danish state-oil company DONG, connects offshore oil fields to an oil terminal located in Frederica. The oil pipeline system was established in the 1980s according to *"The Danish Pipeline Act Concerning the Establishment and Utilization of a Pipeline for Transportation of Crude Oil and Condensate."* The Act requires most oil producers in the

Danish section of the North Sea to use the oil pipeline. The *Netherlands* has two small pipelines connecting to Hoek van Holland and IJmuiden terminals.

Table 57.2

	Proven Oil, Reserves 1/1/03	*Total Oil Production, 2002*	*Net Oil Exports, 2002*	*Proven Natural Gas Reserves, 1/1/03*	*Total Natural Gas, Production 2001*	*Net Natural Gas Exports, 2001*
Norway	10.3 billion barrels	3.3 million bbl/d	3.2 million bbl/d	77.3 Tcf	1.93 Tcf	1.8 Tcf
UK	4.7 billion barrels	2.5 million bbl/d	850,000 bbl/d	24.6 Tcf	3.74 Tcf	461 Bcf
Denmark	1.4 billion barrels	374,750 bbl/d	175,000 bbl/d	3.0 Tcf	296 Bcf	109 Bcf
Netherlands	106 million barrels	99,167d bbl/d	-846,000 bbl/dTcf (net importer)	62.5 Tcf	2.75 Tcf	1
Germany	342 million barrels	146,796 bbl/d	-2.5 million bbl/dTcf (net importer)	11.3 Tcf	783 Bcf	-2.5 Tcf (net importer)

Natural Gas

Unlike North Sea oil production, natural gas production remains on the rise. Energy demand in Europe is growing, and natural gas is expected to meet much of this growth. According to Eurogas, natural gas is expected to cover 28 per cent of the EU's primary energy demand by 2020.

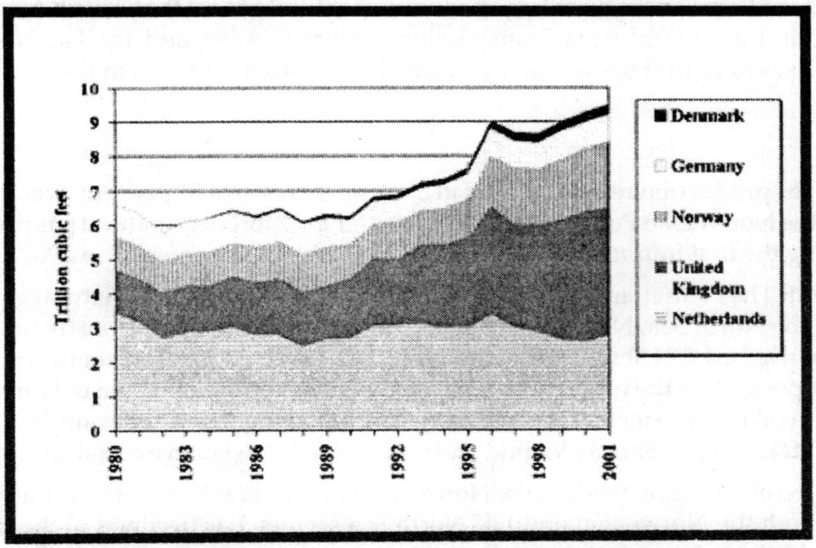

NORTH SEA COUNTRIES NATURAL GAS PRODUCTION, 1980-2001*

* ***Please note that data are for each country's total natural gas production and are not limited to offshore North Sea production***

North Sea natural gas has a geographical advantage over other world natural gas sources, as North Sea natural gas is closer and therefore less expensive to transport to major European markets. Most of continental Europe is already linked, directly or indirectly, to North Sea natural gas sources, and many EU energy companies have large, long-term contracts with major exporters, the Netherlands and Norway. However, the UK, the largest producer of natural gas out of the group, is expected to become a net importer of natural gas by 2005.

The Netherlands

For years, the Netherlands, along with Russia, has been one of the top natural gas suppliers for Western Europe. About two-thirds of the country's current natural gas production is located onshore. According to a report published by Dutch Ministry of Economic Affairs, total net natural gas production decreased by 2.9 per cent in 2002, with offshore production dropping 7.8 per cent. Dutch natural gas reserves decreased by about 1.8 trillion cubic feet (Tcf) for the year, including new finds. The country has made a policy decision to cut back production at its large onshore Groningen field in order to maintain reserves for future use. Most of the declining production is due to this field's intentional decreased production.

The United Kingdom

The UK produces the most natural gas in the North Sea region, with estimated 3.74 Tcf in 2001. Production, however, has declined after peaking in 2000. Although BP received approval in May 2003 to develop the Rhum field, with an estimated reserve of 1.1 Tcf, the field is unlikely to delay the country's need to import natural gas as production declines.

Though most of the UK's natural gas production is in the North Sea, there is a small amount of production onshore and several large fields producing in the Irish Sea. The largest number of non-associated gas fields is located off the English Coast in the Southern Gas Basin, adjacent to the Dutch North Sea sector. However, the three largest sources of natural gas are the Scottish Area Gas Evacuation (SAGE) system, the Central Area Transmission System (CATS), and the Far North Liquids and Associated Gases System (FLAGS), all of which are composed of fields in the central and northern British Sector.

Norway

Natural gas production reached 2.3 Tcf in 2002, an increase of 21 per cent over the previous year, according to the Norwegian Petroleum Directorate (NPD). Moreover, natural gas production is up 18 per cent during the first four months of 2003 campared to the same period in 2002.

Along with The Netherlands, Norway is a key European supplier of natural gas, particularly as new fields in the North Sea, Norwegian Sea, and Barents Sea have come onstream. In 2002, exports from Norway reached 2.24 Tcf, according to NPD. It is expected that total natural gas exports could reach 3.85 Tcf per year in the next 8 -12 years. Along with exports, a portion of Norway's natural gas is re-injected in oil fields to increase oil recovery. The following fields, for example, Oseberg, Åsgard, Statfjord, Gullfaks, Njord, Snorre, Visund and Grane, use Norwegian natural gas for re-injection.

Some of the older, larger fields in the North Sea are in decline. For example, the Frigg field, which is located in both the Norwegian and UK North Sea sectors, has declined to the point that certain associated fields have been shut down. Frigg is expected to cease production this year. New fields, such as Ormen Lange, Norway's second-largest natural gas discovery with estimated reserves of 14.1 Tcf, is expected to help increase production. Other developments include the Halten Bank West, which contains Kristin and the Lavrans, Erlend, Morvin and Ragnfrid fields, with estimated reserves of 1.2 Tcf and the Kvitebjørn field, with estimated reserves of 1.8 Tcf. As the prospects in the Norwegian North Sea decline, operators are looking to expand exploration in the deepwater regions of the Norwegian Sea and the Barents Sea. The government, however, has been hesitant to open more acreage in this region on account of environmental impact concerns.

Denmark and Germany

Denmark and *Germany* are smallest producers of natural gas in the North Sea region. Danish natural gas production has steadily increased over the last three decades, peaking in 2001 at 296 billion cubic feet (Bcf). Similar to Norway, more than a quarter of the country's production in 2002 was re-injected to boost oil production, according to the Danish Energy Authority.

Germany, in contrast, produces more natural gas than Denmark, with an estimated 783 Bcf in 2001. However, nearly all German production is located onshore. Germany's only offshore natural gas field is located in the German North Sea, about 190 miles offshore and began production in September 2000. Known as the Deutsches Nordseekonsortium (German North Sea Consortium), the field is operated by a consortium of Wintershall (operator), BEB Erdgas und Erdoel, BASF, and RWE-DEA. New pipelines will transmit an anticipated 116.5 million cubic feet per day of production. The field is expected to produce for 16 years.

Natural Gas Infrastructure

The already substantial North Sea natural gas infrastructure continues to grow. The Netherlands and the UK have the most extensive pipeline networks in place, while Norwegian export routes are expanding.

The *UK* has many pipeline connections to its prolific southern North Sea, bordering on the Dutch North Sea sector. The major receiving ports for these pipelines include Bacton, Theddlethorpe, and Easington, all located in the middle of England's North Sea coast. One pipeline also connects to Teesside, further north on the English coast. There are many connections from the northern North Sea to Scotland which come onshore at St. Fergus (Vesterland), just north of Aberdeen. The UK is a small gas exporter and has a pipeline connection to Zeebrugge, Belgium (Interconnector) that can be reversed so that Britain is also able to import natural gas. There is also a connection to Ireland off the west coast of Britain.

The Netherlands has natural gas pipelines connecting to terminals located at Uithuizen, Callantsoog, Wijk aan Zee, and Maasvlakte, as well as numerous pipelines which connect among offshore fields. In May 2003, Gasunie subsidiary GasTransport Services plans to construct a new pipeline from Balgzand, near Den Hague to the Bacton terminal on the UK's east coast. If the project proceeds, construction of the pipeline would begin in 2004.

Norway has a number of natural gas trunklines connecting to Europe: the *Europipe I* and *Norpipe* systems to Emden, Germany; the *Zeepipe* trunkline to Zeebrugge, Belgium; the *Franpipe* line to Dunkerque, France; and the *Europipe II* line from Kårstø, Norway to Dornum, Germany. These Norwegian trunklines provide an estimated combined gas transport capacity of 3.3 Tcf per year. The *Vesterled* pipeline transports natural gas from Norway's Frigg field and Heimdal Riser platform, as well as Britain's Galley, Bruce and Alwyn fields to St Fergus in Scotland. Norway signed an agreement with Poland in 1999 whereby Poland would import Norwegian gas from 2001 to 2006. However, construction of this pipeline has been postponed for the indefinite future.

Norway has several pipelines connecting to the country's terminals in Kollsnes and Kårstø. The *Åsgard Transport System* and the *Haltenpipe* transport natural gas from the Norwegian Sea to Kårstø and Tjeldbergodden terminals.

Norway also plans to construct a pipeline linking the Ormen Lange field to the UK's Easington-Dimlington terminal. The pipeline is expected to have a capacity of 2.1 Bcf per day, according to the field's developer, Norsk Hydro. The development of this pipeline depends, however, on whether Norway and the UK can come to an agreement on a treaty (mentioned above). Marathon Oil plans to build the first third-party open access natural gas pipeline in the North Sea–the Symphony pipeline. The 500-mile natural gas pipeline system would connect the Norwegian Heimdal area to the UK's Bacton terminal. Natural gas shipments are expected to begin in 2005, with a capacity of 2.4 Bcf per day.

Germany and *Denmark* have fewer North Sea gas pipelines. Germany has connections to Norway's Ekofisk and Sleipner fields as noted above. Denmark has pipelines connecting its own gas fields to its port of Kaergard in Jutland.

58

Norway

Norway is a major non-OPEC source of oil and was the world's third largest net oil exporter in 2001. Norway is the second-largest natural gas exporter to western Europe.

General Background

Norway's economy is characterized by substantial oil and natural gas revenues, growing government expenditures, a tight labor market, and closer linkage to international oil and gas prices than to the OECD business cycle. Norway is the third largest net oil exporter in the world, and the recent period of high oil prices have made for government budget and current account surpluses and rising disposable income. The petroleum sector represents over 20 per cent of Norway's gross domestic product (GDP). Norway continues to record large trade surpluses, mostly due to hydrocarbon exports. Real GDP growth for 2002 is forecast at 2.3 per cent, a solid rate. Trade surpluses are expected to decline from about 14 per cent to 8 per cent into the later part of this decade. The consumer price growth forecast is at 1.5 per cent in 2002, though the Norwegian Central Bank recently said that core inflation of 2.7 per cent year-on-year in June had been "somewhat higher" than the bank had projected, so the bank raised interest rates slightly. The phasing-in of revenue from the state Petroleum Fund through additional spending and reduced taxation is expected to stimulate consumer spending. As this change is implemented, the central government's non-oil deficit is expected to rise from 2 per cent of mainland GDP in 2001 to 5.5 per cent by 2010, increasing mainland GDP by 0.4 per cent annually.

Norway has a small industrial base apart from its oil and gas, shipping, and fishing industries, and its mainland (*i.e.* excluding oil and natural gas) economy is forecast to grow by 1.2 per cent in 2002. Manufacturing activity was up 1.4 per cent year-on-year for the second quarter of 2002. Norway's government is concerned about its economic welfare once its oil runs out, as is predicted by the end of the first half of the 21st century. Norway makes annual contributions to its Petroleum Fund, a financial safety net for the time when oil revenues decline (and a means of reducing the inflationary impact of

oil revenues). The government was able to pay Norwegian krone (Nkr) 53.5 billion (about $7.1 billion) into the Petroleum Fund in the second quarter of 2002, for a total value of Nkr 605.4 billion.

A new center-right coalition took power in October 2001 after the Labor Party lost seats in the parliamentary election. The coalition consists of Prime Minister Kjell Magne Bondevik's Christian People's Party, the Conservative Party, and the Liberal Party. The government has sought to lessen government involvement in business and to lower taxes, though it remains quite involved in social and environmental policy. The government currently does not have plans to seek membership in the European Union.

Norway is part of the European Economic Area (EEA), but Norwegians have voted in two referenda against joining the European Union (EU). Recent polls have shown some increase in support for joining the EU. Norway has a history of state control over major industry, but this is beginning to change. Norway's reliance on oil revenues in the past resulted in a government preference for keeping Norwegian businesses under Norwegian control.

North Sea Oil and Natural Gas

North Sea oil and natural gas were first discovered in the 1960s. The North Sea did not emerge immediately as a key non-OPEC oil producing area. North Sea production grew as major discoveries continued throughout the 1980s and into the 1990s. Although the region is a relatively high cost oil producer (breakeven is about $12-$14 per barrel, vs. $3-$4 per barrel in Iran, for example), its political stability and proximity to major European consumer markets have allowed it to play a major role in world oil and gas markets.

Many of the world's major crude oil prices are linked to the price of the North Sea's Brent crude oil–about $150 billion in annual petroleum trade. Brent crude is a blend of North Sea crude oils and does not come exclusively from the Brent field. Because Brent crude is traded on the International Petroleum Exchange in London, fluctuations in the market are reflected in the price of Brent. Therefore, all other crude oils linked to Brent can be priced according to the latest market conditions. Brent production is forecast to fall precipitously from its current 400,000 barrels per day (bbl/d) by 2005, making the Brent price marker increasingly dated. Liquidity has fallen to about 10 cargoes per delivery month compared with 300-400 deals per month in the early 1990s. In response to this, pricing service Platts made a change effective July 10, 2002 allowing for substitution–at seller's option–of UK Forties and Norwegian Oseberg for Brent in an attempt to increase potential volumes and reduce volatility resulting from traders "cornering the market." The change has not been universally accepted, and it remains to be seen whether it will be successful. The International Petroleum Exchange (IPE), which runs the Brent futures market, appears to be waiting to see whether the over-the-counter market adopts Brent-Forties-Oseberg (BFO). The first full BFO contract was sold on August 8.

The late 1997-1998 oil price collapse had an adverse effect on North Sea production. In 1997 and 1998, North Sea oil production remained stable, whereas previous years had shown average annual increases of 400,000 bbl/d. The 1999-2000 oil price increase had the opposite effect: North Sea oil and gas production reached new heights in 2000, with oil production exceeding 6 bbl/d for the first time. However, the North Sea area is considered to be increasingly "mature," with few additional large discoveries likely to be made. Some predict that the North Sea will reach peak production of about 7 million bbl/d in the next two or three years, although technology developments could delay this. The average recovery rate for Norwegian fields is expected to eventually reach 44 per cent. Because the North Sea is believed to be nearing its peak production, in both of the major North Sea producing nations, Norway and the United Kingdom (UK), government and industry are taking steps to restructure their oil and natural gas sectors to make them more internationally competitive and also are increasing cooperation between the two countries. On August 28, 2002, Norway and the UK released a joint plan to increase cooperation, cut costs, and raise output, especially on aging fields. However, taxation rates will remain unharmonized. Norway also signed a cooperation agreement with Russia that same day that opens energy dialogue on the Arctic Barents Sea shared by the two countries.

Oil

Norway has proven oil reserves of 9.44 billion barrels. In 2001, Norway was the world's third largest net oil exporter. Norway consumes very little of the oil it produces, and its oil exports are the country's greatest source of revenue. Norway's oil reserves are located exclusively offshore and mostly in the North Sea, with smaller deposits in the Norwegian Sea. The Barents Sea also is being explored. Oil production was about 3.4 million bbl/d in 2001, an increase of about 100,000 bbl/d over 2000. Production in the first half of 2002 was affected by a production cut agreement with OPEC.

In November 2001, Norway's Energy and Oil Minister announced that Norway would cooperate with OPEC and cut crude oil production for the first half of 2002 in an effort to shore up prices in the face of sagging demand. Norway later agreed to cut production by 150,000 bbl/d, with target production at 3.02 million bbl/d. Rather than cutting production steadily across the period, Norwegian production cuts were concentrated in the last month of each quarter, *i.e.*, March and June. A preliminary estimate of crude oil production for the first half of 2002 is 3.06 million barrels per day. In June, it was announced by the Oil and Energy Ministry that "The Norwegian government has decided not to extend the restriction on oil production into the second half of 2002." By this time, Brent prices were some $5 per barrel higher than they had been in November 2001. In late June, Norway informed operators that they could produce at 13 per cent above field production limits in an effort to reach a government target of 3.02 million bbl/d of crude oil for 2002.

Oil service workers struck from July 5 until August 10, when the Norwegian oil industry association (OLF) and oil union Nopef arrived at a new agreement covering 3500 employees in oil service companies The oil industry reportedly suffered a loss of more than Nkr 330 million ($42.2 million) during the strike. A strike began September 10, 2002 by dock workers at Statoil's 205,000-barrels-per-day refinery in Mongstad, Norway, that would curb exports from one of Europe's key gasoline-producing plants. The strike could curb crude production from Norsk Hydro's Troll B and C platforms, which send their output to the terminal at the Mongstad refinery via pipelines.

Oil Sector Restructuring

The Norwegian oil sector has undergone massive restructuring during the past few years. Norway's oil sector had been characterized by extensive public ownership. Its largest oil company, Statoil, was 100 per cent state-owned, while Norsk Hydro, the second largest oil company, was majority state-owned. The only completely privately-held company was Saga Petroleum. In addition, state-owned Statoil managed another entity even larger than itself, the State Direct Financial Interest (SDFI), which represented the state's holdings in 150 offshore oil and natural gas fields and about 40 per cent of total production.

In late 1999, Norsk Hydro completed its acquisition of Saga, reducing its public ownership, originally 51 per cent, to 44 per cent. In April 2001, the Norwegian parliament approved plans to sell between 10 per cent and 25 per cent of Statoil to private investors and to sell 15 per cent of the SDFI to Statoil prior to Statoil's listing on the New York and Oslo stock exchanges. Norsk Hydro (taking the largest share) and eight other Norwegian North Sea operators were sold another 6.5 per cent of the SDFI in March 2002. The remainder of the SDFI (78.5 per cent) was reorganized into a new state company called Petoro. Petoro is the world's fifth largest oil and gas firm in terms of production, with estimated production of 1.4 million bbl/d of oil, though Petoro functions entirely as a management company, having no operations itself. Statoil completed its purchase of 15 per cent of the SDFI in May 2001 for $4.24 billion, and on June 18, 2001, Norway sold 17.5 per cent of its holding in Statoil in an initial public offering for $2.9 billion. These changes should introduce more efficiency into the system, as Statoil was uncompensated for managing the SDFI, and raise more capital for Statoil in order for it to compete globally as the company explores regions such as offshore west Africa and Venezuela.

Norsk Hydro sold two production licenses to Marathon Oil of the United States in July. There is speculation that Norsk Hydro may spin off its oil unit to focus on its aluminum and fertilizers businesses. Statoil is the most likely buyer, which would create a company with production approaching 1 million bbl/d of crude, condensate, and natural gas liquids. Statoil announced in May 2002 that it is selling its 7,000 bbl/d assets in the Danish North Sea to Denmark state oil company DONG for $127 million in order to concentrate on core areas.

Oil Production

Norway's major Norwegian North Sea production areas include: Ekofisk, Sleipner, Frigg, Statfjord, and Oseberg and Troll. There are also five fields producing in the Norwegian Sea. (The 62nd line of latitude separates the North Sea and the Norwegian Sea.)

Norwegian oil investment was about Nkr 56.9 billion ($7.5 billion) in 2001, an increase from the $6.2 billion invested in 2000, but down from the peak of NKr 80 billion ($10.6 billion) in 1998. Investment levels reflect expectations that Norway's oil production will remain roughly constant until 2004, and then begin a gradual decline. Oil fields and projects under development include: Fram West, Grane, Tune, and the Valhall Flanks and water injection. Three new offshore oil fields came on stream in the second half of 2001: Tambar, Glitne, and Huldra. Some important oil discoveries offshore Norway in the past 12 months include: Staerne, near the Norne field, with estimated reserves of 30 million barrels; increased reserves in the Oseberg unit; and additional oil at the Goliat continental oil shelf in the Barents Sea (estimates of 75-107 million barrels increased to 91-250 million barrels). Overall, about 250 million barrels of oil and condensate were added to Norwegian reserves in 2001. A total of 32 blocks were offered at the 17th Norwegian Continental Shelf licensing round in June 2002. Eleven companies will share blocks that comprise six production licenses. This round focused on the Norwegian Sea.

Ekofisk, in the southern North Sea sector, was the first North Sea oil field to be discovered, in the late 1960s, and developed, with production beginning in 1971. Since 1975, oil has been piped from Ekofisk to the UK (Teesside, England). There are currently 29 platforms installed in the area, some of which are in the British North Sea. The most recent phase of development began in 1994, when the Phillips group (the U.S. company that leads the Ekofisk operating consortium, which includes TotalFinaElf, Norsk Agip, Norsk Hydro, and Statoil) installed two new platforms at "Ekofisk II". Ekofisk II came onstream in August 1998. The Phillips license runs through 2028. In December 2001, it was decided by the government that Phillips would remove 14 of the 29 Ekofisk platforms between 2003 and 2018, at an estimated cost of $1 billion (NKr 8 billion). About 10 per cent of the removal cost will be paid by Phillips, 72 per cent by the Norwegian government, and the remainder will be paid by the other members of the consortium. Phillips plans to bring the steel structures and the topside of the concrete Ekofisk tank ashore for recycling, to leave the rest of the concrete tank and barrier wall in place, and also to leave about 150 miles of pipelines buried. Ekofisk's production (including Eldfisk, Embla, and Tor) is expected to be about 381,000 barrels per day of crude oil in 2002. The Valhall field's production continues to decline, with expected production in 2002 at 72,000 bbl/d. However, the recently approved Valhall water injection and the Valhall flanks should improve recovery from the field. The Yme field has ceased production.

Sleipner West was discovered in 1974, but Sleipner East went into production first, in 1993. Sleipner West is tied back into Sleipner East, and the fields share the same operations organization. Sleipner is mostly important for natural gas production, including liquids and condensate (2002 condensate production in East and West is estimated at 3.7 million cubic meters), but the Varg field is estimated to produce 8,300 bbl/d crude oil in 2002. Varg is scheduled to cease production within the next few years.

Moving to the northern North Sea sector, the Frigg-Heimdal area is also mostly important as a natural gas producing area, though the Balder and Jotun fields together are expected to produce about 124,000 bbl/d of crude oil in 2002. Balder was proven as early as 1967, though production did not commence until 1999. Shuttle tankers are loaded from a production ship tied to subsea-completed walls. Several structures close to Balder are being developed by Ringhorne platform. Jotun also commenced production in 1999, from a floating production, storage, and offloading vessel (FPSO) that is serviced by shuttle tankers.

The Statfjord area is one of the largest oil producing areas in the North Sea. The Statfjord field itself was discovered by Mobil in 1974, and it extends into the British North Sea. Production began from Statfjord A in 1979, from Statfjord B in 1982, and from Statfjord C in 1985. Production from the Statfjord North and Stafjord East subsea installations are tied back into Stafjord C. Statoil took over the operations from Mobil in 1987. Three large concrete platforms with storage cells have been installed on Statfjord. Britain's 14.5 per cent share goes by pipeline via the Brent field to Scotland. Statfjord's production has exceeded the most optimistic expectations, but all Statfjord fields are now in decline. Norway's share of Statfjord crude oil production in 2002 (including North and East) is expected to be 205,000 bbl/d. Statfjord should continue producing until 2020.

The Snorre field, with production rising, has become the largest single field in the area, with 2002 production estimated to be 228,000 barrels per day. It was discovered in 1979, and production commenced in 1992. Norway's third largest field is Gullfaks, which, including West and South, is expected to produce 223,000 bbl/d in 2002. Gullfaks (including West) has declined by over 50 per cent since its peak in 1995, but Gullfaks South (including Rimfaks and Gullveig) has had increasing production since it came online in 1998, to 70,000 bbl/d expected for 2002. Vigdis continues to decline from its peak in 1999, but Visund, which is east of Snorre, has had its production increase, with 2002 expected to be 43,000 bbl/d.

The various Oseberg fields (Oseberg, East, South, West) together are the largest oil producing fields in their area, whereas Troll is the largest gas field in the area. Oseberg began production in 1988, and peaked at about 500,000 bbl/d in 1996, and has declined since to about 176,000 bbl/d (including West), far below the capacity of the three platforms there. The surrounding East and South Oseberg fields have come online in 1999 and 2000, respectively, supplementing the declining production at Oseberg with 130,000 bbl/d expected for 2002. Both East and South peaked in 2001. There is a pipeline from Oseberg to the Sture terminal on the Norwegian coast, with tie-backs from East and South to Oseberg. A thin layer of oil underlies the entire Troll field, but it is only sufficiently thick for commercial recovery in the Troll West region. This is where Troll Phase II is expected to produce 316,000 bbl/d in 2002–production has been relatively flat since 2000, though Troll achieved a daily record of 440,000 bbl in May 2002. There is a pipeline from Troll West to the Mongstad crude oil terminal on the Norwegian coast.

The Norwegian Sea has seen production increase at a higher rate than North Sea production in recent years, though it is in an earlier stage of development, the first field having come on stream in 1993. Total production for the area for 2002 is predicted to be 725,000 bbl/d. Much of the increase comes from the new Asgard field, which went into production in 1999, and now produces about 148,000 bbl/d. Norne's production also increased in 2001, though a slight decline is predicted for 2002. Heidrun's production has declined to less than that of the Norne field. Draugen's production has been flat in the past two years, but it is still has the highest production at about 200,000 bbl/d. Shuttle tankers are used to take oil from the platforms or production ships, as there is currently not an oil pipeline from the Norwegian Sea.

Natural Gas

Norway holds 44 trillion cubic feet (Tcf) of natural gas reserves. Norway is not a major natural gas consumer, although its consumption is expected to increase in coming years as natural gas-fired power plants come online. It is estimated that just 16 per cent of Norway's gas reserves have been extracted since output began, though Norway produced more gas than it discovered for the first time in 2001, as the increase in reserves was between 700-900 billion cubic feet (Bcf). Natural gas accounts for about 60 per cent of Norway's overall offshore hydrocarbon reserves and is expected to account for an increasing portion of Norway's energy exports. Norway continues to be the second-largest natural gas exporter in Europe, with its growth outpacing that of Europe's largest natural gas exporter, Russia. Exports are forecast to be between 1.9-2.3 Tcf in 2002. Export volumes peaked at about 6.7 Bcf per day in the second quarter of this year, but will have to decline slightly if the forecast is correct. Norway's sub-sea natural gas export network is capable of transporting about 3 Tcf per year.

Natural Gas Exports

Norway, as a member of the European Economic Area (EEA), is bound by certain EU economic directives, and the EU recently has forced Norway to restructure the way it sells natural gas to other European countries. Prior to June 1, 2001, all Norwegian gas was sold through the Gassforhandlingsutvalget (GFU, meaning Gas sales negotiating committee). Although ownership of Norway's gas is held by many different companies and formerly the SDFI, now Petoro, the GFU consisted of just Norsk Hydro and Statoil. The GFU would set a price for all Norwegian gas available for purchase, instead of letting the various producers compete against each other. The EU claimed that this violated fair trading practices and threatened Statoil and Norsk Hydro with huge fines. In July 2002, Norway and the European Commission resolved this long-running dispute over the legality of long-term contracts negotiated by the defunct Gas Sales Negotiating Committee (GFU). The European Commission had threatened to take legal action against Statoil and Norsk Hydro because long-term contracts already in place that account for about 20 per cent of western Europe's gas imports were negotiated by means of the GFU and because many of these contracts have destination clauses (prohibition of resale). Under the negotiated deal, the Commission relented on its demand to have the long-term contracts revised in return for Statoil and Norsk Hydro agreeing to sell 530 Bcf over a four-year period to new European customers (customers without GFU-era contracts). In May 2002, most natural gas exporting companies agreed to coordinate ownership of their pipeline assets through shares in the new government-backed Gas-Led company. The state-owned company Gassco is the operator on all of Norway's natural gas pipelines (since January 1, 2002), as the partial privatization of the former operator, Statoil, created a conflict of interest.

The effects of all these changes are yet to be seen, though the expectation is that the price of Norwegian natural gas will be reduced, at least in the short to medium run. A major constraint for upstream gas companies competing for sales in the newly deregulated market will be limited infrastructure to take the gas out, because various companies share the same pipeline. Norwegian gas arrives in Europe through the following trunklines: the Europipe I and Statpipe/Norpipe systems to Germany; the Zeepipe trunkline to Zeebrugge in Belgium; the NorFra line to Dunkerque in northern France; and the Europipe II line from Kårstø north of Stavanger to Emden. These Norwegian trunklines provide a combined gas transport capacity of 2.7 Tcf per year. There are also pipelines to the UK, including the new Vesterled pipeline, which opened in October 2001, with volumes at about 138 million cubic feet per day. Marathon is exploring the potential demand for its proposed Symphony natural gas pipeline, which would bring additional Norwegian natural gas to the UK through a link between the Heimdal complex and the Brae/Miller complex in the UK sector.

Statoil expects Norway's share of natural gas deliveries to continental Europe to rise from 14 per cent in 1996 to 20 per cent by 2005. The following companies currently buy Norwegian gas: Ruhrgas, BEB, Meeg, Thyssengas and Verbundnetz Gas (Germany), Gaz de France (France), Gasunie, SEP (the Netherlands), Distrigaz (Belgium), Enagas (Spain), Austria Ferngas, OMV (Austria), Snam (Italy), Energia (Italy), Polish Oil and Gas Company (Poland), Transgas (Czech Republic), and Centrica (UK). Germany is the largest natural gas market in continental Europe, and about 20 per cent of the gas that Germany currently consumes comes from Norway. Ruhrgas expects Norway to supply 30 per cent of Germany's imports. About half of the gas from the NorFra line transits through France to points in

Italy and Spain, while the other half is consumed in France. By 2005, this pipeline is expected to supply one-third of France's total gas consumption.

In July 2001, Stoltenberg and Polish Prime Minister Jerzy Buzak signed a joint declaration for the deliveries of 177 billion cubic feet (Bcf) of natural gas from Norway annually. Existing Polish infrastructure cannot support significant imports from non-Russian sources, so a pipeline across the Baltic through either Sweden or Denmark was being planned, but it now appears unlikely that a natural gas pipeline to Poland will be built because of insufficient demand volumes. There is a competing plan to import liquefied natural gas (LNG) from Norway to a planned LNG terminal on Poland's Baltic Coast. Norway began piping a relatively small amount of gas through Germany in October 2000, based on an earlier contract signed in May 1999, for the delivery of 17.7 Bcf annually, under an agreement between Germany's Ruhrgas and Verbundnetz Gas and Poland's state-held gas monopoly.

The United Kingdom, the largest natural gas market in Europe, will also soon become an importer of Norwegian gas again. Norway had once supplied up to a quarter of British demand in the 1980s, but this dwindled as the Frigg field that supplied the gas was depleted. Vesterled will connect the existing Frigg pipeline with the Heimdale platform, which is already connected by pipeline to the Sleipner gasfields, and from there to other areas of the Norwegian North Sea such as the Ormen Lange gasfield that is scheduled to come on stream in 2006. In July 2001, BP announced a 15-year contract to buy 56.5 Bcf natural gas per year from Statoil. In June 2002, Centrica of the UK signed a 10-year deal with Statoil for the purchase of 483.5 million cubic feet per day, with prices linked to natural gas rather than oil.

Natural Gas Production

The Troll field (East and West) contains over half of Norwegian natural gas reserves and, as Norway's largest natural gas field, Troll produces over 72 Bcf per month. It has a production capacity of 100 million cubic meters (3.5 Bcf) per day, and estimated production in 2002 is expected to be 22.8 billion cubic meters (805 Bcf). The Troll Gas development Phase I in Troll East comprises the Troll A platform, the gas treatment plant at Kollsnes near Bergen, and pipelines linking these two installations. Troll East is thought to contain two-third's of Troll's natural gas reserves. Phase III (under development) will extract gas from Troll West. Troll A is the tallest structure ever moved by humans. Its concrete gravity base structure has been built for a lifetime of 70 years. The division of roles on the field has been controversial. Currently, Statoil and the new Petoro have about three-quarters of the shares and Statoil is the operator, followed by Norsk Hydro, Shell, TotalFinaElf, and Conoco. The gas is taken by the Zeepipe to Zeebrugge and Statpipe/Norpipe to Emden.

Troll is not the only active natural gas field in Norway's North Sea. Gas sales began in 1977 from Ekofisk and Frigg. Ekofisk, in the southern North Sea sector, supplies Ruhrgas, Gaz de France, Gasunie and Distrigaz. Ekofisk has declined from its peak in the late 1970s and a production spike in the 1990s, though it is still expected to produce 5.95 billion cubic meters (210 Bcf) in 2002. Frigg production is sold to British Gas, though Frigg has declined to the point that production is expected to cease sometime this year. Nearby Heimdal's declining production is also set to cease this year. Agreements on selling gas from Statfjord, Gullfaks and Heimdal were signed in 1981 and deliveries began in 1985 to Ruhrgas, BEB, Thyssengas, Gaz de France, Gasunie, Distrigaz, Elf and Meeg. Remaining commitments under these deals average six billion cubic meters per year (212 Bcf). Sleipner, East and West, is expected to produce 13.6 billion cubic meters (479 Bcf) in 2002; this gas is currently sold under the Troll gas sales agreements. Though Sleipner East is declining, most natural gas production is from Sleipner West, which continues to have sharply increasing production. The Norwegian share of gas from the field is piped through the Statpipe/Norpipe system to Emden in Germany via Kårstø, north of Stavanger.

Huldra commenced production with an unmanned platform in November 2001, with natural gas production steadily rising and already at about 350 million cubic feet per day (total expected production 3.19 billion cubic meters or 113 Bcf for 2002). Huldra also produces condensate and about 28,000 bbl/d of crude oil. The crude and condensate are piped to Veslefrikk B, and the gas is piped to Heimdal.

The Åsgard field on the Halten Bank in the Norwegian Sea is one of Norway's most important new projects. The field has been developed as a chain of four interconnected projects: development of Åsgard itself, construction of the Åsgard Transport gas trunkline from the field to the Kårstø gas treatment plant north of Stavanger, the Kårstø development project, and the Europipe II gas trunkline from Kårstø to Dornum in northern Germany. Gas production from the floating platform began in October 2000, and is expected to be 8.9 billion cubic meters (314 Bcf) in 2001. Statoil is the operator of the project, which is one of Norway's giant offshore developments, on par with Ekofisk and Troll. Subsea production installations in the field are planned to be the most extensive in the world, embracing a total of 51 wells grouped in 17 seabed templates. It will link the Halten Bank area to Norway's gas transport system in the North Sea.

Statoil now is developing the Halten Bank South area of the Norwegian Sea, having taken over as operator in January 2000 (Saga had been the operator). Recoverable reserves of the Halten Bank South fields are estimated at 140 billion cubic meters (almost 5 Tcf) of gas and about 440 million barrels of oil and condensate–on par with Åsgard. The Kristin field of the Halten Bank has already secured sales of up to 31 billion cubic meters (1.1 Tcf) from 2005 to 2016. ExxonMobil made the largest discovery of 2000 in this area, the Bella Donna field, with estimated reserves between 60 and 125 billion cubic meters (2.1-4.4 Tcf).

In March 2002, the Norwegian parliament approved Statoil's plans to develop the $5 billion Snohvit project. If it is completed, Snohvit will be the largest sub-sea liquefied natural gas (LNG) project in the world, as well as the most northerly as it is located in the Barents Sea. Approximately 201 Bcf per year of natural gas would be piped to the coast, liquefied, and transported to customers in Spain and the United States by means of four carriers. In June 2002, El Paso of the United States, announced that it had final Norwegian and U.S. government approval for its plans to import 1.8 million metric tons of LNG to the United States from Snohvit. This is over 40 per cent of the project's capacity, and much of the LNG may be delivered to El Paso's Cove Point, Maryland regasification facility. Construction of Snohvit restarted in June as well.

The huge Ormen Lange field in the Norwegian Sea, Norway's second largest natural gas discovery with estimated reserves of of 14.1 Tcf, has its blocks divided into three production licenses, with the unusual characteristic that Statoil/SDFI has only a 30 per cent share of one of the licenses, such that non-Norwegian companies are the majority owners of one of the licenses. Norsk Hydro is the operator in the development phase, and Shell will be the operator in the production phase. Gas production is planned to commence in 2007.

Coal

Norway's coal production occurs on Spitsbergen of the Svalbard Islands, off the country's northern coast. This island also has Norway's only coal-fired power plant. In December 2001, the Norwegian Parliament voted to give a $16.9 million subsidy to state-owned coal monopoly Store Norske Sisbergen

Kulkompani. Mining in Svalbard will be expanded and eventually, coal exports are planned. However, Norway is currently a net importer of coal, though overall consumption is small at 1.57 million short tons in 2000.

Electricity

In 2000, 99 per cent of Norway's electricity generation came from its 27 million kilowatts of installed hydroelectric capacity. Norway has one of the highest rates of per-capita consumption of electricity in the world. In December 2001, state-owned Norwegian electricity company Statkraft purchased independent electricity company Trondheim Energiverk for $483 million. This makes Statkraft the second-largest electricity supplier in Scandinavia and gives the company over 50 per cent of the Norwegian electricity market. Prime Minister Stoltenberg declared, in January 2001, that "the era of large-scale new hydropower is over" and that several large new hydro projects are to be abandoned, including Beiarn, Bjollaga, and Melfjord. A new hydro plant with greater capacity is being constructed to replace the exisiting one at Tyin.

Norway is planning to construct three new natural gas-fired power plants. Construction of two natural gas-fired power plants by Naturkraft appears set to go ahead sometime this year. Naturkraft recently asked the government to extend its license to build these plants beyond 2004. This issue, which has not been completely resolved, is extremely important in Norway, as Prime Minister Bondevik's previous term of office ended in a vote of no confidence that overrode his opposition to the plants. Industrikraft Midt-Norge also plans to build a natural gas-fired plant, and has an allowance to emit 2.2 million tones of carbon dioxide per year. This 2X400 gas-fired combined heat and power plant in Skogn, central Norway is slated to begin construction in 2002. U.S.-based Mirant has bought 40 per cent of five-member industrial consortium IMN, which will build, operate, and own the plant.

Norway has had a surplus of hydroelectric power in the past two years, but in drier years it must import electricity. In January 2001, E.On of Germany, Statkraft, and Elsam of Denmark agreed to free up capacity on key power cables linking the high tension electricity grids of Scandinavian countries to Germany, including the Skaggerrak cable, the only cable connecting western Denmark and Norway.

In May 2002, the European Free Trade Area (of which Norway is a member) informed the government that industry's exemption from taxation on electricity cannot continue. Consumers currently pay a 9 per cent tax on electricity.

Environment

Norway is a proponent of "green power" from renewable sources and has made efforts to make its oil sector as environmentally friendly as possible. Under its Kyoto Protocol commitment, Norway has agreed to limit its carbon emissions to a 1 per cent increase from 1990 levels by the 2008-2012 commitment period. In a dual effort to meet its Kyoto target and to further develop technologies to make oil and gas production less environmentally damaging, Norway has been a leader in alternatives for reducing carbon emissions. As a result of high activity in the oil and gas extraction sectors, Norway is relatively more energy-intensive than most OECD countries, and possesses one of the highest per capita energy consumption levels in the world. Air pollution in Oslo is not as severe as in other major world cities.

Country Overview

Independence: October 26, 1905 (from Sweden)

Population (2001E): 4.5 million

Location/Size: Northern Europe, bordering the North Sea and the North Atlantic Ocean, west of Sweden/123,843 square miles (slightly larger than New Mexico)

Capital City: Oslo

Language: Norwegian (small Lapp- and Finnish-speaking minorities)

Ethnic Groups: Germanic (Nordic, Alpine, Baltic), Lapps (Sami) 20,000

Religions: Evangelical Lutheran 87.8 per cent (state church), other Protestant and Roman Catholic 3.8 per cent, none 3.2 per cent, unknown 5.2 per cent

Defense (8/98): Army, 28,900; Navy, 6,100; Air Force, 6,700 (including 16,500 conscripts)

Economic Overview

Currency: Norwegian Krone (NKr)

Exchange Rate (9/09/02): 1 US Dollar = 7.52 Kroner

Gross Domestic Product (GDP, 2001E): $163.7 billion

Real GDP Growth Rate (2001E): 1.4 per cent *(2002F):* 2.3 per cent

Inflation Rate (consumer prices, 2001E): 3.0 per cent *(2002F):* 1.5 per cent

Unemployment Rate (2001E): 3.6 per cent *(2002F):* 4 per cent

Merchandise Exports (2001E): $58.6 billion

Merchandise Imports (2001E): $33.6 billion

Merchandise Trade Surplus (2001E): $25 billion

*Major Trading Partners:*UK, Germany, Sweden, Denmark, United States, other EU members

Major Exports: Fuels and other energy products; food and beverages; manufactured materials

Major Imports: Machinery and transport equipment, miscellaneous manufactures, food, beverages, and tobacco

Energy Overview

Proven Oil Reserves (1/1/02E): 9.4 billion barrels

Oil Production (2001E): 3.4 million barrels per day (bbl/d), of which 3.1 million bbl/d was crude oil

Oil Consumption (2001E): 0.2 million bbl/d

Net Oil Exports (2001E): 3.3 million bbl/d

Crude Oil Refining Capacity (1/1/02E): 310,000 bbl/d

Natural Gas Reserves (1/1/02E): 44 trillion cubic feet (Tcf)

Natural Gas Production (2000E): 1.81 Tcf

Natural Gas Consumption (2000E): 0.087 Tcf

Net Natural Gas Exports (2000E): 1.7 Tcf

Electrical Generation Capacity (1/1/00E): 27.2 gigawatts

Electricity Generation (2000E): 141 billion kilowatthours (bkwh)

Electricity Consumption (2000E): 112 bkwh

Recoverable Coal Reserves (12/31/99E): 1 million short tons (Mmst)

Coal Production (2000E): 0.55 Mmst

Coal Consumption (2000E): 1.57 Mmst

Major Systems: Statfjord, Oseberg, Gullfaks, Ekofisk

Major Companies: BP, ConocoPhillips, ExxonMobil, TotalFinaElf, Norsk Hydro, Shell, Statoil, Chevron, Petoro

Environmental Overview

Total Energy Consumption (2000E): 1.8 quadrillion Btu* (0.5 per cent of world total energy consumption)

Energy-Related Carbon Emissions (2000E): 10.3 million metric tons of carbon (0.2 per cent of world total carbon emissions)

Per Capita Energy Consumption (2000E): 399.6 million Btu (vs. U.S. value of 348.9 million Btu)

Per Capita Carbon Emissions (2000E): 2.3 metric tons of carbon (vs. U.S. value of 5.7 metric tons of carbon)

Energy Intensity (2000E): 10,619 Btu/$1995 (vs U.S. value of 10,390 Btu/$1996)**

Carbon Intensity (2000E): 0.06 metric tons of carbon/thousand $1995 (vs U.S. value of 0.17 metric tons/thousand $1996)

Sectoral Share of Energy Consumption (1998E): Industrial (52.1 per cent), Residential (21.7 per cent), Transportation (13.1 per cent), Commercial (13.1 per cent)

Sectoral Share of Carbon Emissions (1998E): Industrial (57.0 per cent), Transportation (37.9 per cent), Residential (2.6 per cent), Commercial (2.5 per cent)

Fuel Share of Energy Consumption (2000E): Oil (21.9 per cent), Natural Gas (5.3 per cent), Coal (2.2 per cent)

Fuel Share of Carbon Emissions (2000E): Oil (72.7.1 per cent), Natural Gas (16.7 per cent), Coal (10.6 per cent)

Renewable Energy Consumption (1998E): 1,248 trillion Btu* (5 per cent increase from 1997)

Number of People per Motor Vehicle (1998): 2.0 (vs. U.S. value of 1.3)

Status in Climate Change Negotiations: Annex I country under the United Nations Framework Convention on Climate Change (July 9th, 1993). Signatory to the Kyoto Protocol (signed April 29th, 1998- not yet ratified). Under the Protocol, Norway has agreed to a 1 per cent increase from 1990 emissions levels of a basket of greenhouse gases.

Major Environmental Issues: Water pollution; acid rain damaging forests and adversely affecting lakes, threatening fish stocks; air pollution from vehicle emissions.

Major International Environmental Agreements: A party to Conventions on Air Pollution, Air Pollution-Nitrogen Oxides, Air Pollution-Sulphur 85, Air Pollution-Sulphur 94, Air Pollution-Volatile Organic Compounds, Antarctic-Environmental Protocol, Antarctic Treaty, Biodiversity, Climate Change,

Desertification, Endangered Species, Environmental Modification, Hazardous Wastes, Law of the Sea, Marine Dumping, Nuclear Test Ban, Ozone Layer Protection, Ship Pollution, Tropical Timber 83, Tropical Timber 94, Wetlands and Whaling. Has signed, but not ratified: Air Pollution-Persistent Organic Pollutants.

The total energy consumption statistic includes petroleum, dry natural gas, coal, net hydro, nuclear, geothermal, solar, wind, wood and waste electric power. The renewable energy consumption statistic is based on International Energy Agency (IEA) data and includes hydropower, solar, wind, tide, geothermal, solid biomass and animal products, biomass gas and liquids, industrial and municipal wastes.

59

Oman

With relatively modest oil reserves, Oman is important to world oil markets due to its strategic location overlooking the Strait of Hormuz. Oman also opened a facility for the export of liquefied natural gas (LNG) in April 2000, which is undergoing expansion.

General Background

Oman remains heavily dependent on oil revenues, which account for nearly 80 per cent of the country's export earnings and 40 per cent of gross domestic product (GDP). Real GDP growth was 3.3 per cent for 2001 and is projected at 3.0 per cent for 2002. Inflation is mild, projected at 0.9 per cent for 2002.

Oman has made privatization and diversification of its economy top policy priorities. Expanded utilization of natural gas is central to Omani diversification plans, for export as well as for domestic use. Natural gas projects include a liquefied natural gas (LNG) project in Sur, which began exports in April 2000, an aluminum smelter and petrochemicals plant in Sohar, and a urea fertilizer plant in Sur. These projects are intended not only to help alleviate Oman's dependence on oil, but also reduce dependence on governmental spending and employment.

In recent years, Oman also has put forth great efforts to attract foreign investments, particularly in light industry, tourism, and electric power generation. Foreign investment incentives include a 5-year tax holiday for companies in certain industries, an income tax reduction for publicly held companies with at least 51 per cent Omani ownership, and soft loans to finance new and existing projects. Oman became a member of the World Trade Organization (WTO) in October 2000.

Constitutional reforms in Oman have been part of an ongoing process of modernization. The Basic Law, announced in late 1996, institutionalized a new structure for the country's political institutions, creating an upper legislative chamber, the Majlis al-Dawla (Council of State), alongside

the Majlis al-Shura, for which elections were held in 1997. The two chambers now form the Council of Oman.

Oil

In many ways, Oman is atypical of Persian Gulf oil producers. Oil was not discovered in commercial quantities until 1962–decades after most of Oman's neighbors. Oman's oil fields also are generally smaller, more widely scattered, less productive, and more costly per barrel than in other Persian Gulf countries. The average well in Oman produces about one-tenth the volume per well compared to neighboring countries. Oman continues to use a variety of enhanced oil recovery (EOR) techniques in order to minimize the costs of exploration and further development at new and existing oil fields. Using these technologies, Oman has succeeded in bringing down the cost of oil production to $3 per barrel in some fields and $4 per barrel in others–but these figures, while still low by world standards, are substantially above most other Persian Gulf oil fields. Oman is not a member of the Organization of Petroleum Exporting Countries (OPEC) or the Organization of Arab Petroleum Exporting Countries (OAPEC).

Most of Oman's 5.5 billion barrels in proven oil reserves are located in the country's northern and central regions. In the North, the Yibal, Natih, Fahud, al-Huwaisah and Lekhwair fields combined account for almost half of total Omani oil production. Yibal, which produces around 180,000 bbl/d, is the largest oil field in the country. Crude oil found in this region is mainly medium or light, with gravities in the 32°-39° API range. Northern oil is mostly found along with natural gas. Heavier oil is found in southern Oman, particularly in the Nimr and Amal fields, with gravities averaging 20° API, and normally not associated with natural gas. Oman's main oil export blend is a medium sour crude.

Petroleum Development Oman (PDO), the country's second-largest employer after the government, holds over 90 per cent of the country's oil reserves, and accounts for about 94 per cent of production. PDO is a consortium comprised of the Omani government (60 per cent), Shell (34 per cent), Total (4 per cent), and Partex (2 per cent). However, Shell operates most of Oman's key fields, including Yibal and Lekhwair. As part of a strategy to increase its oil reserves, PDO has set out to develop additional exploration and recovery techniques. PDO' s stated aim is to double its average recovery rate to 50 per cent (a goal known as "Target 50") by investing over $1 billion in enhanced recovery projects over the next five years, but it remains to be seen if this will be possible. Use of enhanced recovery technologies also increases production costs substantially. Despite these efforts, Oman's crude oil production has fallen in the first half of 2002, averaging 918,425 bbl/d, down from an average of 959,816 bbl/d for 2001. Oman had pledged a 40,000 bbl/d cut in output effective January 1, 2002 in support of OPEC production cuts, though the decline may also reflect declining production at mature fields.

Yibal, discovered in 1962, is Oman's largest producing oil field, supplying around one-quarter of PDO's total production. In 1986, the field's output was boosted from 120,000 bbl/d to more than 140,000 bbl/d with the installation of water injection facilities. Production was increased further following the completion of a $200-million development project, called Yibal Shusiba Phase II, in 1994. The project involved drilling 96 wells, mostly horizontal, and modifications to production stations B, C, and D, which included the installation of gas injection facilities. Yibal currently produces around 180,000 bbl/d. Oman's second largest oil field, Nimr, was discovered in 1980 and is located in the southern part of the country. Nimr currently produces about 178,000 bbl/d from more than 307 wells.

Foreign companies recently awarded concessions for exploration include France's TotalFinaElf, which signed a deal for a 100 per cent stake in Block 34 in southern Oman in March 2002, and Hunt Oil, which which was awarded a concession in May 2002 for Block 51 in the Sharqiyah region. China's CNPC also has acquired a foothold in Oman in 2002–a 50 per cent stake in Block 5 which it acquired after it was relinquished by the Japanese firm Japex.

Most of Oman's crude oil exports go to East Asia, with China, Japan, and South Korea the largest importers. India also is a significant importer of Omani crude oil.

Refining and Petrochemicals

In 1982, Oman constructed its first refinery, at Mina al-Fahal. Subsequently, the 50,000-bbl/d plant was expanded to 85,000 bbl/d. SK Engineering of South Korea was awarded a contract in June 2002 for the construction of a new desulfurization unit at Mina al-Fahal. Output from the facility, which is operated by the state-owned Oman Refinery Company (ORC), is used to meet local product demand.

A second refinery is planned, near the northern city of Sohar. Bids for construction of the project were solicited in March 2002. A final selection of a construction contractor has not been made.

As part of its effort to diversify its economy and to develop value-added industries, Oman has begun to invest in petrochemical production. Oman is moving ahead with plans to contruct a large-scale joint-venture petrochemical project in Sohar for the production of polyethylene and fertilizer.

Natural Gas

Oman has made natural gas–both for export as well as for domestic gas-intensive industries–the cornerstone of its diversification and economic growth strategy. Through an extensive exploration program, Oman has consistently increased its natural gas reserves in recent years. As of January 1, 2002, Oman's estimated proven natural gas reserves were approximately 29.3 trillion cubic feet (Tcf), up from only 12.3 Tcf in 1992, largely of associated gas. A significant gas discovery by PDO was reported in March 2001 in the Dakhiliya region. Most of these reserves are located in areas owned by PDO, which produces the majority of Oman's natural gas. More than 10 Tcf of Oman's non-associated natural gas is located in deep geological structures, many of which are beneath active oil fields.

Oman is extending its existing natural gas pipeline network. Two contracts were awarded to separate firms for projects connecting natural gas deposits in central Oman to the coastal cities of Sohar in the north and Salalah in the south. India's Dodsal Company completed construction of a $124-million line to Sohar in August 2002. A consortium including Italy's Snamprogetti and Saipem completed the $180-million line to Salalah in August 2002. In April 2001, Oman awarded a contract to operate the country's natural gas transportation and distribution infrastructure for the next five years to Canada's Enbridge and BC Gas. The contract includes a provision for technology transfer and training, so operation can be transferred to Omani staff after five years.

Oman also is one of the potential participants in the Dolphin Project, which will provide gas from Qatar's North Dome field to the United Arab Emirates (UAE). Initially, the project will link gas consumers in the United Arab Emirates with gas supplies from Qatar. While Oman originally was discussed as an importer, recent indications are that Oman might export a small quantity of gas to the UAE. Later, if Omani demand for petrochemical plants is sufficient, the flow might be reversed. Recent discussions have focused on possible exporta of Omani natural gas to Fujairah from late 2003 until 2006, when Qatari supplies are to become available, via a relatively inexpensive 30-mile spur pipeline. Oman could also serve as a transit corridor for gas exports from the Dolphin Project to Pakistan through a subsea pipeline, but sufficient demand and financing to make this possible are not expected in the next several years.

Oman made an agreement with Iran in 1997 for the development of the Hengam/Bukha offshore field in the Strait of Hormuz, which straddles the line between the two countries territorial waters. The field is currently producing 40 million cubic feet per day (Mmcf/d) of gas. In May 2000, the two governments agreed that the Iranian share of the gas would be exported to Oman once it is developed.

Liquefied Natural Gas (LNG) Exports

Oman began exports of LNG in early 2000, with the completion of a 6.6-million-ton-per-year liquefaction plant is located at Qalhat, near Sur. The project was developed by the Oman Liquefied Natural Gas Company (OLNGC), a joint venture of the Omani government (51 per cent), Shell (30 per cent), Total (5.54 per cent), Korea LNG (5 per cent), Mitsubishi (2.77 per cent), Mitsui & Co. (2.77 per cent), Partex (2 per cent), and Itochu (0.92 per cent). The LNG plant consists of two, 3.3-million-ton-per-year, LNG trains supplied by non-associated natural gas from the Saih Nihayda, Saih Rawl, and Barik fields. The second train became operational in May 2000, and the government of Oman approved plans for a third train in May 2002.

Korea Gas Corporation (Kogas) is the anchor customer for the LNG project, having contracted for 4.1 million tons per year of Omani LNG over a 25-year period. Japan's Osaka Gas Company is another client, and will receive 700,000 tons per year for 25 years. The third client, the Dabhol power project in

India, may not take its volume due to financial problems. The spot market for LNG relatively strong, so it is not expected that Oman LNG will have trouble maintaining operation at full capacity. New sales agreements have been concluded in the first half of 2002 with Union Fenosa of Spain and Gaz de France for LNG supplies which will become available with the completion of the third train.

In an effort to enhance prospects for future gas-based export projects, Oman has signed a number of joint venture agreements to carry out exploration and development activities. Canada-based Gulfstream Resources Ltd. is planning to invest more than $60 million over the next eight years to develop a gas field in Haffar Block 30. In November 1998, Occidental, Amoco and Neste Oy of Finland announced the establishment of a joint venture firm to develop, explore and produce natural gas reserves in northern Oman. The area consists of five blocks, covering approximately 3,150 square miles. Target markets include the city of Sohar and the northern UAE through the large natural gas supply hub located in Sharjah, UAE. The project will include development, well drilling, construction of a gas collection system and processing plant, as well as building gas transmission lines to Sohar and the Sharjah gas hub.

Electricity

Oman's government is undertaking a major reform of the country's electric power system. Demand is growing rapidly, at about 4-5 per cent per year, and the Omani government is seeking to attract foreign investment to ensure adequate generating capacity. The 90-megawatt (MW) al-Manah power plant became the Persian Gulf's first independent power project (IPP) in 1996, and an additional 180-MW of capacity was added there in early 2000. The Omani government currently is working on a new legal framework for a privatized power sector, which is to be enacted by the end of 2002. It is to provide for the sale of existing assets, including transmission and distribution to private owners, beginning in 2003.

Three new IPPs are planned for the near future. The 280-MW Al-Kamil power project, which reached financial close in March 2001, is being built by International Power PLC. Arab International Contractors of Egypt also in involved in the project. It is expected to be operational by late 2002. AES has been selected for the other IPP project, the 430-MW Barka power plant. It is expected to be completed by April 2003. Both plants will run on natural gas. PSEG of the United States has been chosen to build an integrated power project to supply the Dhofar region. The 200-MW generating plant is under construction. A contract for another 140-MW plant at Qarn Alam in southern Oman was awarded in May 2002 to Bharat Heavy Electricals of India. The facility is scheduled for completion by mid-2004.

Country Overview

Independence: 1650 (end of Portuguese rule)

Population (2001E): 2.6 million

Location/Size: Southeast Arabian Peninsula/82,030 sq. mi. (about the size of Kansas)

Major Cities: Muscat (capital), Salalah, Sur, al-Khasab

Languages: Arabic (official), English

Ethnic Groups: Arab, Baluchi, South Asian (Indian, Pakistani, Sri Lankan, Bangladeshi), African

Religion: Muslim (Ibadi–75 per cent, Sunni, Shi'a), Hindu

Defense (8/98): Army (43,500), Navy (4,200), Air Force (4,100), Royal Household (6,500)

Economic Overview

Currency: Omani Rial

Exchange Rate (9/02): $1 = 0.386 Omani Rial

Nominal Gross Domestic Product (GDP) (2001E): $20.9 billion *(2002E):* $21.9 billion

Real GDP Growth Rate (2001): 3.3 per cent *(2002E):* 3.0 per cent

Inflation Rate (consumer prices)(2001E): -1.1 per cent *(2002E):* 0.9 per cent

Major Trading Partners: Japan, United Arab Emirates, South Korea, United Kingdom, United States, Thailand

Merchandise Trade Balance (2001E): $3.1 billion *(2002E):* $2.8 billion

Major Export Products: Petroleum, fish, processed copper, textiles

Major Import Products: Machinery, transportation equipment, manufactured goods, food, livestock, lubricants

Monetary Reserves (2002E, non-gold): $2.4 billion

Total External Debt (2002E): $2.9 billion

Energy Overview

Proven Oil Reserves (1/1/02E): 5.5 billion barrels

Oil Production (2001E): 963,816 barrels per day (bbl/d), of which 959,816 bbl/d is crude oil

Oil Consumption (2001E): 55,000 bbl/d

Net Oil Exports (2001E): 908,816 bbl/d

Crude Oil Refining Capacity (1/1/02E): 85,000 bbl/d

Oil Export Customers (2001): China, Japan, South Korea, Thailand, Singapore, Taiwan, India

Natural Gas Reserves (1/1/02E): 29.3 trillion cubic feet

Natural Gas Production (2000E): 320 billion cubic feet (Bcf)

Natural Gas Consumption (2000E): 221 Bcf

Electric Generation Capacity (1/1/00E): 2.1 gigawatts

Electricity Production (2000E): 8.1 billion kilowatthours

Environmental Overview

Total Energy Consumption (2000E): 0.34 quadrillion Btu* (<0.1 per cent of world total energy consumption)

Energy-Related Carbon Emissions (2000E): 6.0 million metric tons of carbon (<0.1 per cent of world total carbon emissions)

Per Capita Energy Consumption (2000E): 135.3 million Btu (vs. U.S. value of 351.0 million Btu)

Per Capita Carbon Emissions (2000E): 2.4 metric tons of carbon (vs. U.S. value of 5.6 metric tons of carbon)

Energy Intensity (2000E): 20,599 Btu/$1995 (vs U.S. value of 10,918 Btu/$1995)**

Carbon Intensity (2000E): 0.36 metric tons of carbon/thousand $1995 (vs U.S. value of 0.17 metric tons/thousand $1995)

Sectoral Share of Energy Consumption (1998E): Industrial (44.1 per cent), Residential (20.0 per cent), Transportation (24.3 per cent), Commercial (11.6 per cent)

Sectoral Share of Carbon Emissions (1998E): Industrial (41.7 per cent), Residential (19.6 per cent), Transportation (27.4 per cent), Commercial (11.2 per cent)

Fuel Share of Energy Consumption (2000E): Natural Gas (67.6 per cent), Oil (32.4 per cent), Coal (0.0 per cent)

Fuel Share of Carbon Emissions (2000E): Natural Gas (63.3 per cent), Oil (36.7 per cent), Coal (0.0 per cent)

Renewable Energy Consumption (1998E): 0 Btu*

Number of People per Motor Vehicle (1998): 6.6 (vs. U.S. value of 1.3)

Status in Climate Change Negotiations: Non-Annex I country under the United Nations Framework Convention on Climate Change (ratified February 8th, 1995). Not a signatory to the Kyoto Protocol.

Major Environmental Issues: Rising soil salinity; beach pollution from oil spills; very limited natural fresh water resources.

Major International Environmental Agreements: A party to Conventions on Biodiversity, Climate Change, Desertification, Hazardous Wastes, Law of the Sea, Marine Dumping, Ship Pollution and Whaling.

The total energy consumption statistic includes petroleum, dry natural gas, coal, net hydro, nuclear, geothermal, solar, wind, wood and waste electric power. The renewable energy consumption statistic is based on International Energy Agency (IEA) data and includes hydropower, solar, wind, tide, geothermal, solid biomass and animal products, biomass gas and liquids, industrial and municipal wastes.

Oil and Gas Industries

Organizations: Petroleum Development Oman Ltd. (PDO) controls all oil resources. PDO is a partnership between the Omani government (60 per cent), Royal Dutch/Shell (34 per cent), Total (4 per cent), and Partex (2 per cent). Oman Oil Company (OOC) is the overseas investment arm of the Ministry of Petroleum.

Major Foreign Oil Company Involvement (non-PDO): BP, CNPC, IPC, Itochu, Japex, Occidental, Phillips

Major Oil Fields: Roughly 1.8 billion barrels in reserves are located in the large northern structure containing the Yibal, Natih, Fahud, al-Huwaisah, Lekhwair, and Shibkah fields. Other key fields are the southern Marmul and Nimr fields as well as Occidental's 120-million barrel Safah field and the estimated 400-million barrel Amal Eastern High field, which contains heavy crude oil.

Major Refinery: Mina al-Fahal (85,000 bbl/d)

Major Oil Terminal: Mina al-Fahal

60

Pakistan

Opportunities exist for foreign direct investment in Pakistan's energy sector, though some foreign investors have encountered problems in recent years.

General Background

Recent economic developments in Pakistan have been dominated by the military campaign in neighboring Afghanistan beginning in October 2001. The United States permanently lifted sanctions against both India and Pakistan in September 2001: sanctions were imposed in 1998 in the wake of the nuclear tests, but a substantial aid program now has resumed. International donor organizations such as the International Monetary Fund (IMF) and the World Bank are concerned that Pakistan's reforms are losing momentum due to the difficulty in building a consensus in Pakistan's Parliament. While some reforms have already been carried out, it was noted that difficult reforms, such as raising natural gas prices, have been put on hold. The two institutions previously agreed to provide substantial amounts of additional credit, totalling nearly $10 billion over the next three years.

Pakistan's critical textile industry has been adversely affected since September 2001, and agricultural production was already suffering from a severe drought in 2001, but an inflow of aid has improved Pakistan's short-term financial situation considerably. The United States also has agreed to reduce or suspend some tariffs on imports of Pakistani textile products, which should help boost Pakistani export earnings.

Pakistan's real gross domestic product (GDP) growth rate was 3.6 per cent in 2002, falling well short of earlier forecasts. Real GDP growth for 2003 is expected to reach 4.1 per cent. Pakistan's external debt is equal to around 60 per cent of its annual GDP, and its debt payments due each year exceed its receipts from exports. The government also suffers from a relatively ineffective system for tax collection, with only 1 per cent of the population paying income taxes, though the Musharraf

government has begun a program to increase tax collection rates, and the effort is showing some signs of success.

While formal legal protections for foreign investment in Pakistan generally are good, inadequate infrastructure, a poorly educated workforce, sectarian and ethnic violence, and a slow-moving judicial system have proven to be obstacles to attracting foreign investment. While some of the well-publicized disputes between Pakistani state entities and the country's Independent Power Producers (IPP's) have been resolved, the dispute has been a major blow to foreign investor confidence in Pakistan. Foreign Direct Investment (FDI) in Pakistan in recent years has been only a small fraction of the comparable figures for the mid-1990s. Pakistan's often tense relationship with India also has been a limiting factor. In April 2003, India and Pakistan moved resume full diplomatic relations, which had been severed in December 2001.

Oil

Pakistan produced 53,000 barrels per day (bbl/d) of oil in 2002 (of which 50,000 bbl/d was crude oil and the rest refinery gain and other liquids), and consumed 365,000 bbl/d of petroleum products. Net oil imports were 312,000 bbl/d. While there is no prospect for Pakistan to reach self sufficiency in oil, the government has encouraged private (including foreign) firms to develop domestic production capacity. Pakistani domestic oil production centers on the Potwar Plateau in Punjab and lower Sindh province.

Most of the foreign firms active in Pakistan in the oil exploration and production sector are small independent firms. The two most significant foreign oil firms in Pakistan are BP and Eni (which acquired British independent Lasmo in early 2001). State-owned Oil and Gas Development Corporation (OGDC) also is a major player. Malaysia's Petronas has acquired a stake in an exploration block in Sindh province, in cooperation with Eni. In November 2000, the Pakistani government awarded two exploration blocks: the onshore Mehran Block 2467-4, to a team including Union Texas Pakistan, a subsidiary of BP, and Occidental Petroleum; and an offshore block to Ocean Energy. American independent Orient Petroleum, which also hold concessions in Pakistan, is investing approximately $70 million in seismic surveys and exploratory drilling over an initial period of three years.

In September 2002, Pakistan awarded an exploration license to Eni SPA for Manchar Block 2667.5, which is located in Sindh Province.

The Pakistani government had planned to move forward with the sale of a significant number of the production assets of OGDC in late 2001, but postponed the sale due to investor concerns about the country's stability. Current plans call for the sale of nine currently producing oil and gas fields, with bids to be solicited after the bidding for PSO is completed.

Refining/Downstream

Pakistan's net oil imports are projected to rise substantially in coming years as demand growth outpaces increases in production. Demand for refined petroleum products also greatly exceeds domestic oil refining capacity, so nearly half of Pakistani imports are refined products. Pakistan's Pak-Arab Refinery (PARCO) became operational in late 2000, adding 100,000 bbl/d to the country's refining capacity, and alleviating refined product import dependence.

A small, 30,000 bbl/d refinery is being set up by private Bosicor Pakistan near Karachi and is slated to begin commercial operation in June 2003. A refurbished unit previously owned by Tesoro Petroleum in the United States is being used in the construction of the refinery.

Another major planned project is the "Iran-Pak" refinery, which would have a capacity of 130,000 bbl/d. The refinery would be located near the border with Iran in Baluchistan province and would be a 50:50 partnership between Pakistan's Petroleum Refining and Petrochemical Corporation (PERAC) and the National Iranian Oil Company (NIOC). Oil processed at the Iran-Pak refinery would come almost exclusively by sea from Iran, and would be unloaded at a terminal to be built for the refinery. The project has failed to reach financial closure, however, and seems unlikely to be built as NIOC's demand for a guaranteed rate of return is at odds with Pakistan's policy against such guarantees.

Oil sector reforms in Pakistan are said to be on track, including the privatization of Pakistan State Oil (PSO). On April 26, 2003, the Pakistani government will accept bids for a 51 per cent stake in PSO, which holds a 70 per cent domestic market share and has more than 3,800 retail outlets. Deregulation of prices for petroleum products is being pursued in parallel with the privatization of PSO. Final approval for the privatization was granted in January 2001.

As part of the privatization process, the government of Pakistan is setting up the Gas Regulatory Authority (GRA) and the Petroleum Regulatory Board (PRB), which will separate out the government functions from the state-owned companies to be privatized. Pakistan's government hopes to reap significant revenues from these privatizations over the next several years.

Natural Gas

Pakistan has 25.1 trillion cubic feet (Tcf) of proven gas reserves, and currently produces around

0.8 Tcf of natural gas per year, all of which is consumed domestically. Natural gas producers include Pakistani state-owned companies Pakistan Petroleum Ltd. (PPL) and Oil and Gas Development Corporation (OGDC), as well as BP, Eni, OMV, and BHP. The largest currently productive fields are Sui, by far the largest at 650 million cubic feet per day (Mmcfd), Adhi and Kandkhot (120 Mmcfd), Mari, and Kandanwari.

Pakistan's demand for natural gas is expected to rise substantially in the next few years, with an increase of roughly 50 per cent by 2006, according to Pakistan's oil and gas ministry. Pakistan also plans to make gas the "fuel of choice" for future electric power generation projects. This will necessitate a sharp rise in production of natural gas, and also has generated interest in Pakistan in pipelines to facilitate imports from neighboring countries.

Much of Pakistan's increased natural gas demand will be satisfied by increased domestic production. Austrian company OMV's 1998 discovery at Sawan is expected to produce 340 Mmcfd by 2003. Lasmo (now Eni) reported a new find in March 1999, in western Sindh province, which is expected to produce 20 Mmcfd. Hardy Oil (UK) also reported a new discovery in 1999, in the Middle Indus region of Sindh, which tested at an initial 58 Mmcfd. Petronas reported a new discovery in September 2001 near Sukkur. Offshore exploration concessions have also been granted to Eni, Shell, OMV, and others.

Development of new natural gas fields is proceeding, with Pakistan's government expecting recently discovered fields to add about 1 billion cubic feet per day (Bcf/d) to Pakistan's natural gas production. The Zamzama field in Sindh province came onstream in early 2001, and produces 60 Mmcfd. Pakistan has a contract with Eni for the development of the Bhit gas field, which came onstream in March 2003 and has reached peak production of 235 Mmcfd.

Several import schemes also have been under discussion in recent years, though recent finds now under development have made it unlikely that Pakistan will need to import natural gas within the next few years. In mid-2000, Pakistan's government stated that it would permit a natural gas pipeline linking Iran's massive reserves to rival India across Pakistani territory. Pakistan would earn transit fees for Iranian gas supplied to India and also would be able to purchase some gas from the pipeline when and if its own demand was sufficient. While Iran and Pakistan have shown great interest in the project, India has been reluctant to move forward as long as political and military tensions with Pakistan over Kashmir persist. The recent escalation of tensions between the two countries has made any movement on the project even more unlikely, though a feasibility study is still underway.

Another natural gas import possibility is an eventual link with the Dolphin Project, a scheme to supply gas from Qatar's North Dome gasfield to the United Arab Emirates and Oman, via a subsea pipeline from Oman. Even though Pakistan has signed a preliminary agreement to eventually purchase natural gas from Qatar, it seems increasingly unlikely that Pakistan will be included in the project in the near-term, due to its financial weakness and uncertainty about whether there will be sufficient domestic natural gas demand growth.

A third possible natural gas pipeline would link gas-rich Turkmenistan with Dalautabad in central Pakistan via Afghanistan. Unocal had been the main foreign backer of the plan until August 1998, when it withdrew from the project after the U.S. strikes against terrorist training camps associated with Osama bin Laden in Afghanistan. Subsequently, the governments of Pakistan and Turkmenistan held talks with the Afghan Taliban authorities about continuing the project without Unocal. The new Afghan transitional government of Hamid Karzai has endorsed the pipeline idea, but it seems unlikely to be implemented due to lack of interest by potential investors.

Some independent observers of the Pakistani natural gas market believe that increases in domestic production, coupled with a slower growth in demand than projected by the Pakistani government, will render the gas pipeline projects economically unviable. In addition, Pakistan's weak financial position makes it difficult to secure financing for such ambitious projects. A pipeline from Iran to India would make sense in financial terms, as its primary justification would be sales to India, with Pakistan as only a secondary customer, but the energy security issues it raises for India make it unlikely to proceed.

Electricity

Pakistan has 18 gigawatts (GW) of electric generating capacity. Thermal plants using oil, natural gas, and coal account for about 68 per cent of this capacity, with hydroelectricity making up 28 per cent and nuclear plants 2.6 per cent. Pakistan's total power generating capacity has increased rapidly in recent years, due largely to foreign investment, leading to a partial alleviation of the power shortages Pakistan had faced earlier. Rotating blackouts ("load shedding") are, however, necessary at times in some areas. Transmission losses are about 30 per cent, due to poor quality infrastructure and a significant amount of power theft. Seasonal reductions affect the availability of hydropower, and the country faced a severe shortage in 2001. With much of the Pakistan's rural areas yet to receive electric power, and less than half of the population connected to the national grid, significant power demand growth is expected in the long term, though in the short term, Pakistan has some excess generation capacity.

The electric power sector in Pakistan is still primarily state-owned, but a privatization program is underway. The main state-owned utilities are the Water and Power Development Authority (WAPDA), and the Karachi Electricity Supply Corporation (KESC), which serves only Karachi and surrounding areas. WAPDA, which is made up of eight regional electricity boards, is to be split up for privatization. One regional entity, the Faisalabad Area Electricity Board, has begun the privatization process, which was scheduled to conclude in late 2002, but has been delayed for unspecified reasons.

Growth in power generation in recent years has come primarily from new independent power producers (IPP's), some of which have been funded by foreign investors, and a few WAPDA hydroelectric dam projects. The two largest private power plants in Pakistan are the Hub power company (HUBCO) and the Kot Addu power company. HUBCO is owned by a consortium of National Power (UK), Xenal (Saudi Arabia), and Mitsui Corporation, and has a 1,300-MW capacity. Kot Addu, with a 1,500-MW capacity, was privatized in 1996 (from WAPDA), and is owned by National Power. Both of these plants, as well as a few other small private operators, sell power to the national grid currently run by WAPDA.

In April 2003, the Ministry of Industries and Production announced that it was planning to build coal-fired power-generation plants in export processing zones and in special industrial states to provide a less expensive source of energy. Officials hope to exploit the large, untapped coal reserves in Tharkparkar. At present, coal makes up less than a 5 per cent share in overall energy production.

Two power projects involving U.S.-based companies (Babcock and Wilcox, and General Electric) received $293 million from the U.S. Export-Import Bank in early 1998. These projects involve equipment and services for Uch Power Ltd., and the Saba power plant. Uch Power became operational in December 2000, with a capacity of 550 MW. Private sector projects will rely primarily on increased use of natural gas. New WAPDA projects are confined to hydropower, including projects such as the 1,425-MW Ghazi-Barotha plant which takes advantage of the enormous untapped potential of the Indus River. The dam is under construction, and a payment dispute with construction firms involved in the project was resolved in February 2002. Construction had been halted in September 2001 due to the payment dispute and withdrawal of expatriate staff due to security concerns.

IPP's have been involved in disputes and litigation with the government over the rates set in their Power Purchase Agreements (PPA's) with the national WAPDA grid. Under Prime Minister Nawaz Sharif, Pakistan's government had charged that the IPP's had engaged in price fixing and had paid bribes to officials of the previous Benazir Bhutto government. The Sharif government's main demand was for a reduction in rates to 4.5 cents per kilowatt hour (kwh), from the 6.6 cents per kwh which most of the IPPs had in their original contracts. Both of the largest IPP's, HUBCO and Kot Addu, were targeted, as well as the Malaysian-owned Dharki power plant. In response to the Pakistani government's demands for a rate reduction, the IPP's demanded that prices for fuels be lowered, in particular oil, which is supplied by a state controlled monopoly. HUBCO's dispute with WAPDA was settled by an agreement on a new price of 5.6 cents per kwh in December 2000, and Kot Addu agreed to a rate reduction in October 2000. HUBCO has recovered reasonably well from the dispute, and reported a profit for 2001. Libery Power, a new 235-MW IPP backed by Tenaga Nasional of Malaysia, began operation in August 2001, after resolving its tariff dispute with WAPDA.

Power theft is a pressing issue in Pakistan. While it is impossible to precisely measure theft (as opposed to line loss), it is obvious that it constitutes a sizable proportion of Pakistan's overall 30 per cent loss rate. The situation was so severe by early 1999 that the Pakistani government assigned army units to look for illegal connections to transmission lines and rigged meters. Power theft is just one part of the financial problems for WAPDA, however. WAPDA is at the center of a public sector "circular debt" problem, in which state firms and government ministries have failed to pay power bills, and WAPDA has failed to meet obligations to them and to private sector creditors.

Coal

Coal currently plays a relatively minor role in Pakistan's energy mix, but the discovery of large volumes of low-ash, low-sulfur lignite in the Tharparkar (Thar) Desert in Sindh province could increase its importance. Thar reserves are being developed under the jurisdiction of the provincial Sindh Coal Authority and have enormous economic potential. The Authority's policy is to develop the reserves primarily to fuel large electric power plants to be built in tandem with the coal mines. A feasibility study recently was carried out for the construction of a coal-fired power plant near the Thar coal mines, and President Musharraf has stated that coal should make up more than the current 1 per cent of electric power generation in Pakistan.

Environment

Pakistan's attempt to raise the living standards of its citizens has meant that economic development has largely taken precedence over environmental issues. Unchecked use of hazardous chemicals, vehicle emissions, and industrial activity has contributed to a number of environmental and health hazards, chief among them being water pollution. Much of the country suffers from a lack of potable water due to industrial waste and agricultural runoff that contaminates drinking water supplies. Poverty and high population growth have aggravated, and to a certain extent, caused, these environmental problems.

In February 2003, the Asian Development Bank (ADB) offered Pakistan $50 million to assist its Rawalpindi Environment Improvement Project. The project includes provisions for augmenting the water supply system; rehabilitating and extending the sewerage and drainage network; building a sewerage treatment plant; supporting a solid waste management system; administering environmental improvement work on the upstream of Rawal Lake and Nullah Leh (natural storm water drain) and strengthening the Tehsil Municipal Administration (TMA) and related civic agencies. ADB officials will visit Pakistan later this year to further study the project.

Although Pakistan is renowned for its mountain ranges and areas of untouched wilderness, the country has passed legislation to protect its environment only in the past 10 years. Environmental groups have questioned the country's commitment to environmental protection, pointing to the decision in August 1999 to allow oil and gas exploration in Kirthar National Park, the country's oldest national wildlife park, by a multi-national company.

In the cities, widespread use of low-quality fuel, combined with a dramatic expansion in the number of vehicles on Pakistani roads, has led to significant air pollution problems. Although Pakistan's energy consumption is still low by world standards, lead and carbon emissions are major air pollutants in urban centers such as Karachi, Lahore, and Islamabad.

Theft or diversion of electricity in transmission, as well as a lack of energy efficiency standards, have contributed to Pakistan's high energy and carbon intensities. To increase energy efficiency, the country is stepping up its use of renewable energy sources to bring electricity to rural areas. As urbanization continues and the population grows at a rapid rate, in the 21st century Pakistan will need to confront its environmental problems in order to safeguard the health of it citizens.

COUNTRY OVERVIEW

Independence: August 14, 1947 (from UK)

Population (7/01E): 144.6 million

Location/Size: Southern Asia/310,500 square miles (about twice the size of California)

Major Cities: Islamabad (capital), Karachi, Lahore, Faisalabad

Languages: Urdu (national/official), English (official), Punjabi, Sindhi, Pashtu, Baloch

Ethnic Groups: Punjabi, Sindhi, Pashtun (Pathan), Baloch, Muhajir (immigrants from India and their descendants)

Religions: Muslim, 97 per cent (Sunni 77 per cent, Shia 20 per cent); Christian, Hindu, and other, 3 per cent

Defense (8/98): Army (520,000); Air Force (45,000); Navy (22,000); Paramilitary Forces (247,000)

Economic Overview

Currency: Pakistani Rupee

Average Exchange Rate (04/24/03): U.S.$1 = 57.7 rupees

Gross Domestic Product (GDP, market exchange rates, 2002E): $59.8 billion

Real GDP Growth Rate (2002E): 3.6 per cent *(2003F):* 4.1 per cent

Inflation Rate (2002E): 3.7 per cent *(2003F):* 5.3 per cent

Current Account Balance (2002E): -$1.4 billion

Merchandise Imports (2002E): $11.8 billion

Merchandise Exports (2002E): $9.9 billion

Merchandise Trade Balance (2002E): -$1.9 billion

Total External Debt (2001E): $34.1 billion

Major Trading Partners: United States, Japan, Germany, United Kingdom, and Saudi Arabia

Major Export Products: Raw cotton and textiles; rice; leather manufactures

Major Import Products: Petroleum; machinery and transport equipment; food

Energy Overview

Proven Oil Reserves (1/1/03E): 310 million barrels

Oil Production (2002E): 53,000 barrels per day (bbl/d), of which 50,000 bbl/d was crude oil

Oil Consumption (2002E): 365,000 bbl/d

Net Oil Imports (2002E): 312,000 bbl/d

Crude Oil Refining Capacity (1/1/03E): 238,850 bbl/d

Natural Gas Reserves (1/1/03E): 26.4 trillion cubic feet (Tcf)

Natural Gas Production (2001E): 0.8 Tcf

Natural Gas Consumption (2001E): 0.8 Tcf

Coal Production (2001E): 3.5 million short tons (Mmst)

Coal Consumption (2001E): 4.6 Mmst

Net Coal Imports (2001E): 1.1 Mmst

Recoverable Coal Reserves (1/1/03E): 2.5 billion short tons

Electric Generation Capacity (1/1/01E): 18.0 gigawatts (68 per cent thermal, 28 per cent hydro, 2.6 per cent nuclear)

Electricity Generation (2001E): 67 billion kilowatthours

Environmental Overview

Total Energy Consumption (2001E): 1.9 quadrillion Btu* (0.46 per cent of world total energy consumption)

Energy-Related Carbon Emissions (2001E): 29.2 million metric tons of carbon (0.44 per cent of world total carbon emissions)

Per Capita Energy Consumption (2001E): 12.9 million Btu (vs. U.S. value of 341.8 million Btu)

Per Capita Carbon Emissions (2001E): 0.2 metric tons of carbon (vs. U.S. value of 5.5 metric tons of carbon)

Energy Intensity (2001E): 26,229 Btu/$1995 (vs U.S. value of 10,736 Btu/$1995)**

Carbon Intensity (2001E): 0.41 metric tons of carbon/thousand $1995 (vs U.S. value of 0.17 metric tons/thousand $1995)

Fuel Share of Energy Consumption (2001E): Oil (42.2 per cent), Natural Gas (41.2 per cent), Coal (4.8 per cent)

Fuel Share of Carbon Emissions (2001E): Oil (54.6 per cent), Natural Gas (38.1 per cent), Coal (7.3 per cent)

Status in Climate Change Negotiations: Non-Annex I country under the United Nations Framework Convention on Climate Change (ratified June 1st, 1994). Not a signatory to the Kyoto Protocol.

Major Environmental Issues: Water pollution from raw sewage, industrial wastes, and agricultural runoff; limited natural fresh water resources; a majority of the population does not have access to potable water; deforestation; soil erosion and desertification.

Major International Environmental Agreements: A party to Conventions on Biodiversity, Climate Change, Desertification, Endangered Species, Environmental Modification, Hazardous Wastes, Law of the Sea, Nuclear Test Ban, Ozone Layer Protection, Ship Pollution and Wetlands. Has signed, but not ratified, Marine Life Conservation.

The total energy consumption statistic includes petroleum, dry natural gas, coal, net hydro, nuclear, geothermal, solar, wind, wood and waste electric power. The renewable energy consumption statistic is based on International Energy Agency (IEA) data and includes hydropower, solar, wind, tide, geothermal, solid biomass and animal products, biomass gas and liquids, industrial and municipal wastes.

Energy Industries

Organization: Oil and Gas Development Corporation (OGDC), a state company, handles oil and gas exploration and development; Water and Power Development Authority (WAPDA) supplies electricity to most of the country; Karachi Electric Supply Corporation Limited (KESC) serves the greater Karachi metropolitan area; Pakistan Atomic Energy Commission (PAEC) operates one nuclear power plant

Major Foreign Energy Company Involvement: AES, Atlantic Richfield, British National Power, Coastal Power, Gaz de France, Total, General Electric, Lasmo Oil (U.K.), Marubeni (Japan), ExxonMobil, Monument Oil & Gas, Premier Oil, Royal Dutch Shell, Xenal (Saudi Arabia)

Major Ports: Gwadar, Karachi, Muhammed bin Qasim, Ormaro

Major Gas Fields: Bhit, Dhodak, Kadanwari, Mari, Prikoh, Qadipur, Sawan, Sui

Major Oil Fields: Dhurnal, Fimkasser, Liari, Mazari, Thora

Major Pipelines: Sui Northern Gas Pipeline; Sui Southern Gas Pipeline; Pak-Arab Refinery Company (PARCO) petroleum product pipeline

Major Refineries (Capacity): Pak-Arab Refinery near Multan (100,000 bbl/d); Attock Refinery in Rawalpindi (35,000 bbl/d), National Refinery in Korangi (62,050 bbl/d), Pakistan Refinery Ltd. in Karachi (46,300 bbl/d)

61

Panama

Panama is important to world energy markets because the Panama Canal is a major transit center for oil shipments and a potential choke point. Panama is key to plans to connect the electricity and natural gas grids of North and South America.

General Background

Notwithstanding its small population and area, (2.9 million and 30,193 square miles respectively) Panama is an important center for international trade in the Western Hemisphere as both a major shipping thoroughfare and a regional economic power. Approximately two million long tons of cargo pass through the Panama Canal annually. Panama is also a financial and communications hub that sits at the crossroads of five international fiber-optic networks and hosts 110 international banks.

The Panamanian economy is one of Latin America's most stable. The Panamanian Balboa has been fixed to the dollar since 1903, and the country has consistently maintained Central America's second highest per capita gross domestic product (after Costa Rica). Panama's Colon free trade zone is the largest in the Western Hemisphere, generating approximately $11 billion in export and re-export business annually; well above the average Central American gross domestic product (GDP) in 2000 of $8.3 billion.

Recently, after maintaining healthy economic growth throughout the 1990s, the global economic slowdown has stunted Panamanian development. Panama's real GDP growth has fallen from 2.5 per cent in 2000 to only 0.3 per cent in 2001. Economic malaise continues in 2002, as toll revenues at the Panama Canal were reportedly $12.3 million under budget as of June 2002. Annual real GDP growth for 2002 is forecast at only 0.7 per cent.

The economic policies of Panamanian President Mireya Moscoso, elected in 1999 to a five-year term, have focused on both increased international investment and development of the country's latent tourism potential. In 1999, the Moscoso administration implemented a domestically controversial

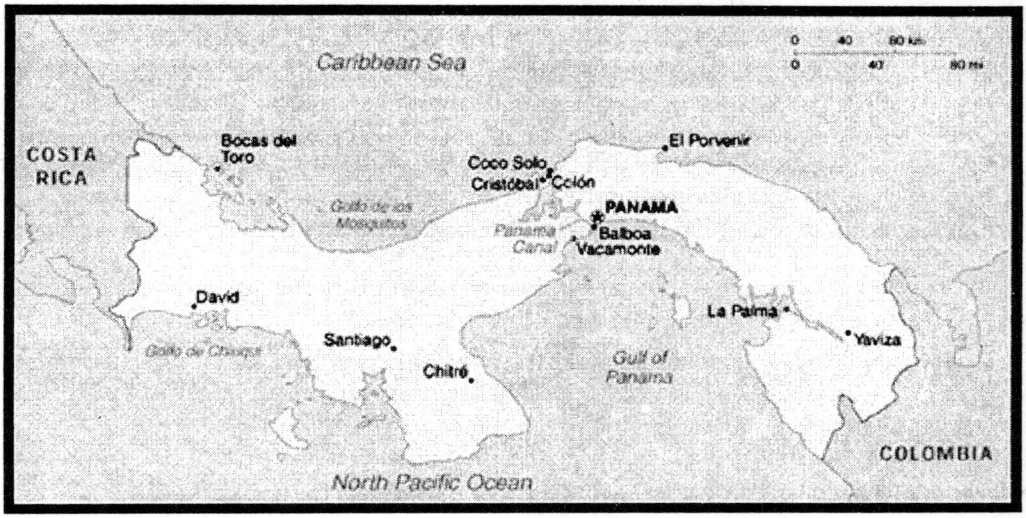

protocol agreement with the U.S. Overseas Private Investment Corporation (OPIC), which offers, for the first time, political risk insurance to American investors looking at Panama. The country's tourism institute, IPAT, is currently implementing a $10 million public relations program designed to make Panama's rainforests and tropical wildlife an international eco-tourism attraction.

Energy

Panama has negligible hydrocarbon energy reserves and imports over 70 per cent of its energy. Virtually all oil is imported, and the country neither produces nor consumes natural gas.

Electricity generation accounts for most of Panama's domestic energy production; hydroelectricity alone accounts for 70 per cent of the country's total energy production. In 1998, the state-owned electricity company, IRHE, was broken up. There are currently eight electricity generators and three distributors operating in Panama. Foreign company involvement in the Panamanian electricity sector includes U.S.-based Coastal, AES, Illinova, and Noresco, Canada's Hydro Quebec, and Spain's Union Fenosa. EGE Fortuna, owned by Coastal and Hydro Quebec (51 per cent) and the Panamanian government (49 per cent), is the country's largest generator, with 300 megawatts (MW) of hydropower. Transmission remains in the hands of the government through the publicy-owned Etesa company. Also in 1998, a regulatory body was created, *Ente Regulador*, to oversee the electric, telecommunications, and water sectors.

Panamanian electricity demand has grown by about 40 per cent since 1991, and is expected to continue growing significantly in the coming few years. Numerous new projects are planned to help meet growing demand. In May 2000, the Inter-American Development Bank (IDB) Private Sector Department approved its first project in Panama. The IDB will loan $59.8 million to IGC/ERI Pan Am Thermal Generating Limited (PATG) for the construction and operation of a 96-MW, $92 million thermal electric plant near Panama City. Construction of AES's $190 million, 132-MW Esti hydroelectric project in Panama's Chiriqui province began in early August 2000 and is expected to reach completion in late 2002. Esti is made up of two hydroelectric plants, Guasquitas and Canjilones, on the Chiriqui river.

Central America has been discussing plans to link the region's electricity grids. The *Sistema de Interconexion Electrica para America Central* (SIEPAC) project calls for the construction of transmission lines connecting 37 million consumers in Panama, Costa Rica, Honduras, Nicaragua, El Salvador, and Guatemala. In April 2002, a transmission line between Honduras and El Salvador was opened, marking the complete interconnection of all six SIEPAC countries. Endesa (Spain) is currently planning a high-voltage, 230-kilovolt (kV), 1,140-mile transmission line that will extend from Guatemala to Panama with connections to the grids and substations of all of the member countries to be completed by 2006. This overarching power line is designed to mitigate the poor quality of existing interconnections, making regional transactions possible.

Panama has become increasingly concerned about its environment. With the 1998 electricity restructuring, the government also restructured its environmental regulatory framework. *Autoridad National del Ambiente* (ANAM) became the government agency responsible for the implementation and administration of environmental law. ANAM holds significant enforcement authority, in contrast to the regulatory body that preceded it.

The Panama Canal

The Panama Canal extends approximately 50 miles from Panama City on the Pacific Ocean to Colon on the Caribbean Sea. It is widely considered to be one of the world's great engineering achievements. The United States is the largest user of the Canal in terms of cargo tonnage, as either port of origin or destination, although Asian countries are beginning to close the gap. Ships bound for Japan from the East Coast of the United States save about 3,000 miles by going through the Canal; ships sailing from Ecuador to Europe save about 5,000 miles.

Treaties

In 1903, the Republic of Panama and the United States signed the original Panama Canal Treaty, which allowed the United States to build and operate a canal connecting the Pacific Ocean with the

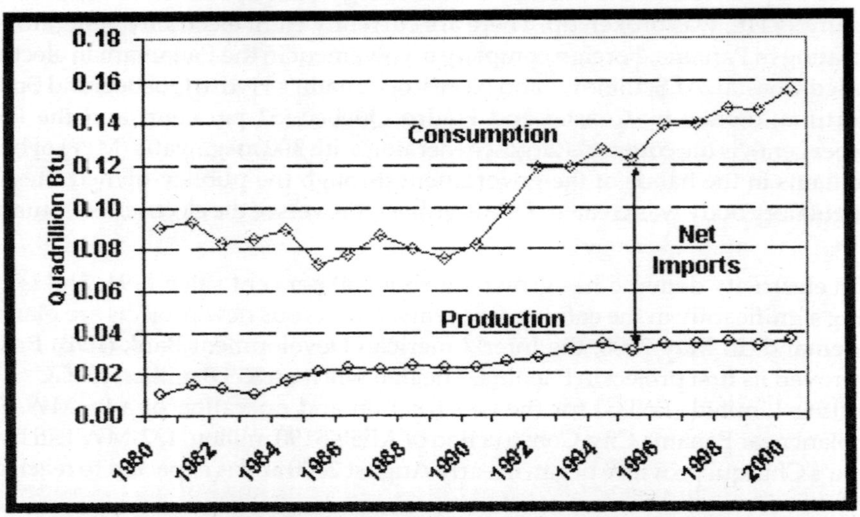

PANAMANIAN ENERGY PRODUCTION AND CONSUMPTION, 1980-2000

Caribbean Sea through the Isthmus of Panama. The Treaty granted the United States the use, occupation, and control of a Canal Zone, approximately 10 miles wide, in which the United States possessed full sovereign rights. In return, the United States guaranteed the independence of Panama and paid the government of Panama $10 million, as well as an annuity of $250,000, which each year increased at a rate far beyond that of inflation.

On September 7, 1977, a new Panama Canal Treaty was signed by President Torrijos of Panama and President Carter of the United States that transferred full control of the Canal to Panama on December 31, 1999. Under this Treaty, the Panama Canal Company, the Canal Zone, and its government were disenfranchised on October 1, 1979, and replaced by the Panama Canal Commission that operated the Canal during the 20-year transition period that began with the Treaty. The Panama Canal Commission has now been replaced by a new Panamanian entity, the Panama Canal Authority. The treaty guarantees permanent neutrality of the Canal. Control over U.S. military facilities in the former Panama Canal Zone has reverted to Panamanian authority. The U.S. Southern Command and U.S. Army South troops moved out of Panama at the end of 1999. A Hong Kong-based company, Hutchison-Whampoa, now operates the ports at both entrances to the Canal. This has been a cause for security concerns among some lawmakers in the United States, although the United States is legally entitled to intervene to maintain the neutrality of the Canal.

Canal Traffic

The U.S. East Coast-Asia route is the dominant trade route for the Panama Canal, and is boosted by increasing U.S.-China trade. Movement between Europe, the U.S. West Coast and Canada comprises the waterway's second major trade route. Recently, North-South trade has been increasing, as Latin America evolves in to an increasingly important trading partner of North America.

The Canal is designed to accommodate about 50 ships per day (the maximum has been 65 transits per day). On average, it takes one ship 24 hours to pass completely through the canal. Oceangoing vessel transits totaled 12,197 in fiscal year 2001, or an average of 33 vessels per day, an almost 1.0 percent decline from 2000's total of 12,303, or 34 vessels per day. Transit by the 50,000-ton Panamax class ships, the largest vessels that the waterway can accommodate, represents about 30 per cent of total oceangoing transit through the Canal.

Petroleum is one of the largest commodities (by tonnage) shipped through the Canal, and accounted for approximately 16 per cent of total canal shipments during 2001. Also in 2001, approximately 613,000 barrels per day (bbl/d) of crude oil and petroleum products passed through the Canal. Around 60 per cent of total oil shipments went from the Atlantic to the Pacific, with petroleum products dominating this traffic. Crude oil accounted for the majority of Pacific to Atlantic oil traffic, with much of this coming from the Alaskan North Slope. Overall, petroleum products far outweigh crude oil, accounting for almost 70 per cent of all petroleum shipments through the Canal. Some coal is shipped through the canal as well, accounting for about 3 per cent of total Canal traffic. Over six million short tons of coal passed through the canal in 2001, with over 80 per cent going from the Pacific to the Atlantic.

Trans-Panama Pipeline

If transit were halted through the Canal, the Trans-Panama pipeline (Petroterminal de Panama, S.A.) could be re-opened to carry oil in either direction. This pipeline is located outside the former Canal Zone near the Costa Rican border, and runs from the port of Charco Azul on the Pacific Coast (near Puerto Armuelles, southwest of Dàvid) to the port of Chiriqui Grande, Bocas del Toro on the

Caribbean. It was opened in October 1982 as an economical alternative to the Panama Canal for transporting Alaskan oil across Panama en route to Gulf Coast ports. Transit time from Alaska to the U.S. Gulf Coast via Panama is about 16 days, whereas a tanker would take 40 days to reach the Gulf Coast from Alaska if rerouted around Cape Horn (the southern tip of South America). To date, more than 2.2 billion barrels of Alaskan crude oil has been transported through the 81-mile pipeline. However, the 860,000-bbl/d pipeline was closed in April 1996 after Alaskan oil shipments to the Gulf Coast declined with falling Alaskan oil production (Alaska now produces about 1 million bbl/d) and increased oil consumption on the west coast of the United States, especially in California. In addition, the decision to allow Alaskan oil to be exported outside the United States reduced the incentives to ship Alaskan oil to the Gulf Coast. There has been some discussion of reversing the direction of the pipeline to allow Caribbean oil producers a less expensive outlet to Pacific destinations.

U.S. Petroleum Imports and the Panama Canal

The United States is not heavily reliant on the Panama Canal for its petroleum imports. In 2001, only 1.7 per cent of total U.S. petroleum imports (crude oil plus petroleum products) transited the Canal en route to American ports (see table). On the whole, much more crude oil destined for U.S. shores passes through the Canal than do petroleum products. However, as a share of U.S. imports, the Canal is far more important for imported petroleum products. In 2001 only 0.8 per cent of imported crude oil came to the United States through the Panama Canal, compared to approximately 8 per cent of imported petroleum products. The United States imports slightly less oil through the Canal than 20 years ago, and no longer relies heavily on the Canal to move oil from one coast to the other. In 1979, 265,768 bbl/d of oil originating in and destined for the United States went through the Canal; this number fell to approximately 21,000 bbl/d in 2001. Proportionally, the countries of South and Central America are much more dependent on the Canal to support their economies than are the United States and Canada.

Canal Expansion and Modernization

In 1996, the Canal Commission Board of Directors initiated a $1 billion modernization and improvement program designed to expand capacity. One of the key programs is the widening of the narrow Gaillard Cut, an 8-mile incision through the rock and shale of the continental divide. Before expansion, the passage was only large enough for one Panamax-sized ship to pass through at a time. In August 2002, after five years of operations, expansion of the Galliard Cut was completed, allowing two Panamax-sized vessels to pass through simultaneously. Other on-going improvement programs include purchasing new hydraulic machinery to replace the existing electro-mechanic rising valves, and installing a new computer traffic monitoring system at the locks. The Panama Canal Authority is also investing in more powerful tugboats and towing locomotives. The modernization and improvement program aims to increase overall Canal capacity by 20 per cent.

Country Overview

Independence: 1903 (from Colombia)

Population (2001E): 2.9 million

Location/Size: Middle America, bordering both the Caribbean Sea and the North Pacific Ocean, between Colombia and Costa Rica/30,200 square miles

Capital City: Panama City

Languages: Spanish (official), English (14 per cent ; many Panamanians are bilingual)

Ethnic Groups: Mestizo (mixed Amerindian and white) 70 per cent, Amerindian and mixed (West Indian) 14 per cent, White 10 per cent, Amerindian 6 per cent

Religions: Roman Catholic 85 per cent, Protestant 15 per cent

Economic Overview

Currency: Balboa

Market Exchange Rate: US $1=1 Balboa (fixed)

Nominal Gross Domestic Product (GDP, 2002E): $10.2 billion *(2003F)*: $10.6 billion

Real GDP Growth Rate (2002E): 0.7 per cent *(2003F):* 2.6 per cent

Inflation Rate (2002E): 0.7 per cent *(2003F):* 1.4 per cent

Annual Toll Collection at Panama Canal (FY 2001): $578,358,220 (6 per cent of GDP)

Major Exports: Bananas, Shrimp, Sugar

Main Destinations of Exports: United States, Seeden, Costa Rica

Major Imports: Capital Goods, Oil, Food Products

Main Origins of Imports: United States, Ecuador, Venezuela

Foreign Debt (2001): $7.6 billion

Energy Overview

Oil Production (2001E): 1,000 barrels per day (bbl/d)

Oil Consumption (2001E): 58,000 bbl/d

Net Oil Imports (2001E): 57,000 bbl/d

Crude Refining Capacity (1/1/02): 60,000 bbl/d

Coal Consumption (2000E): 70,000 short tons (all imported)

Electric Generation Capacity (2000): 1.35 million kilowatts (45 per cent hydroelectric)

Electricity Generation (2000E): 4.9 billion kilowatthours

Electricity Consumption (2000E): 4.7 billion kilowatthours

Environmental Overview

Total Energy Consumption (2000E): 0.16 quadrillion Btu* (<0.1 per cent of world total energy consumption)

Energy-Related Carbon Emissions (2000E): 2.36 million metric tons of carbon (<0.1 per cent of world total carbon emissions)

Per Capita Energy Consumption (2000E): 54.4 million Btu (vs U.S. value of 351.0 million Btu)

Per Capita Carbon Emissions (2000E): 0.8 metric tons of carbon (vs U.S. value of 5.6 metric tons of carbon)

Energy Intensity (2000E): 16,602 Btu/$1995 (vs U.S. value of 10,918 Btu/$1995)**

Carbon Intensity (2000E): 0.25 metric tons of carbon/thousand $1995 (vs U.S. value of 0.17 metric tons/thousand $1995)

Sectoral Share of Energy Consumption (1998E): Industrial (16.2 per cent), Transportation (46.5 per cent), Residential (36.3 per cent), Commercial (1.0 per cent)

Sectoral Share of Carbon Emissions (1998E): Industrial (24.7 per cent), Transportation (41.0 per cent), Residential (32.2 per cent), Commercial (2.1 per cent)

Fuel Share of Energy Consumption (2000E): Oil (75 per cent), Hydro (23 per cent), Coal (1.0 per cent), Natural Gas (0.0 per cent)

Fuel Share of Carbon Emissions (2000E): Oil (98.5 per cent), Coal (1.5 per cent), Natural Gas (0.0 per cent)

Number of People per Motor Vehicle (1998): 9.8 (vs U.S. value of 1.3)

Status in Climate Change Negotiations: Non-Annex I country under the United Nations Framework Convention on Climate Change (ratified on May 23rd, 1995). Ratified the Kyoto Protocol on March 5th, 1999.

Major Environmental Issues: water pollution from agricultural runoff threatens fishery resources; deforestation of tropical rain forest; land degradation.

Major International Environmental Agreements: Biodiversity, Climate Change, Climate Change-Kyoto Protocol, Desertification, Endangered Species, Hazardous Wastes, Law of the Sea, Marine Dumping, Nuclear Test Ban, Ozone Layer Protection, Ship Pollution, Tropical Timber 83, Tropical Timber 94, Wetlands, Whaling.

62

Paraguay

Paraguay has no known oil or gas reserves but is a major producer and exporter of hydroelectric power. Paraguay is part of the Southern Cone and a member of MERCOSUR, the Southern Common Market.

General Background

The Republic of Paraguay plays an important role in the economies of neighboring MERCOSUR members, Brazil and Argentina. Paraguay's two large hydroelectric plants, Itaipu and Yacyreta, meet approximately 25 per cent of Brazilian electricity demand and 40 per cent of Argentine electricity demand respectively. Paraguay also exports soybeans and cotton to its neighbors and many Paraguayans work outside of the country, sending their remittances home. Accordingly, Paraguay's economy is highly sensitive to regional trends. For instance, Argentina's financial crisis, has adversely affected Paraguayan economic growth. Paraguay's real gross domestic product (GDP) grew by only 1.0 per cent in 2001 and is expected to grow by 1.8 per cent in 2002.

Paraguay's President, Luiz Gonzalez Macchi, has been in office since March 1999. President Macchi (formerly President of the Chamber of Senators) constitutionally succeeded Raul Cubas Grau, who had resigned. The country is scheduled to hold Presidential elections on April 27, 2003.

Oil

Oil accounted for slightly less than half of Paraguay's total energy consumption in 2000, with the balance attributed to hydropower. The country relies entirely on imports to meet its crude oil demand, which totaled 29,000 barrels per day (bbl/d) in 2001, a 48 per cent increase since 1991.

Imported crude oil comes primarily from Argentina's Palmar Largo field in Formosa Province and is processed at the country's sole refinery, Villa Elisa. Crude oil trade between Paraguay and Argentina began in October of 1998, when Petroleos Paraguayos (Petropar, the state oil company)

began to purchase its crude oil exclusively from Argentina. Prior to 1998, Paraguay imported oil from Algeria, as that country's generally lighter crude was well suited to the Villa Elisa refinery.

To-date, no oil has been discovered in Paraguay. There has been some exploration by both local and North American companies, encouraged by the discovery in 1984 of substantial deposits in the Argentine province of Formosa, close to the Paraguayan border, and by exploration in the Bolivian Chaco.

Natural Gas

Although Paraguay neither consumes nor produces natural gas, the country could become a transit point for Bolivian natural gas in the near future. In March 2002, the Foreign Ministers of Paraguay and Bolivia signed a preliminary agreement regarding the construction of a pipeline to transport Bolivian natural gas to Brazil via Paraguay. Construction of the pipeline, known as the Integration Gas Pipeline (GASIN), is scheduled to begin in 2003, with completion anticipated in 2005. The pipeline will stretch 3,300 miles, through Bolivia, Argentina, and Paraguay en route to Brazil, which is currently experiencing a crippling energy shortage. Securing access to Bolivian natural gas could also allow Paraguay to incorporate natural gas into its national fuel mix, which currently relies heavily upon hydroelectric generation and imported petroleum.

Electricity

Paraguay is a co-owner of two of the world's largest hydroelectric dams. The Itaipu hydroelectric

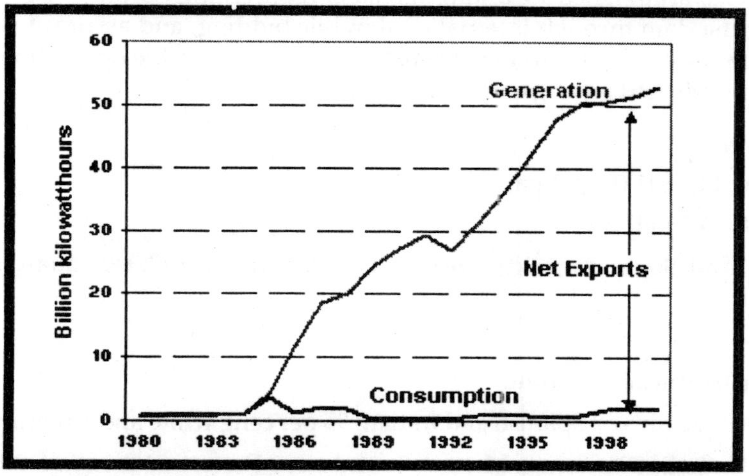

ELECTRICITY GENERATION AND CONSUMPTION IN PARAGUAY, 1980-1998

power plant, located on the Parana River along the country's border with Brazil, supplies 95 per cent of Paraguay's relatively tiny electricity demand. The remainder of Paraguay's electricity needs are supplied primarily by the Yacyreta hydroelectric power plant, located on the southern border with Argentina. Accordingly, Paraguay is a significant net electricity exporter. In 2000, Paraguay exported approximately 49.5 billion kilowatthours (bkwh) of electricity, approximately 90 per cent of its national electricity generation.

Itaipu, the world's largest hydroelectric dam, was completed in 1982 and has 18 functional 700-megawatt (MW) turbines, which combined constitute a total installed capacity of 12,600 MW. The plant is jointly owned and operated by Paraguay and Brazil, which divide Itaipu's electricity generation evenly, according to the 1973 Treaty of Itaipu. In 2001, a prolonged drought caused production at Itaipu to decline by almost 15 per cent. Itaipu produced 79 million megawatt-hours (MWh) in 2001, compared to 93 million MWh in 2000.

Paraguay has plans to add two more 700-MW units to Itaipu's existing 18 units. The first should begin service in 2003, while the second is expected to follow in 2004. The project will increase peaking capacity of the Itaipu station to help meet demands on the Brazilian and Paraguayan grids. It also will allow maintenance outages to be performed on two units at a time, while keeping 18 units operating year-round.

Paraguay's other major power plant, the Yacyreta hydroelectric plant, is equipped with 20 turbines and has an installed capacity of 3,200 MW. Paraguay and co-owner Argentina share the electricity generated at Yacyreta evenly, with most of Paraguay's electricity being exported to Argentina. Paraguay and Argentina signed the Treaty of Yacyreta in 1973, but the project suffered numerous delays, and the initial phase of construction was not completed until September of 1994. The twentieth turbine was finally set in to motion at Yacyreta in July 1989.

Paraguay and Argentina are planning to build another hydroelectric dam on the Paraná, the 3,000-MW Corpus Critsi dam. The dam would be built about 124 miles upstream from Yacyreta. In May 2000, Paraguayan Foreign Minister Juan Esteban Aguirre and Argentine Foreign Minister

Adalberto Rodriguez Giavarini signed a working agreement concerning Corpus Cristi. The agreement calls for constructing the dam through international public bidding, and assures that the project's construction will be environmentally sound. There has been considerable opposition to the project among the local community and among environmentalists.

Country Overview

Independence: May 14, 1811 (from Spain)

Population (2001E): 5.7 million

Location/Size: Central South America, northeast of Argentina/157,000 square miles–slightly smaller than California

Capital: Asunción

Languages: Spanish (official), Guarani

Ethnic Groups: Mestizo (mixed Spanish and Indian) 95 per cent, white and Amerindian 5 per cent

Religion: Roman Catholic 90 per cent, Mennonite and other Protestant denominations 10 per cent

Defense (8/97): Army: 14,900; Navy: 3,600, including 900 Marines; Air Force: 1,700; paramilitary police force: 14,800

Economic Overview

Currency: Guarani

Market Exchange Rate (9/18/02): US$1=6,385.60 Guaranis

Gross Domestic Product (2001E): $7.4 billion

Real GDP Growth Rate (2001E): 1.0 per cent *(2002F):* 1.8 per cent

Inflation Rate (2001E): 6.8 per cent *(2002F):* 7.3 per cent

Current Account Balance (2001E): -$220.2 million (-3 per cent of GDP)

Foreign Reserves (2001E): $713.5 million

Foreign Debt (2001E): $3.32 billion

Major Export Products: Hydroelectricity, cotton, soybeans, meat products

Major Export Destinations: Brazil, Argentina, Uruguay

Major Import Products: Consumer goods, intermediate goods, capital goods, fuels

Main Origins of Imports: Brazil, Argentina, Uruguay

Energy Overview

Oil Production: None

Oil Consumption (2001E): 29,000 barrels per day (bbl/d)

Crude Oil Refining Capacity (12/24/01): 7,500 bbl/d

Natural Gas Production/Consumption: None

Coal Consumption (2000E): 70,000 short tons (all imported)

Electricity Generation Capacity (2000E): 7.4 gigawatts, of which 99.5 per cent was hydroelectric

Electricity Generation (2000E): 51.5 billion kilowatthours (bkwh)

Electricity Consumption (2000E): 2.0 bkwh

Electricity Net Exports (2000E): 49.5 bkwh

Environmental Overview

Total Energy Consumption (2000E): 0.12 quadrillion Btu* (<0.1 per cent of world total energy consumption)

Energy-Related Carbon Emissions (2000E): 1.0 million metric tons of carbon (<0.1 per cent of world total carbon emissions)

Per Capita Energy Consumption (2000E): 21.3 million Btu (vs. U.S. value of 351.0 million Btu)

Per Capita Carbon Emissions (2000E): 0.20 metric tons of carbon (vs. U.S. value of 5.6 metric tons of carbon)

Energy Intensity (2000E): 12,552 Btu/$1995 (vs U.S. value of 10,918 Btu/$1995)**

Carbon Intensity (2000E): 0.12 metric tons of carbon/thousand $1995 (vs U.S. value of 0.17 metric tons/thousand $1995)

Sectoral Share of Energy Consumption (1998E): Residential (39.4 per cent), Industrial (37.9 per cent), Commercial (13.5 per cent), Transportation (9.2 per cent)

Sectoral Share of Carbon Emissions (1998E): Transportation (90.2 per cent), Residential (6.4 per cent), Industrial (3.4 per cent), Commercial (0.0 per cent)

Fuel Share of Energy Consumption (2000E): Oil (48.5 per cent), Coal (1.3 per cent), Natural Gas (0.0 per cent)

Fuel Share of Carbon Emissions (2000E): Oil (96.5 per cent), Coal (3.5 per cent), Natural Gas (0.0 per cent)

Renewable Energy Consumption (1998E): 627 trillion Btu* (0 per cent increase from 1997)

Number of People per Motor Vehicle (1998): 41.7 (vs. U.S. value of 1.3)

Status in Climate Change Negotiations: Non-Annex I country under the United Nations Framework Convention on Climate Change (ratified February 24th, 1994). Paraguay ratified the Kyoto Protocol on August 25th, 1999.

Major Environmental Issues: Deforestation (an estimated 2 million hectares of forest land have been lost from 1958-85); water pollution; inadequate means for waste disposal present health risks for many urban residents.

Major International Environmental Agreements: A party to Conventions on Biodiversity, Climate Change, Desertification, Endangered Species, Hazardous Wastes, Law of the Sea, Ozone Layer Protection and Wetlands. Has signed, but not ratified, the Nuclear Test Ban.

The total energy consumption statistic includes petroleum, dry natural gas, coal, net hydro, nuclear, geothermal, solar, wind, wood and waste electric power. The renewable energy consumption statistic is based on International Energy Agency (IEA) data and includes hydropower, solar, wind, tide, geothermal, solid biomass and animal products, biomass gas and liquids, industrial and municipal wastes.

Energy Industries

Organization: Oil refining–Petroleos Paraguayos (Petropar, the state oil refinery) ; *Electric power*–Administracion Nacional de Electricidad (ANDE), the state electricity corporation

Port: Asunción

Major Oil-Producing Fields: none

Refinery (12/24/01 Capacity): Petroleos Paraguayos–Villa Elisa (7,500 bbl/d)

63

Persian Gulf Oil and Gas Exports Fact Sheet

In 2002, the Persian Gulf countries (Bahrain, Iran, Iraq, Kuwait, Qatar, Saudi Arabia, and the United Arab Emirates) produced about 25 per cent of the world's oil, while holding nearly two-thirds (674 billion barrels) of the world's crude oil reserves. OECD gross oil imports from Persian Gulf countries averaged about 10.6 million barrels per day (bbl/d) during 2002, accounting for 27 per cent of the OECD's total gross oil imports. Besides oil, the Persian Gulf region also has huge reserves (1,923 trillion cubic feet–Tcf) of natural gas, accounting for 36 per cent of total proven world gas reserves.

General Background

The Persian Gulf, also known as the Arabian Gulf, is a 600-mile-long body of water which separates Iran from the Arabian Peninsula, and one of the most strategic waterways in the world due to its importance in world oil transportation. At its narrowest point (the Strait of Hormuz), the Gulf narrows to only 34 miles wide.

There have been, and continue to be, significant territorial disputes between Persian Gulf countries. Besides the Iraqi invasion of Kuwait in August 1990, and before that the Iran-Iraq War from 1980 to 1988, another important dispute is between the UAE and Iran over ownership of three islands–Abu Musa, Greater Tunb Island, and Lesser Tunb Island, all strategically located in the Strait of Hormuz. The three islands were effectively occupied by Iranian troops in 1992. In 1995, the Iranian Foreign

Ministry claimed that the islands are "an inseparable part of Iran." Iran rejected a 1996 proposal by the Gulf Cooperation Council (GCC) for the dispute to be resolved by the International Court of Justice, an option supported by the UAE. In early 1996, Iran took further moves to strengthen its hold on the disputed islands. These actions included starting up a power plant on Greater Tunb, opening an airport on Abu Musa, and announcing plans for construction of a new port on Abu Musa. In September 2000, Iran stated its willingness to resume talks with the UAE on the dispute. In March 2000, *Jane's Defence Weekly* reported that satellite images of Abu Musa and the Tunbs did not show any evidence that Iran had fortified the islands militarily, or turned them into "unsinkable aircraft carriers capable of closing the (Hormuz) Strait during a crisis." On December 31, 2001, the GCC issued a statement reiterating its support for the UAE's sovereignty over Abu Musa and the Tunbs, declared Iran's claims on the islands as "null and void," and backed "all measures.by the UAE to regain sovereignty on its three islands peacefully."

In February 1991, Iraqi troops, before being expelled from Kuwait by coalition forces, dumped millions of barrels of oil into the Persian Gulf, creating an environmental crisis and also threatening desalination plants in the region. During the Iran-Iraq War, oil tankers were attacked in the Persian Gulf by both Iraq and Iran, leading in part to the U.S. decision in 1987 to "reflag" Kuwait tankers and also to increase U.S. naval forces in the Persian Gulf.

On March 20, 2003, a U.S.-led coalition began attacks on Iraqi targets, followed by a ground invasion. By mid-April, U.S. forces entered Baghdad, Mosul, Kirkuk, and other Iraqi cities, while securing important oilfields in the northern and southern parts of the country. The war in Iraq began just over a year after President Bush, in his January 2002 State of the Union address, labeled Iraq (along with Iran and North Korea) as members of an "axis of evil" that supported terrorism and were developing weapons of mass destruction. This speech came five months after the terrorist attacks of September 11, 2001, the worst such attack ever on U.S. soil.

Oil and Gas Reserves, Production, Capacity

The Persian Gulf contains around 674 billion barrels of proven oil reserves, representing approximately two thirds of proven, conventional world oil reserves, and 1,923 Tcf of natural gas reserves (35 per cent of the world total). Also, at the end of 2002, Persian Gulf countries maintained

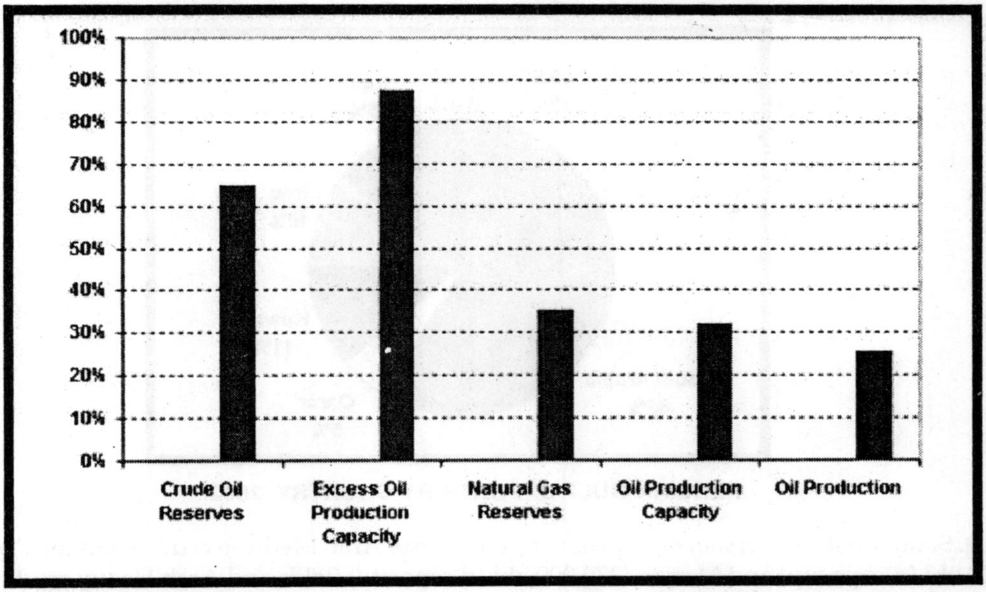

PERSIAN GULF AS A PER CENT OF WORLD (2002)

about 22.3 million bbl/d of oil production capacity, or 32 per cent of the world total. Perhaps even more significantly, the Persian Gulf countries normally maintain an overwhelming share (around 90 per cent) of the world's *excess* oil production capacity (note: as of April 2003, following the Iraq war, excess world oil production capacity was only around 0.7-1.2 million bbl/d, all of which was located in the Persian Gulf region). Excess production capacity is important because, in the event of an oil supply disruption, such as the recent Venezuela, Iraq, and Nigeria situations, this oil can be brought online to compensate.

In 2002, Persian Gulf countries had estimated net oil exports of 15.5 million bbl/d of oil (see pie chart). Saudi Arabia exported the most oil of any Persian Gulf country in 2002, with an estimated 7.0 million bbl/d (45 per cent of the total). Also in 2002, Iran had estimated net exports of around 2.3 million bbl/d (15 per cent), followed by the United Arab Emirates (2.1 million bbl/d–13 per cent), Kuwait (1.7 million bbl/d–11 per cent), Iraq (1.6 million bbl/d–10 per cent), Qatar (0.8 million bbl/d–5 per cent), and Bahrain (0.01 million bbl/d–0.1 per cent).

According to the Energy Information Administration's *International Energy Outlook 2002*, Persian Gulf oil production is expected to reach about 30.7 million bbl/d by 2010, and 42.9 million bbl/d by 2020, compared to about 21.7 million bbl/d in 2000. This would increase Persian Gulf oil production capacity to 35 per cent of the world total by 2020, up from 28 per cent in 2000.

Offshore Persian Gulf Oil Development

Major offshore Persian Gulf oil fields include Khafji and Hout, both of which are connected to Saudi Arabia's Safaniyah, the world's largest offshore oilfield (with estimated reserves of 19 billion

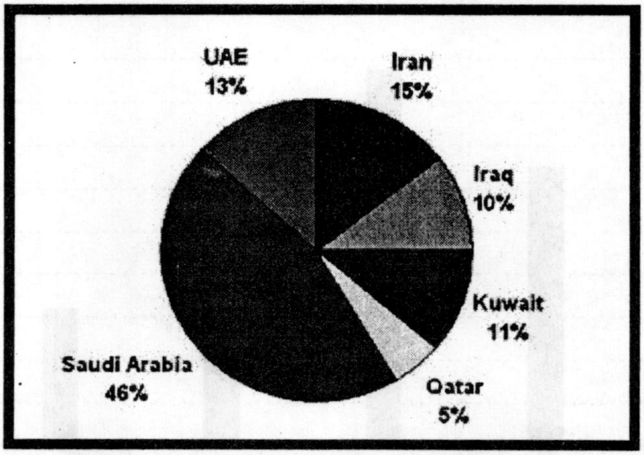

PERSIAN GULF EXPORTS BY COUNTRY–2002

barrels). Saudi offshore Persian Gulf production includes Arab Medium crude from the Zuluf (over 500,000 bbl/d capacity) and Marjan (270,000 bbl/d capacity) fields and Arab Heavy crude from the Safaniya field. The Doroud 1&2, Salman, Abuzar, Foroozan, and Sirri fields comprised the bulk of Iran's offshore output, all of which is exported. Iran plans extensive development of existing offshore fields and hopes to raise its offshore production capacity sharply. Iran's national oil company (NIOC) has expressed interest in developing five oil and gas fields in the Hormuz region (Henjam A–HA, HB, HC, HD, and HE), which, according to NIOC, hold an estimated 400 million barrels of liquids (oil, natural gas condensates, etc.) and have production potential of 80,000 bbl/d.

Offshore Persian Gulf Natural Gas Development

Besides oil, the Persian Gulf region also is important because it contains huge reserves (1,923 Tcf) of natural gas, with Iran and Qatar holding the world's second and third-largest reserves (behind Russia), respectively. This likely will become increasingly important in coming years, as both domestic gas consumption and gas exports (by pipeline and also by liquefied natural gas–LNG–tanker) increase. In late 2000, Saudi Arabia resolved a long-standing offshore Persian Gulf border dispute with Kuwait, opening the door to development of the huge (13-Tcf) Dorra gas field, which lies in waters straddling Iranian, Saudi, and Kuwaiti territories. Most of Qatar's gas is located in the North Dome Field, which contains 380 Tcf of in-place and 239 Tcf of recoverable reserves, making it the largest known non-associated gas field in the world. The Qatari government believes that the country's economic future lies in developing this vast gas potential. Currently, Qatar has two LNG exporters: Qatar LNG Company (Qatargas); and Ras Laffan LNG Company (Rasgas). The $10 billion Dolphin Project is expected to supply gas, beginning in 2005, from Qatar's North Dome to the United Arab Emirates, and Oman. Pakistan also could be supplied by Dolphin at some point in the future, although at present this seems unlikely.

Another major Persian Gulf offshore gas project is Iran's huge South Pars field. Current estimates are that South Pars contains 280 Tcf of gas (some estimates run as high as 500 Tcf), of which a large fraction will be recoverable, and over 17 billion barrels of liquids. Development of South Pars is Iran's largest energy project, and already has attracted around $20 billion in investment. Natural gas from

South Pars largely is slated to be shipped north via the planned 56-inch, $500 million, IGAT-3 pipeline (a section of which is now being built by Russian and local contractors), as well as a possible IGAT-4 line, and then reinjected to boost oil output at the mature Aghajari field (output peaked at 1 million bbl/d in 1974, but has since fallen to 200,000 bbl/d), and possibly the Ahwaz and Mansouri fields (which make up part of the huge Bangestan reservoir in the southwest Khuzestan region). South Pars natural gas also could be exported, both by pipeline and possibly by liquefied natural gas (LNG) tanker.

In February 2003, Oil Minister Zanganeh officially inaugurated Phases 2 and 3 of South Pars development, which began to come onstream in September 2002. Already, Phases 2 and 3 reportedly are producing around 2 Bcf per day of natural gas, and 85,000 bbl/d of condensates. On September 29, 1997, Total (now TotalFinaElf) had signed a $2 billion deal (along with Russia's Gazprom and Malaysia's Petronas) to explore South Pars and to help develop the field during Phase 2 and 3 of its development. In July 2000, Italian firm ENI had signed a $3.8 billion deal with Iran to develop the South Pars region for gas. The deal reportedly was the largest between Iran and a foreign company since the 1979 Islamic Revolution.

In addition to South Pars, Iran aims to develop the 6.4-Tcf, non-associated Khuff (Dalan) reservoir of the Salman oil field, which straddles Iran's maritime border with Abu Dhabi, where it is known as the Abu Koosh field. NIOC is seeking to develop the Khuff reservoir, which could lead to the production of 500 Mmcf/d of non-associated gas, along with the 120,000 bbl/d of crude oil that is now being produced from a shallower reservoir. Also, the 47-Tcf North Pars development will be integral to Iran's long-term gas utilization plans. Development plans call for 3.6 Bcf/d of gas production, of which 1.2 Bcf/d would be re-injected into the onshore Gachsaran, Bibi Hakimeh, and Binak oil fields. The other 2.4 Bcf/d would be sent to the more mature Agha Jari oil field.

Oil Flows

Strait of Hormuz

In 2002, the vast majority (around 88 per cent) of oil exported from the Persian Gulf transited by tanker through the Strait of Hormuz, located between Oman and Iran. By far the world's most important oil "chokepoint," accounting for transit of around two-fifths of all world traded oil, the Strait consists of 2-mile wide channels for inbound and outbound tanker traffic, as well as a 2-mile wide buffer zone. Closure of the Strait of Hormuz would require use of longer alternate routes (if available) at increased

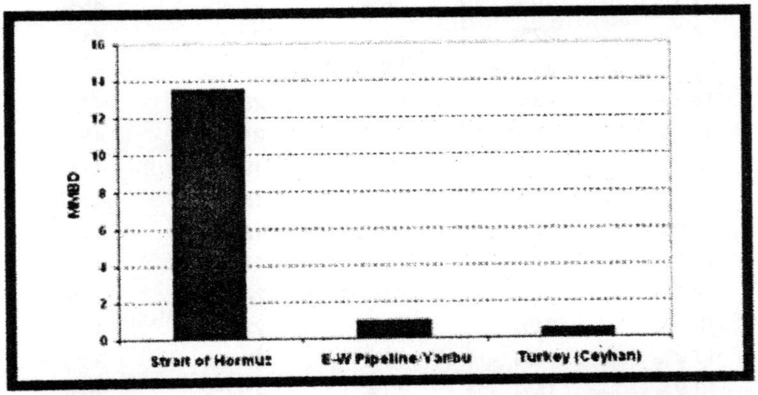

PERSIAN GULF OIL EXPORTS BY ROUTE–2002 (MILLION BARRELS PER DAY)

transportation costs. Such routes include the 4.8-million-bbl/d-capacity East-West Pipeline across Saudi Arabia to the port of Yanbu, and the Abqaiq-Yanbu natural gas liquids line across Saudi Arabia to the Red Sea. The 13.6 million bbl/d or so of oil which transit the Strait of Hormuz goes all over the world, eastwards to Asia (especially Japan, China, and India) and westwards (via the Suez Canal, the Sumed pipeline, or around the Cape of Good Hope in South Africa) to Western Europe and the United States. Another route for Saudi oil exports which reportedly has been under consideration is through Yemen to the Gulf of Aden.

Bab al-Mandab

Oil heading westwards by tanker from the Persian Gulf towards the Suez Canal or Sumed pipeline must pass through the Bab al-Mandab. Located between Djibouti and Eritrea in Africa, and Yemen on the Arabian Peninsula, the Bab al-Mandab connects the Red Sea with the Gulf of Aden and the Arabian Sea. Any closure of the Bab al-Mandab could keep tankers from reaching the Suez Canal/ Sumed Pipeline complex, diverting them around the southern tip of Africa. This would add greatly to transit time and cost, and effectively tie up spare tanker capacity. In December 1995, Yemen fought a brief battle with Eritrea over Greater Hanish Island, located just north of the Bab al-Mandab. The Bab al-Mandab could be bypassed by utilizing the East-West oil pipeline. However, southbound oil traffic, which totaled about 1 million bbl/d in 1997, would still be blocked. In addition, closure of the Bab al-Mandab would effectively block non-oil shipping from using the Suez Canal, except for limited trade within the Red Sea region.

Table 63.1: Net Oil Imports from the Persian Gulf Region

	As % of Demand			*As % of Total Net Oil Imports*		
	US	*W. Europe*	*Japan*	*US*	*W. Europe*	*Japan*
1982	10.4%	NA	58%	16.1%	NA	60%
1983	10.4%	NA	60%	10.1%	NA	60%
1984	10.4%	NA	61%	10.6%	NA	61%
1985	10.3%	NA	58%	7.1%	NA	59%
1986	9.7%	NA	58%	16.7%	NA	58%
1987	8.8%	NA	59%	18.1%	NA	60%
1988	8.7%	NA	57%	23.2%	NA	58%
1989	9.4%	NA	64%	25.8%	NA	63%
1990	11.3%	29%	66%	27.4%	45%	65%
1991	11.0%	27%	64%	27.7%	41%	64%
1992	10.4%	26%	66%	25.6%	42%	66%
1993	10.3%	29%	69%	23.3%	47%	69%
1994	9.7%	25%	70%	21.4%	45%	69%
1995	8.8%	23%	70%	19.8%	44%	70%
1996	8.7%	21%	69%	18.8%	41%	70%
1997	9.4%	23%	75%	19.1%	44%	75%
1998	11.3%	26%	75%	21.8%	47%	77%
1999	12.6%	22%	73%	24.8%	43%	74%
2000	12.6%	21%	75%	23.8%	42%	75%
2001E	13.9%	18%	76%	25.0%	33%	76%
2002E	13.9%	15%	73%	21.7%	29%	75%

Suez/Sumed Complex

After passing through the Bab al-Mandab, oil en route from the Persian Gulf must pass either through the Suez Canal or the Sumed Pipeline complex in Egypt. Both of these routes connect the Red Sea and Gulf of Suez with the Mediterranean Sea. Over 3 million bbl/d of Persian Gulf oil exports transit the Suez Canal/Sumed complex, destined mainly for Europe and the United States. Any closure of the Suez Canal and/or Sumed Pipeline would divert tankers around the southern tip of Africa (the Cape of Good Hope), adding greatly to transit time and effectively tying up tanker capacity.

Other Export Routes

In 2002, around 1.9-2.2 million bbl/d (12 per cent -14 per cent) of oil from the Persian Gulf was exported via routes besides the Strait of Hormuz. This oil was exported mainly: 1) via the Saudi East-West pipeline to the port of Yanbu on the Red Sea (about 1 million bbl/d); 2) via pipeline from Iraq's Kirkuk oil region to the Turkish port of Ceyhan (about 0.5-0.8 million bbl/d); 3) by pipeline via Syria (around 0.2 million bbl/d); 4) by various means (smuggling by truck and small boat, for instance) to a variety of destinations, including Kurdish areas of northern Iraq, Turkey, Jordan, Iran, India, and Pakistan; and 5) by truck to Jordan.

OECD Oil Imports from the Persian Gulf

U.S. gross oil imports from the Persian Gulf fell during 2002, to around 2.3 million bbl/d (almost all of which was crude), from 2.8 million bbl/d in 2001. The vast majority of Persian Gulf oil imported by the United States came from Saudi Arabia (69 per cent), with significant amounts also coming from Iraq (20 per cent), Kuwait (10 per cent), and small amounts (less than 1 per cent total) from Qatar and the United Arab Emirates. Iraqi oil exports to the United States fell sharply in 2002, to around 442,000 bbl/d, compared to 795,000 bbl/d in in 2001. Saudi exports fell from 1.66 million bbl/d in 2001 to 1.55 million bbl/d in 2002. Overall, the Persian Gulf accounted for about 22 per cent of U.S. *net* oil imports, and 11 per cent of U.S. oil demand, in 2002.

Western Europe (defined as European countries belonging to the Organization for Economic Cooperation and Development–OECD) averaged 2.3 million bbl/d of oil imports from the Persian Gulf during 2002. This represented a decrease of about 0.4 million bbl/d from the same period in 2001. The largest share of Persian Gulf oil exports to Western Europe came from Saudi Arabia (51 per cent),

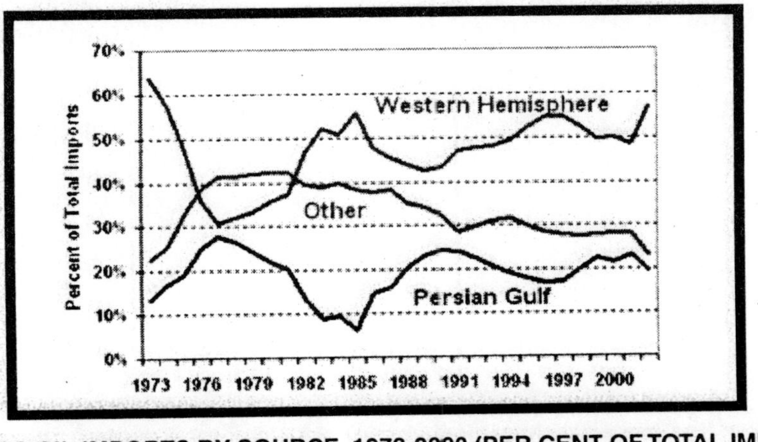

U.S. GROSS OIL IMPORTS BY SOURCE, 1973-2000 (PER CENT OF TOTAL IMPORTS)

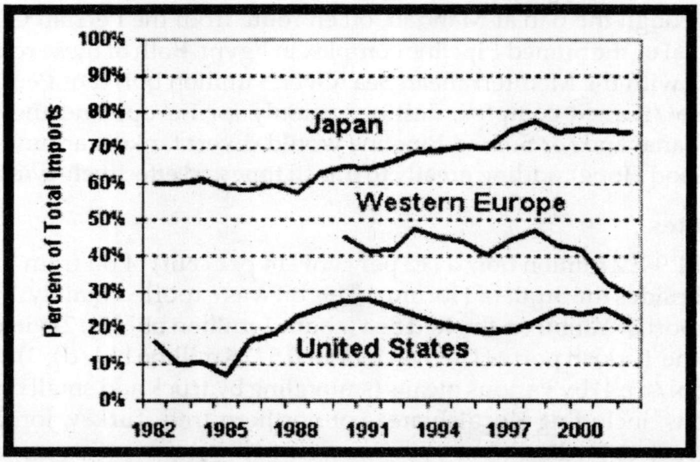

NET OIL IMPORTS FROM THE PERSIAN GULF AS A PER CENT OF TOTAL NET OIL IMPORTS

with significant amounts also coming from Iran (27 per cent), Iraq (13 per cent), and Kuwait (6 per cent).

Japan averaged 3.9 million bbl/d of net oil imports from the Persian Gulf during 2002. Japan's oil imports from the Persian Gulf as a percentage of demand were down just slightly relative to 2001, at about 75 per cent. Japan's dependence on the Persian Gulf for its oil supplies has increased sharply since the low point of 58 per cent in 1986. During 2002, around 31 per cent of Japan's Persian Gulf imports came from Saudi Arabia, 30 per cent from the United Arab Emirates, 14 per cent from Iran, 13 per cent from Kuwait, 11 per cent from Qatar, and around 1 per cent from Bahrain and Iraq combined.

64

Peru

Peru represents a potentially expanding market for U.S. and other foreign energy companies, although privatization is proceeding more slowly than in the 1990s. The development of Peru's huge Camisea natural gas field project, which was on hold for several years, is expected to be delivering natural gas to Lima by August 2004.

General Background

After former President Fujimori fled Peru to seek refuge in Japan in November 2001, an interim administration under Valentin Paniagua led the country, helping to put in place a policy of fiscal restraint and deficit reduction. In July 2001, Alejandro Toledo assumed the presidency after national elections, pledging to boost the economy, reduce unemployment, create 400,000 jobs per year, and significantly reduce poverty in the Andean country.

Since taking office, President Toledo's administration has coordinated an economic recovery, underway since mid-2002. In effort to redistribute political and economic power, President Toledo is attempting to integrate 25 regional governments, created following regional elections in November 2002. The central government still retains the right to decide which state assets it would maintain control and which it would give to regional governors' control.

Over the past several years, Peru's economy has weathered global economic difficulties better than many of its neighbors. Throughout the 1990s, Peru maintained relatively positive economic indicators, such as gross domestic product (GDP) growth, despite fallout from the Asian financial crisis, El Nino weather-related problems, and low prices for mining exports, a major revenue source for Peru. In 2001, world economic slowdown (especially after the September 11 terrorist attacks in the United States), lower prices for its commodity exports (which make up the majority of Peru's export earnings), and domestic political uncertainty, battered Peru's economy. As a result, real GDP grew by only 0.2 per cent in 2001.

For 2002, however, the economy bounced back, with an estimated real GDP growth of about 4.5 per cent (some sources estimated 5.2 per cent). This economic turnaround resulted from increased copper and zinc exports, particularly from the new Antamina mine, which has been in operation for one year, as well as from complying to the IMF standby accord, which requires the country to keep its fiscal deficit at 1.9 per cent of GDP for 2003. The IMF reported in February 2003 that Peru passed a second review of its two-year standby accord. The IMF also stated that it supports its plans to increase tax collection through reforms.

Peru is still grappling with other problems, including: high unemployment; a credit crunch; stagnating foreign investment; a slow-moving privatization program; high levels of private and public sector debt; and rampant poverty. Inflation, on the other hand, remains low, due to tight fiscal and monetary policies, depressed domestic demand, and a relatively stable exchange rate for the country's currency, the Nuevo sol. Peru's former Economy and Finance Minister, Pedro Pablo Kuczynski, has introduced several tax reforms, including a reduction in the "solidarity payroll tax" (IES) from 5 per cent to 2 per cent, and an increase in the selective fuel consumption tax (ISCC).

Peru's privatization program has slowed after a decade of liberalization. In 2002, the Toledo administration managed to raise only half of the expected $700 million in the face of stiff popular opposition. In June 2002, after the Toledo administration agreed to sell two electric power plants, Egasa and Egesur, to Belgium's Tractebel, a six-day protest broke out in Arequipa. The government was forced to suspend the sale in order to restore calm in Arequipa. In January 2003, the government also had to postpone the privatization of the Yuncan hydro power plant, because the regional government did not approve the sale.

Peru is a member of the Andean Community (ANCOM), set up in March 1996 by leaders of Bolivia, Colombia, Ecuador, Peru, and Venezuela. At that time, the five national leaders expressed their intent to move towards a single market along the lines of the European Union, although significant policy differences need to be worked out. In November 1997, Peru joined the Asia Pacific Economic Cooperation (APEC) forum. Peru also receives trade preferences from the United States under the 1991 Andean Trade Preferences Act, which the US House of Representatives (on November 16, 2001) voted to extend through 2006.

Oil

Peru has estimated proven oil reserves of 323 million barrels as of January 2003. Peru's crude oil production fell sharply in the late 1980s and early 1990s, before a modest and short-lived recovery began in 1993. Since 1996, however, Peru's oil production has fallen from 122,000 barrels per day (bbl/d) to about 97,666 bbl/d in 2002. Peru's oil consumption in 2002 was 160,000 bbl/d and net imports were 83,000 bbl/d, mainly from Colombia, Ecuador, and Venezuela.

Production and Exploration

The major oil producing companies in Peru are Pluspetrol and Pecom of Argentina (through Perez Companc del Peru), Peru's Petro-Tech and Aguaytia and China's Sapet. Pluspetrol, which acquired Occidental Petroleum's 100 per cent interest in this Block 1-AB in 2000, was the country's leading producer in 2002, extracting 22.8 million barrels of oil, roughly 40 per cent of Peru's production.

Current oil activities are mainly in the northern jungle along the Ecuador border, northeastern and central Peru and offshore areas stretching from the northern city of Tumbes to Pisco, south of Lima. Peru's 37 million acres of offshore basins are largely unexplored. Most of Peru's ten coastal basins are entirely offshore, but at both ends of the 1,500-mile coastline, the sedimentary basins

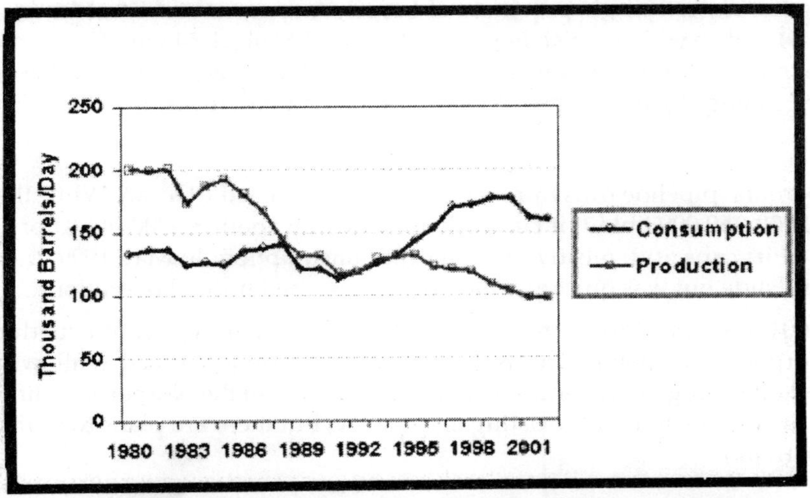

PERU'S OIL PRODUCTION AND CONSUMPTION, 1980-2001

stretch from onshore areas across the continental shelf and down the continental slope. Two of these basins located in the north have commercial production: The Tumbes-Progreso basin in the north at the Ecuador border; and the Talara basin just south of Tumbes. The Pisco Basin, located south of Lima, is going through technical evaluations. The Marañon and Ucayali basins, which stretch from the northern jungle to northeast section of Peru, are also oil producing areas. (Perupetro provides a detailed map of the Peru's oil block distribution).

Since 1986, no major hydrocarbon finds have been made in Peru. Oil exploration efforts in Peru ran into problems in 1999. Shortly before an August 1999 bidding round was to begin, Repsol revealed that it had drilled two dry wells in Block Z-29 offshore in the Trujillo Basin (in July 2000, Repsol returned the block to Perupetro). The U.S. company Anadarko withdrew from Peru after spending three years exploring its Block 84. Pan Energy Exploration, associated with Peruvian companies, withdrew from Block 85 after drilling dry wells. Coastal Peru, which was a partnership including the U.S.-based Hunt Corporation, is no longer operating after three test wells failed to produce any oil. This has discouraged companies in investing in Peru. During 2000 and 2001, oil companies drilled 63 wells and in 2002, only 10.

The Toledo government has been attempting to attract foreign oil investment in part by making terms for investment more attractive, and also by setting up a better legal framework for investors. In early January 2002, President Toledo signed two bills aimed at encouraging oil and mining exploration. Toledo stated that it was "of critical importance" to reverse the several-year decline in Peru's oil production and the corresponding increase in Peru's net oil imports. In 2002, however, the state-owned investment company, Perupetro, which is responsible for promoting the investment of hydrocarbon exploration and exploitation activities in the country, signed only two exploration and production (E&P) contracts with Burlington Resources and Petrotech. In response, the Peruvian government in October 2002 reduced royalties on crude and natural gas exploration and production contracts by up to 30 per cent with the hope of attracting investment and boosting oil production.

A number of companies are conducting exploration in Peru. Repsol-YPF announced in January 2002 that it planned to invest $500 million during 2001-2004 in the Peruvian hydrocarbons sector. In December 2002, Burlington Resources received permission to explore for hydrocarbons in block 70, a region covering the eastern jungle provinces of Loreto, Amazonas and San Martin. Burlington is exploring blocks 34 and 35 with Spanish Repsol-YPF, as well as block 64 with Occidental Petroleum. Perupetro is also in the process of finalizing five 7-year E&P contracts with U.S. based Occidental, Hunt Oil and BPZ Energy, Spain's Repsol-YPF and two Canadian firms.

Pipeline

Peru's Norperuano pipeline runs from the Amazon to the Pacific Ocean. While the pipeline has a capacity of 200,000-250,000 bbl/d, it currently carries only about 80,000 bbl/d of crude oil (less than 40 per cent of its capacity), mostly for domestic consumption. In May 1999, the pipeline was ruptured by a mudslide but was quickly repaired and resumed normal operations.

In August 2001, the first shipment of oil from the Amazonian region of Ecuador was sent by river to Peru's Norperuano pipeline. This is the first stage in a project aimed at allowing Ecuador to export landlocked oil through Peru, and also to more fully utilize the Norperuano line. For now, oil is being sent via river barge to Peru's existing oil pipeline, but there are plans to build a connecting pipeline to Norperuano.

Refining/Downstream

Peru has six refineries. The largest, La Pampilla, with a crude oil capacity of 100,000 bbl/d, underwent a 60 per cent privatization in 1996 and is controlled by Repsol-YPF. La Pampilla handles over half the total production of refined products. Petroperu operates 4 refineries: the 62,000-bbl/d-capacity Talara refinery; the 10,500-bbl/d Iquitos Loreto refinery in the northeast, the 6,500-bbl/d Conchan refinery; and the1,700-bbl/d El Milagro refinery. In 1994 the U.S.- based Maple Gas Corporation entered into a long-term lease agreement for the Pucallpa refining (3,250 bbl/d) and distribution facilities of Petroperu.

Peru's government was to begin privatizing state-owned refineries in the late 1990s; however, this was delayed despite considerable interest from international companies. The government renewed it efforts to privatize the four refineries–Talara, Iquitos, El Milagro, and Conchan–in 2002. These refineries, particularly Talara, need investment to expand and upgrade their facilities, which the state is unable to finance. But Peruvian public opinion continues to oppose privatization. Meanwhile, the government's recent transfer of power from the capital to the provinces is likely to complicate negotiations.

Natural Gas

Peru has proven natural gas reserves of about 8.7 trillion cubic feet (Tcf) as of January 2003. Peru produced around 13.1 billion cubic feet (Bcf) in 2001. Production is likely to increase sharply in coming years with development of the giant Camisea natural gas field, the largest in South America. Peru's demand is expected to increase as well, as the government plans to implement two measures to encourage the use of natural gas: 1) phasing out of the use of 84-octane gasoline, which has already been banned under international environmental agreements; and 2) substituting diesel-2 fuel in the transportation and industrial sector with natural gas. In 2001, natural gas made up only 2.2 per cent of total energy consumption in Peru.

Camisea Project

The Camisea natural gas field contains an estimated 13 trillion cubic feet (Tcf) of natural gas and 600 million barrels of natural gas liquids. Camisea gas reserves are found in two reservoirs, San Martín and Cashiriari, lying on either side of the Camisea River in the Ucayali basin, in the remote jungle east of the Andes, about 300 miles from Lima.

In 1986, Shell discovered the two natural gas fields in the Camisea area. In 2000, after much delay, the Special Committee of the Camisea Project (CECAM) finalized contracts for the development of Camisea. CECAM divided the $2.6 billion Camisea project into a 40-year contract covering exploration and production (upstream) and a 33-year contract dealing with transportation and distribution (downstream).

In February 2000, the government awarded Pluspetrol (operator), Hunt Oil, SK Group, and Tecpetrol (which is owned by Techint) the upstream portion of the contract. The consortium holds the rights for the extraction of natural gas and liquids hydrocarbon in Block-88 for 40 years. As of March 30, 2003, the consortium had drilled three wells (San Martín 1001,1002 and 1004). The fourth well, 1004, will be finished by May 2003.

The downstream portion contained three subcontracts: 1) transportation of gas from Camisea to Lima; 2) transportation of natural gas liquids from Camisea to the coast; and 3) distribution of gas in Lima and Callao. In October 2000, the Peruvian government awarded Transportadora de Gas del

Peru (TGP) consortium the first two contracts of the "downstream" portion of the Camisea project. Argentina's Techint and Pluspetrol, U.S.-based Hunt Oil, Peru's Graña y Montero, Algeria's Sonatrach, and South Korea's SK Corporation and Belgium's Tractebel make up TGP. The Project includes the construction and operation of two pipelines, one for natural gas (446 miles) and one for natural gas liquids (338 miles). Commercial operation is set to begin in August 2004.

In May 2002, the government of Peru awarded Tractebel the third portion of the downstream contract, a 30-year concession for the construction and operation of a gas distribution network in Lima and adjacent port Callao. Tractebel will build a 37.5-mile main distribution pipeline that will deliver natural gas to some of the larger industries and power generators around Lima.

Natural gas from Camisea is slated to fuel power plants in Lima and north-central Peru, as well as large industrial customers. The initial capacity of the Camisea pipeline is expected to be 250 million cubic feet per day (MMcf/d) of natural gas, expandable up to 729 MMcf/d and 70,000 bbl/d of condensate. First deliveries should begin in August 2004 to Lima. Natural gas is to be either re-injected or transported across the Andes to Lima (via Pisco on the southern coast) by pipeline. In the Pisco area, the fish industry and an iron-carbide plant represent potential customers. In Lima, the cement industry and Lima's private electric generating company, Etevensa, are probable main customers for Camisea gas. Etevensa runs the country's largest power plant at Ventanilla, just north of Lima.

Another possibility for Camisea natural gas production is to export it as liquefied natural gas (LNG), most likely to the U.S. West Coast or Mexico. President Toledo has proposed that Peru and Bolivia (which has the second-largest natural gas reserves in Latin America, after Venezuela) combine their individual natural gas development efforts, including LNG exports and construction of a petrochemical complex (fed by Peruvian and Bolivian natural gas) at the Peruvian port of Ilo. This plan also would require construction of a pipeline connecting landlocked Bolivian gas reserves (in the Margarita fields) to Ilo. Pacific LNG, a consortium comprising Repsol-YPF, BG, and Pan American Energy, is developing the project and favors a Chilean port, as it would be cheaper. However, a land dispute between Bolivia and Chile has made the decision politically sensitive (in the 1879-1883 War of the Pacific, Bolivia lost its only outlet to the sea to Chile). A majority of Bolivians prefer a Peruvian port to a Chilean. The Bolivian government hopes to make a decision by late spring 2003.

Hunt Oil, along with SK Corporation, is leading a separate LNG project. Hunt has already conducted a feasibility study to determine the prospects for exporting excess Camisea gas to Mexican and United States markets, as well as environmental impact study. Hunt plans to call for bids in April 2003 on the construction of a liquefaction plant. The project, however, will only go forward if Hunt can secure LNG buyers. Hunt Oil has been in negotiations with Sempra Energy International, which is planning to construct a regasification plant in Mexico. Hunt Oil has also indicated that it is willing to consider a joint venture with Pacific LNG consortium.

Camisea is expected to generate \$5-\$6 billion in royalties and tax revenues for Peru over the next 30+ years. Government officials also say that the project will lower energy costs, replace dirtier fuels and create jobs. Nonetheless, the Camisea project has encountered stiff opposition since its inception. Environmental groups argue that extraction processes and pipeline will cause irreversible damage to the rainforest and to the lifestyle of indigenous people living in the region. The Peruvian government and Camisea companies have been recently trying to secure \$350 million in loans from the Inter-American Development Bank (IDB) to help finance the remaining work of the project. IDB will review an environmental impact study, which was completed in December 2002, before it makes a decision. The IDB indicated recently that a decision on an initial loan of $75 million would be made in May 2003.

Aguaytia Project

In 1993, Maple Gas was awarded the concession for the Aguaytia gas field, the first natural gas field to be commercially developed in Peru. The Aguaytia natural gas project accounts for about 42 per cent of Peru's total gas output. Aguaytia has produced an average of about 55 MMcf/d of natural gas since it came onstream in June 1998. Natural gas from Aguaytia fires a 160-MW power plant that feeds Peru's North Interconnected Power grid. Maple attracted the following companies to participate in operating the project: PanEnergy Corporation (now part of Duke Energy Corporation); El Paso Energy Corporation; Illinova Corporation (now part of Dinergy Corporation); and Pennsylvania Power and Light; as well as the Scudder Latin American Power Fund. The project is being reorganized into separate units: natural gas (Aguaytia Energy del Peru); electricity generation (Termoselva); and power transmission (Eteselva).

Electricity

Peru has about 6.1 million kilowatts of installed electric generation capacity, about half of which is hydroelectricity, with the rest mainly diesel and fuel oil. Peru generated 20.6 billion kilowatthours (kwh) of electricity in 2001, up from 19.5 billion kwh in 2000. Peru consumed 19.1 billion kwh in 2001, of which over 90 per cent was supplied by hydropower. Besides being the main source of electric power in Peru, hydroelectricity also supplies about one-third of Peru's total energy requirements. Because hydroelectric power output fluctuates due to rainfall patterns, Peru is attempting to reduce its dependence on hydropower and replace it with natural gas-fired plants.

Sector Organization

In the 1990s, Peru's government approved plans for liberalizing the country's electricity sector. In 1992, an Electricity Concessions Law was passed, which allowed for private generation, transmission, and distribution of electricity in Peru. The state utility ElectroLima and most of ElectroPeru's assets were divested as a result of the law. In November 1997, Peru's Congress approved legislation, limiting firms to a 15 per cent market share in electricity generation, transmission or distribution. The law also allows the government to block any acquisition giving a private company more than a 5 per cent market share in more than one electric power sector. Finally, the legislation gives Peru's government the right to veto any acquisitions deemed contrary to the "national interest."

Portions of Peru's electric sector, particularly electricity generators, remain wholly or partially state-owned, including ElectroPeru (the largest generator), Edegel, Egasa, and Egesur. Privatization of generation plants, which began in 1995, was seen as a means of rapidly increasing capacity, but the sales have stalled in the face of public opposition. Edegel, the second largest electricity generator after ElectroPeru, was privatized in 1995. In 1996, Electroperu's Egenor was privatized and is now controlled by Duke Energy. A consortium led by the Spanish-owned Endesa subsidiary, Endesar, bought 63 per cent of Empresa Electrica de Piura. In December 2001, PSEG Global acquired of 100 per cent of Empresa de Electricidad de los Andes S.A. (ElectroAndes). Electroandes main assets are four hydroelectric plants: Yaupi (108 MW), Malpaso (54 MW), Pachachaca (12 MW), and Oroya (9 MW).

In June 2002, the government faced riots in Arequipa, Peru's second largest city, after the Toledo administration agreed to privatize two electric power plants (Egasa and Egesur) to Belgium's

Tractebel. The government was forced to suspend the sale in order to restore calm in Arequipa. Tractebel eventually backed out of the deal. In March 2003, Peru's Constitutional Tribune began hearing arguments for and against the privatization. If the Tribune votes in favor of the state, the government could call for another bidding round later this year. Peru's privatization agency, ProInversion, has also delayed the auctioning of another power plant, Yuncan, in January 2003, in the face of potential riots.

In September 1998, the Fujimore government froze concessions for all hydroelectric projects as part of push to maximize the potential value of the Camisea natural gas project. In December 2000, Peru's Ministry of Mines and Energy lifted restrictions on new hydroelectric power projects, resulting in a number of hydro dam construction approvals: Chevez (560 MW); Platanal (270 MW); San Gaban I (100 MW initially, with possible expansion to 460 MW); Tarucani (100 MW); Maranon (100 MW); and Huanza (86 MW); Poeches I and II (27 MW each); Quitarasca I (112 MW), Pias (20 MW); Ocona (140 MW); and Yuncan (130 MW). The Peruvian government has tried to privatize the Yuncan hydro project, but the Pasco regional government managed to halt the privatization of Yuncan, planned originally for January 17, 2003. The Pasco government had demanded a stake in the Yuncan plant. According to recent reports, it appears that the Pasco government has reached an agreement and will allow the Yuncan project to proceed. US-based Public Service Enterprise Group, Volcan, the Peruvian mining company, and Belgium's Tractebel are all interested in the project. The government has not yet committed to the sale of the $1billion Mantaro hydroelectric complex.

Power Grid

In September 2002, Colombian companies, Interconexion Electrica (ISA), Traselca and Empresa de Energia de Bogota (EEB), took over management of the Peruvian electricity distribution network. The Colombia firms created Red de Energia del Peru to operate the system. The company is responsible for the work previously maintained by Peru's state-owned Etecen and Etesur electric companies.

Peru has also been in the process of integrating its power grid with those of Colombia and Ecuador. The three countries have signed two preliminary agreements to this end in September 2001 and April 2002. These agreements are a part of a larger movement by the Andean Community to create a common electricity market. Once the integrated market is operating, the transmission lines will allow for a permanent flow of power between each one of the electric power systems, allowing each country to purchase power under the most favorable conditions

In November 2002, Peru and Ecuador signed a bilateral electricity integration agreement with Ecuador, under which Peru is to begin exporting electricity to Ecuador in September 2004. ISA of Colombia is responsible for constructing the Peruvian part of a $15 million 35-mile transmission line, which will run from Zorritos to Zarumilla. In the first stage, the line will have a capacity of 90 MW. The second stage of the project will involve the construction of $30 million 'back-to-back' continuous AC substation, allowing the transmission line to transport 150 MW in both directions. The third stage includes building a second substation to raise the line's capacity to 250 MW. The goal of the project is to allow Peru to transport excess hydropower during its rainy season to Ecuador, when its hydro plants experience low capacity.

In 1998, the Peruvian government awarded the Transmantaro Consortium, which is controlled by Hydro-Québec International, to build a 605-kilometer north-south transmission line (Mantaro-Socabaya), which unified Peru's central-northern and southern power grids. Since October 2000, Hydro- Québec TransÉnergie has been operating the line, under a 33-year contract.

Country Overview

Independence: July 28, 1821 (from Spain)

Population (2003E): 27 million

Location/Size: Western South America, between Chile and Ecuador; 796,836 square miles (slightly smaller than Alaska)

Major Cities: Lima (capital)

Languages: Spanish (official); Quechua (official); Aymara

Ethnic Groups: Indian (45 per cent); mestizo (37 per cent); white (15 per cent); black, Japanese, Chinese, and other (3 per cent)

Religions: Roman Catholic

Defense (8/98): Army (85,000), Navy (25,000), Air Force (15,000), Paramilitary Police (78,000)

Economic Overview

Currency: 1 Nuevo sol (Ns) = 100 centimos

Market Exchange Rate (4/04/03): US$1 = NS 3.60

Gross Domestic Product (2002E): $57.4 billion

Real GDP Growth Rate (2002E): 4.5 per cent *(2003F):* 4.2 per cent

Inflation Rate (consumer prices, 2002E): 0.2 per cent

Unemployment Rate (2002E): 8.4 per cent *(2003F):* 8.4 per cent

Merchandise Exports (2002E): $7.7 billion

Merchandise Imports (2002E): $7.3 billion

Merchandise Trade Balance (2002E): $0.4 billion

Major Export Products (2002): Gold, copper, fish and fish products, textiles, zinc

Major Import Products (2002): Intermediate goods; capital goods; consumer goods

Major Trading Partners (2002): United States, Japan, Germany, other Latin America (Brazil, Colombia, Chile, etc.)

Foreign Exchange Reserves (non-gold; 2002E): $9.5 billion

Total Foreign Debt (2002E): $28.3 billion

Energy Overview

Proven Oil Reserves (1/1/03): 323 million barrels

Oil Production (2002E): 97,666 barrels per day (bbl/d), of which 93,666 bbl/d was crude oil

Oil Consumption (2002E): 160,000 bbl/d

Net Oil Imports (2002E): 83,000 bbl/d

Crude Oil Refining Capacity (1/1/03): 190,950 bbl/d

Natural Gas Reserves (1/1/03): 8.7 trillion cubic feet (Tcf)

Natural Gas Production (2001E): 13.1 billion cubic feet (Bcf)

Natural Gas Consumption (2001E): 13.1 Bcf

Recoverable Coal Reserves (2001E): 1.17 billion short tons

Coal Production (2001E): 50,580 short tons

Coal Consumption (2001E): 920,000 short tons

Electric Generation Capacity (2001E): 6.1 million kilowatts (kw), of which 53 per cent was thermal, 47 per cent hydroelectric

Electricity Generation (2001E): 20.6 billion kilowatthours (kwh)

Environmental Overview

Total Energy Consumption (2001E): 0.55 quadrillion Btu* (0.14 per cent of world total energy consumption)

Energy-Related Carbon Emissions (2001E): 7.2 million metric tons of carbon (0.11 per cent of world total carbon emissions)

Per Capita Energy Consumption (2001E): 20.9 million Btu (vs. U.S. value of 341.8 million Btu)

Per Capita Carbon Emissions (2001E): 0.27 metric tons of carbon (vs. U.S. value of 5.51 metric tons of carbon)

Energy Intensity (2001E): 9,038 Btu/$1995 (vs U.S. value of 10,736 Btu/$1995)**

Carbon Intensity (2001E): 0.12 metric tons of carbon/thousand $1995 (vs U.S. value of 0.17 metric tons/thousand $1995)

Fuel Share of Energy Consumption (2001E): Oil (60.0 per cent), Coal (4.4 per cent), Natural Gas (2.2 per cent)

Fuel Share of Carbon Emissions (2001E): Oil (97.7 per cent), Coal (8.9 per cent), Natural Gas (3.5 per cent)

Status in Climate Change Negotiations: Non-Annex I country under the United Nations Framework Convention on Climate Change (ratified June 7th, 1993). Signatory to the Kyoto Protocol signed November 13th, 1998 and ratified September 12, 2002.

Major Environmental Issues: Deforestation; overgrazing of the slopes of the cost and sierra leading to soil erosion; desertification; air pollution in Lima; pollution of rivers and coastal waters from municipal and mining wastes.

Major International Environmental Agreements: A party to the Antarctic-Environmental Protocol, Antarctic Treaty, Biodiversity, Climate Change, Desertification, Endangered Species, Hazardous Wastes, Nuclear Test Ban, Ozone Layer Protection, Ship Pollution, Tropical Timber 83, Tropical Timber 94, Wetlands and Whaling.

The total energy consumption statistic includes petroleum, dry natural gas, coal, net hydro, nuclear, geothermal, solar, wind, wood and waste electric power. The renewable energy consumption statistic is based on International Energy Agency (IEA) data and includes hydropower, solar, wind, tide, geothermal, solid biomass and animal products, biomass gas and liquids, industrial and municipal wastes.

Oil and Gas Industries

Organization: Perupetro, which started operating in 1993, is the state company responsible for overall regulation and licensing of the country's oil and gas industries. Perupetro also negotiates oil and gas contracts with companies to explore and/or produce in Peru. Petroperu is the state oil company and Electroperu is the state electric power company. Regional state-owned electric company Egesur (for the south), as well as state mining company Centromin, are also slated for privatization.

Ports: Callao, Chimbote, Ilo, Iquitos, Matarani, Paita, Pucallpa, Salaverry, San Martin, Talara, Yurimaguas

Major Natural Gas Field: Camisea and Aguaytia

Foreign Energy Company Involvement: Barrett Resources, Burlington Resources, Coastal, Duke Energy, Elf Aquitaine, Empresa de Energia de Bogota, Enterprise Oil, ExxonMobil, Hanwha Energy, Hunt Oil, Hyundai, Interconexion Eletrica, Korean Petroleum Development Corp., Maple Production, Mosbacher Energy, Occidental Petroleum, Pangaea Peru Energy, Perez Companc, Petro-Tech, Phillips Petroleum, Pluspetrol, Quintana Minerals, Repsol-YPF, Sapet, Shell, SK Corp., Sonatrach, Technit, Tractebel, Traselca.

Pipelines: 500 miles for crude oil; 40 miles for natural gas (which will increase to 486 miles once Camisea project is completed).

Refineries (crude oil capacity): La Pampilla Lima (100,000 bbl/d); Talara (62,000 bbl/d); Iquitos Loreto (10,500 bbl/d); Conchan (6,500 bbl/d); Pucallpa (3,250 bbl/d); El Milagro (1,700 bbl/d)

65 Philippines

The Philippines is important to world energy markets because it is a growing consumer of energy, particularly electric power, and a potential market for foreign energy firms. It also may become a significant producer of natural gas.

General Background

Under the leadership of President Gloria Macapagal-Arroyo, the Philippines has undertaken an economic transformation, deregulating its energy sector and offering new incentives for foreign investment. President Macapagal-Arroyo, a trained economist, came into power when former President Joseph Estrada was forced to resign in 2001. But while a certain degree of success has been achieved, the country's fiscal deficit and declining currency value are still problematic.

Real gross domestic product (GDP) grew by 4.4 per cent in 2002. This increase exceeded both Philippine and international expectations. Much of the country's renewed economic vibrancy results from improved agricultural yields, as well as from an increase in domestic consumption brought about by reduced inflation, even though exports of consumer electronics have still been relatively weak. Real GDP growth for 2003 is projected at 3.7 per cent.

U.S. troops have been invited into the Philippines to work with local armed forces on the Southern island of Basilan. Basilan is home to the Abu Sayyaf, a violent Philippine rebel group. As of mid-2003, approximately 2,000 American military personnel were stationed in the Philippines.

The Philippines is one of the claimants, along with China, Taiwan, Malaysia, and Vietnam, to the Spratly Islands, located in the South China Sea. Potential oil and natural gas reserves surrounding the islands have sparked the interest of all the littoral states, though no exploratory drilling has been carried out due to the dispute.

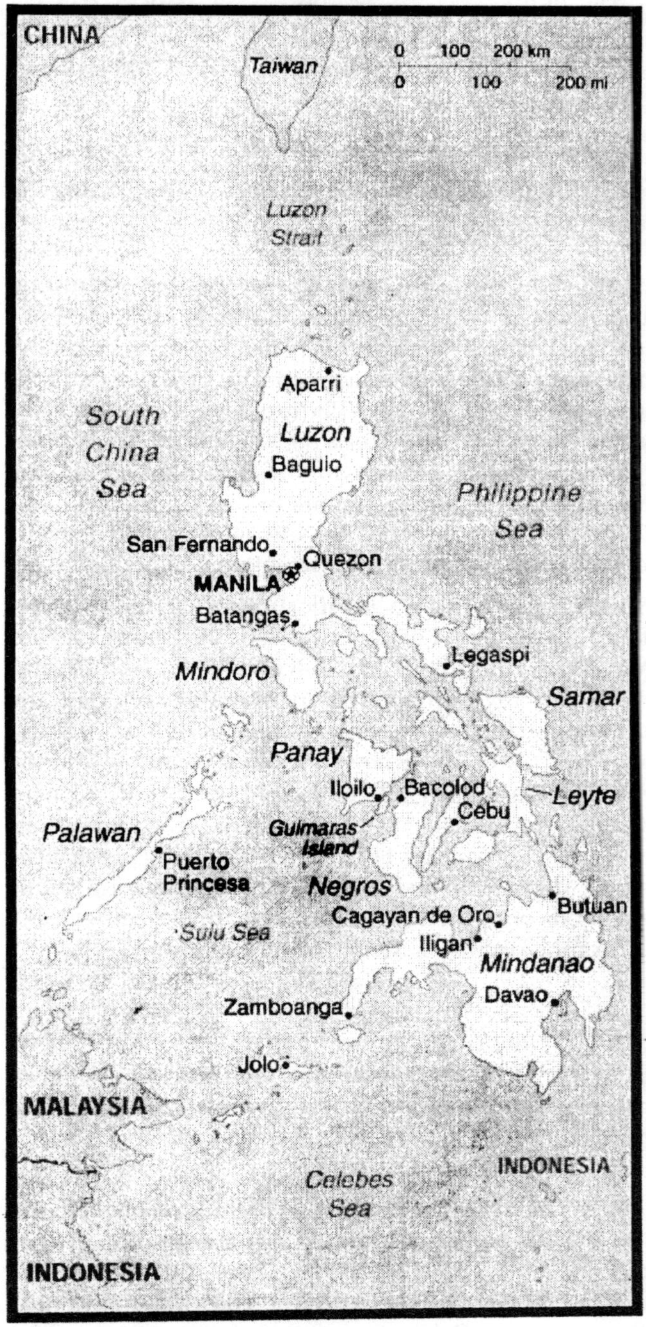

Oil

The Philippines began 2001 producing an average of only 1,000 barrels per day (bbl/d) of crude oil. In 2002, however, crude oil production averaged 23,512 bbl/d. This dramatic increase was due primarily to the development of new deep-sea oil deposits beneath the natural gas-bearing structures in the Malampaya field. The increased production volume is still modest, however, in relation to the country's needs. The Philippines consumed 342,000 bbl/d on average in 2002, resulting in net oil imports of 318,488 bbl/d.

This dependence on imported oil makes the Philippine economy vulnerable to sudden spikes in world oil prices. Oil consumption is expected to increase by over 5 per cent annually over the next several years as economic growth increases demand in most sectors. Oil demand for power generation, however, is expected is declining sharply, as many aging oil-fired electric power plants are shut down or converted to burn natural gas.

Despite small proven oil reserves, the Philippines has enjoyed a recent wave of optimism amongst domestic and foreign drillers. In October 2001, exploration underneath the Malampaya gas field revealed an estimated 85 million barrels of oil condensate. Shell Philippines Exploration (SPEX) has committed about $2 billion to the upstream components of the combined oil/natural gas project and currently operates the joint venture with partners Texaco Philippines and the Philippines National Oil Company. Production currently is around 25,000 bbl/d. In addition, six new offshore exploration projects have commenced in the Malampaya basin, led by Nido Petroleum, Philippines National Oil Company Exploration Corp., Trans-Asia Oil, Unocal Corp., and Philodril. Also, Trans-Asia has conducted exploratory drilling at the San Isidro well in the East Visayan Basin. The Philippine government issued a solicitation for bids in early August 2003 under a new licensing system, the First Petroleum Public Contract Round (PCR-1), covering 46 additional exploration blocks. The closing date is March 2, 2004. PCR-1 replaces the old system of negotiations with individual companies.

Refining and Downstream

The Philippines' downstream oil industry is dominated by three companies: Petron; Pilipinas Shell (Royal Dutch/Shell's Philippine subsidiary); and Caltex (Philippines). Petron is the Philippines' largest oil refining and marketing company. The company was a wholly owned subsidiary of the state-owned Philippine National Oil Company (PNOC) until 1994. Currently, the Philippine government and Saudi Aramco each own 40 per cent of the company, with the remaining 20 per cent held by portfolio and institutional investors, making it the only publicly listed firm amongst the three oil majors. Petron's Limay, Bataan refinery has a crude processing capacity of 180,000 bbl/d. Petron's market share as of mid-2003 is around 39 per cent. Caltex (Philippines), a subsidiary of Caltex, the Texaco-Chevron joint venture based in Singapore, operates a 86,500-bbl/d refinery, two import terminals, and more than 1,000 retail gasoline stations throughout the Philippines. Pilipinas Shell has a 153,000-bbl/d refinery, one of the largest foreign investments in the Philippines, and operates some 1,000 Shell gasoline stations. Overall, Philippine refineries run at around 80 per cent of capacity, and there is not a great deal of demand for new refinery construction.

Oil market deregulation, beginning in 1998, continues to have a significant effect on the industry. Since deregulation started, 62 new firms, including TotalfinaElf, Flying V, SeaOil (Philippines), Eastern Petroleum, Trans-Asia Energy and Unioil Petroleum Philippines Inc., have invested heavily and built several hundred new retail stations. While the three original companies still dominate the market, these firms have captured a steadily growing share of the petroleum products market, rising from around 10 per cent in 2000 to 17 per cent by mid-2003. These new entrants have organized the "New Players Petroleum Association of the Philippines" (NPPAP), and have been credited with putting significant downward pressure on retail fuel prices in the country. Currently, the Philippines enjoys the lowest fuel prices of any non oil-exporting Asian country. However, price swings associated with deregulation and higher world oil prices have angered many Filipinos. Despite recurring public calls for price controls, the government has remained committed to deregulation. In December 1999, the Supreme Court upheld the constitutionality of the country's deregulation program.

Natural Gas

The Philippines has 3.8 trillion cubic feet (Tcf) of proven natural gas reserves, but had no significant production until late 2001. While in the past the country's natural gas sector has not been developed extensively, the government has made expanding gas use a priority, particularly for electric power generation, in an effort to cut oil import expenses.

The impetus for the dramatic change in the country's natural gas sector is the Malampaya offshore field. Malampaya is the largest natural gas development project in Philippine history, and one of the largest-ever foreign investments in the country. Shell Philippines Exploration (SPEX, operator, with a 45 per cent stake), Texaco (45 per cent), and the PNOC (10 per cent) have come together to form the $4.5 billion Malampaya Deepwater Gas-to-Power Project. The Malampaya field is located in the South China Sea, off the northern island of Palawan, and contains an estimated 2.6 Tcf of natural gas. A 312-mile (504-kilometer) pipeline links the field to three power plants in Batangas. The pipeline is among the longest deep-water pipelines in the world, with half of its length more than 600 feet deep. With completion of the sub-sea pipeline and conversion of the first of three power stations, (San Rita, operated by British Gas and Philippines 1st Gas Corp.), the Malampaya project was officially inaugurated on October 16, 2001. Natural gas from Malampaya eventually will fuel three power plants with a combined 2,700-megawatt (MW) capacity for the next twenty years and will displace 26 million barrels of fuel oil, according to the Philippine government. The BG/Philippines 1st Gas Corporation partnership has announced that it expects to have a second station, the San Lorenzo facility, converted for natural gas use by 2003. The government has publicly considered selling its10 per cent share in the Malampaya project to the public; however no date has yet been set for the IPO.

A $100 million expansion pipeline from Batangas to Metro Manila ("Bat-Man") has been considered by numerous investors including PNOC, Shell, Brunei's Mashor Group, First Gas, and Sumitomo. This pipeline would supply gas to additional power plants as well as the industrial and commercial sectors. Negotiations on the financial aspects of the project are ongoing, and construction is expected to begin in 2004, with the pipeline commencing operation in 2006.

Exploration continues is other parts of the country, but no major discoveries have been reported. Three small natural gas fields were closed down in 2001. Fields in the Tukankuden and the Cotabato Basin were shut down due to security problems, while another field in Victoria, Tarlac, was closed because the natural gas discovered was too saturated with water for commercial production.

The Philippine government is developing a policy framework for the emerging natural gas industry that foresees the government's role as that of facilitator while attempting to ensure competition. Domestic development is to be encouraged, but competition from imported gas also is to be allowed. Gas supply to wholesale markets will have market-set prices, while prices for captive markets and small consumers will be regulated.

Liquefied natural gas (LNG) has begun to receive added attention as a potential source of natural gas supplies. PNOC has been considering the construction of an LNG regasification terminal in Bataan, which would serve the Manila area. A letter of intent has been signed for natural gas imports into the Philippines from BP's Tangguh LNG project in Indonesia.

Coal

Development of new natural gas projects in the Philippines has come largely at the expense of the country's struggling coal industry. PNOC's coal mining subsidiary produced 1.5 million short tons of coal in 2001. While coal is a declining share of the Philippines fuel mix, there are still new small coal mines under development, mainly on the southern island of Mindanao. The country has decided to restructure the use of its 366 million short tons of estimated coal reserves, which is mostly low-rank lignite, for processing in smaller "clean coal" plants, for eventual end-use as household fuel, and briqueting.

The Philippines consumed 9.3 million short tons of coal in 2001, 7.8 million short tons of which were imported. Indonesia, China, and Australia are major exporters of coal to the Philippines.

World Trade Organization (WTO) regulations require that the Philippines lift import restrictions on coal. Since the 1970s, when the National Coal Authority was created, Philippine coal importers have been required to obtain a government certificate of compliance before importing coal, allowing the authorities to force importers to buy domestic coal each time they purchased coal from abroad. President Macapagal Arroyo has committed to honoring the international coal supply contracts approved by the previous government.

Electricity

Energy production in the Philippines is concentrated in the electricity sector. Geothermal power accounts for the country's largest share of indigenous energy production, followed by hydropower, natural gas, coal, and oil. The Philippine government has made shifting from reliance on imported oil a major goal, and is pushing the current boom in natural gas-fired electricity development.

The most significant event in the Philippine energy industry in recent years was the Power Industry Reform Act (PIRA) of 2001. After seven years of congressional debate and litigation, the Act came into force on June 26, 2001. The act has three main objectives: 1) to develop indigenous resources; 2) to cut the high cost of electric power in the Philippines; and 3) to encourage foreign investment. Passage of the Act set into motion the deregulation of the power industry and the breakup and eventual privatization of state-owned enterprises.

PIRA required the state-owned utility National Power Corporation (Napocor) to break up its vertically integrated assets into smaller sub-sectors such as generation, transmission, distribution and supply in order to prepare for eventual privatization. The result will be a system in which privatized generators would sell directly to private distribution companies. Working with consultants from the law firm of Hunton and Williams (U.S.), the government has designated two new entities designed solely for the eventual privatization of state assets. These two concerns, Transco and PSALMcorp, will entail the state's high voltage transmission infrastructure, and power plants, respectively. The government also will sell off its share of Meralco, a vital distribution utility on the island of Luzon that serves Manila and the immediate surrounding area by buying power from various Independent Power Producers (IPPs).

Napocor will need to transfer its existing power purchase obligations to private distributors, and also to renegotiate high-priced contracts. The cost savings lie in the fact that private distributors will likely be unwilling to enter into agreements that are above market rates. There are other financial incentives for the government as well. Napocor's huge debt and $9 billion in power purchase agreements are unsustainable, and the government must already contribute $300 million per year to keep Napocor afloat.

In order to make the sale of Napocor more attractive to investors, the government has absorbed a significant amount of Napocor's debt. In addition, the $9 billion in power purchase agreements with IPPs also will be sold off. The transmission system is to be transferred to an independent company, Transco, which is to be privatized. According to deregulation laws, no single potential buyer will be allowed to own more than 30 per cent of the Philippines' generating assets. Privatization of Transco has been delayed, though, due to the fact that only one offer was received during the first two rounds of bidding. It is possible that the Philippine government will pursue negotiated contract with Singapore Power, the lone bidder.

Electricity demand in the Philippines is expected to grow by around 9 per cent per year through the end of the decade, necessitating as much as 10,000 MW of new installed electric capacity. Current contracts will provide about half of that amount, with the remainder expected to be filled once the market deregulates. Medium-term increases in power demand are to be satisfied largely by the three gas-fired plants (Ilijan, Santa Rita, and San Lorenzo) that will be linked to the Malampaya natural gas field. The Korea Electric Power Corporation (KEPCO) began commercial operation of the 1,200-MW Ilijan plant in June 2002. KEPCO will run the plant under a build-operate-transfer scheme for 20 years, after which ownership will revert to Napocor. Minority stakeholders in the plant are Southern Energy of the United States (20 per cent) plus Mitsubishi (21 per cent) and Kyushu

Power (8 per cent) of Japan. First Gas Power completed its 1,020-MW plant at Santa Rita in August 2000, and it switched to natural gas as a fuel in January 2002.

There are two new projects in Luzon. The CE Casecnan Water and Energy Company (a subsidiary of California Energy International) is constructing a multipurpose irrigation and 150-MW hydroelectric facility. Also, the Washington Group completed the 350-MW San Roque multipurpose hydro project, which began commercial operation in May 2003.

Mirant is the Philippines' largest IPP, operating five power plants in the country. Mirant's coal-fired Sual plant began commercial operation in late 1999. The 1,218-MW plant is located about 130 miles north of Manila, and is the nation's largest and lowest-cost electricity producer. Napocor is the sole purchaser of power from Sual.

Several power-generating facilities also are under extensive rehabilitation. The 100-MW Binga hydroelectric plant in Itogon, Benguet has been under renovation since 1993 following damage from a 1990 earthquake. After years of delays, it resumed operation in July 2002. A larger project is the $470 million contract with Argentine firm IMPSA (*Industrias Metalurgicas Pescarmona Sociedad Anonima*) to rehabilitate and operate the 750-MW Caliraya-Botocan-Kalayaan (CBK) power complex in Laguna, south of Manila. The CBK complex is the grid regulator in Luzon, and as such is able to transmit power to other plants on the grid in the event of breakdowns. IMPSA, in conjunction with new partner Edison Mission Energy of the United States, was able to get a performance undertaking guarantee despite Napocor's and some government officials' objections, facilitating long-delayed financing of the project.

The Philippines, due to its geography, has problems linking all of its larger islands together into one grid and ensuring availability of electric power in rural areas. The government has set a target date of 2006 for electrification of all these villages, and also is taking steps to link together the country's three major power grids (Luzon, Visayas, and Mindanao). Where it is not economical to link small islands' grids into the national grid, separate local systems are being established around small generating plants.

Renewables

The Philippines is the world's second largest producer of geothermal power, with an available capacity of 1,931 MW, according to the Philippine government. The government would like to add another 990 MW, bringing capacity to 2,921MW, and exceeding the U.S. capacity of 2,775 MW. Geothermal power currently makes up around 16 per cent of the Philippines' installed power generation capacity, most of which has been developed by the PNOC–Energy Development Corporation (PNOC-EDC). Privatization of PNOC-EDC is planned, though as with other generation assets, the process has progressed much slower than originally planned. Kyushu Electric company is in a joint venture with PNOC-EDC to develop a 40-MW geothermal plant in Sorsogon, Albay province, and Marubeni of Japan has expressed its intent to build the 100-MW Cabalian geothermal plant in Leyte. California Energy's Philippine unit is working with PNOC to develop three new geothermal power plants in Leyte, producing a total of 540 MW of electricity. Plans are underway to develop nine new facilities in Luzon, ranging from 20 MW to 120 MW, that will eventually bring a total of 440 MW of geothermal energy to the grid. By 2005, the new 40-MW Mambucal and 40-MW Rangas power stations in Dauan, Negros Oriental are expected to come online. Financing for the projects was secured from the Development Bank of the Philippines (DBP) in June 2003.

Besides geothermal, the Philippines also is exploring the use of other renewables for electricity generation, particularly in the country's unelectrified villages. In March 2001, the Philippine and Spanish governments, in conjunction with BP, agreed to a $48 million contract to bring solar power to 150 villages. BP and the government of Australia also have partnered with the Philippines to supply solar power to rural villages, bringing 1,145 solar-powered systems to 52 new municipalities.

The Philippines appears to have a strong potential for wind generation. The United States Department of Energy wind mapping survey estimates that wind resources in the Philippines have a power generation potential of as much as 70,000 MW, seven times the country's current power demand. The 40-MW, PNOC-EDC, Northern Luzon project in Ilocos Norte began operation in late 2002. A contract for a second, 40-MW phase of the project was signed with Aboitiz Power in March 2003.

Country Overview

Independence: July 4, 1946 (from United States)

Population (2003E): 84.6 million

Location/Size: Southeast Asia/115,830 sq. mi. (slightly larger than Arizona)

Major Cities: Manila (capital), Quezon City, Cebu, Davao

Languages: Pilipino (official; based on Tagalog), English (official)

Ethnic Groups: Christian Malay (91.5 per cent), Muslim Malay (4 per cent), Chinese (1.5 per cent), other (3 per cent)

Religions: Roman Catholic (83 per cent), Protestant (9 per cent), Muslim (5 per cent), Buddhist and other (3 per cent)

Defense (8/98): Army (74,500), Navy (25,900), Air Force (17,400), Reserves (131,000)

Economic Overview

Currency: Philippine peso

Market Exchange Rate (8/18/03): $1 = 55.15 pesos

Gross Domestic Product (GDP, 2002E): $78.0 billion

Real GDP Growth Rate (2002E): 4.4 per cent *(2003E):* 3.7 per cent

Inflation Rate (consumer prices, 2002E): 4.9 per cent *(2002F)* 4.9 per cent

Current Account Balance (2002E): $4.2 billion

Major Trading Partners: United States, Japan, EU, Singapore, Hong Kong

Merchandise Exports (2002E): $35.2 billion

Merchandise Imports (2002E): $35.4 billion

Major Export Products: Electronic equipment, machinery, garments, coconut oil

Major Import Products: Machinery and equipment, fuel products, textile yarns, chemicals

Total External Debt (2002E): $53.7 billion

Energy Overview

Proven Oil Reserves (1/1/03E): 178 million barrels (Oil and Gas Journal)

Oil Production (2002E): 23,512 bbl/d

Oil Consumption (2002E): 342,000 bbl/d

Net Oil Imports (2002E): 318,488 bbl/d

Crude Oil Refining Capacity (1/01/03E): 419,500 bbl/d

Natural Gas Reserves (1/1/03E): 3.772 trillion cubic feet

Natural Gas Production and Consumption: (2001E): 353 million cubic feet (negligible)

Recoverable Coal Reserves (2001E): 366 million short tons

Coal Production (2001E): 1.5 million short tons

Coal Consumption (2001E): 9.3 million short tons

Electric Generation Capacity (2001E): 13 gigawatts (GW)

Electricity Generation (2001E): 45.2 billion kilowatthours (bkwh) (55.4 per cent thermal, 17.5 per cent hydro, and 27.0 per cent geothermal)

Electricity Consumption (2001E): 42.0 bkwh

Environmental Overview

Total Energy Consumption (2001E): 1.25 quadrillion Btu* (0.3 per cent of world total energy consumption)

Energy-Related Carbon Emissions (2001E): 18.6 million metric tons of carbon (0.3 per cent of world total energy-related carbon emissions)

Per Capita Energy Consumption (2001E): 16.3 million Btu (vs. U.S. value of 341.8 million Btu)

Per Capita Carbon Emissions (2001E): 0.2 metric tons of carbon (vs. U.S. value of 5.5 metric tons of carbon)

Energy Intensity (2001E): 13,748 Btu/$1995 (vs U.S. value of 10,736 Btu/$1995)**

Carbon Intensity (2001E): 0.20 metric tons of carbon/thousand $1995 (vs U.S. value of 0.17 metric tons/thousand $1995)

Fuel Share of Energy Consumption (2001E): Oil (56.8 per cent), Coal (16.3 per cent), Natural Gas (0.03 per cent)

Fuel Share of Carbon Emissions (2001E): Oil (72.6 per cent), Coal (27.4 per cent), Natural Gas (0.03 per cent)

Renewable Energy Consumption (2001E): 0.25 quadrillion Btu*

Status in Climate Change Negotiations: Non-Annex I country under the United Nations Framework Convention on Climate Change (ratified August 2nd, 1994). Signatory to the Kyoto Protocol (signed April 15th, 1998).

Major Environmental Issues: Uncontrolled deforestation in watershed areas; soil erosion; air and water pollution in Manila; increasing pollution of coastal mangrove swamps which are important fish breeding grounds.

Major International Environmental Agreements: A party to Conventions on Biodiversity, Climate Change, Endangered Species, Hazardous Wastes, Law of the Sea, Marine Dumping, Nuclear Test Ban, Ozone Layer Protection. Tropical Timber 83, Tropical Timber 94, Wetlands and Whaling. Has signed, but not ratified, Desertification.

The total energy consumption statistic includes petroleum, dry natural gas, coal, net hydro, nuclear, geothermal, solar and wind electric power. The renewable energy consumption statistic is based on EIA data and includes geothermal, solar, wind, wood and waste electric power consumption.

Oil and Gas Industries

Organization: The Philippine National Oil Company (PNOC) is the country's state-owned energy company responsible for oil and development of local energy resources. Petron, privatized in 1994, is considered to be the country's largest oil refining company, with Shell and Caltex also significant. National Power Corporation (NPC) is the state-owned electric company.

Major Foreign Energy Company Involvement: Caltex, Royal-Dutch Shell, Petroleum Authority of Thailand, TotalFinaElf

Major Natural Gas Fields: Malampaya-Camago

Major Oil Refineries (capacity–bbl/d): Petron–Limay, Bataan (180,000 bbl/d); Pilipinas Shell–Tabangao (153,000) bbl/d); Caltex–Batangas (86,500 bbl/d)

66

Portugal

Portugal's energy imports are expected to increase significantly in coming years, as the country lacks domestic resources for energy production.

General Background

Portugal is a member of the European Union (EU) and the common European currency, the euro. Portugal is a relatively poor country economically in comparison to its neighbors, with the lowest nominal per-capita GDP of all EU member states in 2002. However, since joining the EU, the country's economy has expanded, with an average real economic growth rate of 3.7 per cent per year from 1986 to 2000. Funds to invest in the country's infrastructure received from the European Commission have helped contribute to Portugal's successful economic performance.

Before Portugal's newly-elected prime minister, José Manuel Durão Barroso of the center-right Social Democratic Party (PSD), entered office in 2002, Portugal had received a formal reprimand from the European Commission (EC) for the size of the country's budget deficit, which reached 4.1 per cent in 2002, breaching the 3 per cent ceiling allowed by the EU's Stability and Growth Pact. Although the Portuguese government has succeeded in reducing the budget deficit to 2.7 per cent of GDP in 2002, its austere fiscal policy, in conjunction with weaker global growth, has made it difficult for Portugal's economy to recover. Real GDP for 2002 was estimated at 0.5 per cent. The budget deficit issue remains a problem, and Portugal could once again breach the 3 per cent ceiling in 2003. Portugal's unemployment rate also increased to 5 per cent in 2002 from 4.1 per cent at the end of 2001.

Energy

Portugal has extremely limited domestic energy resources, and therefore imports about 90 per cent of its energy needs. Because of Portugal's geographic location, bordering only Spain and the

Atlantic Ocean, much of its energy imports are transported through Spain. The Iberian peninsula has an extensive natural gas network that links the two countries with Algeria via Morocco. The two Iberian countries signed an agreement in November 2001 to integrate their electricity markets completely by 2003. Both governments recently decided to extend the completion of the integrated market until 2006.

Energy Sector Restructuring

On April 3, 2003, the Portuguese government approved a plan to restructure the energy sector. The government plans to split the state-controlled oil and gas holding company GalpEnergia into three sectors. Galp's main natural gas business, Gas de Portugal, will merge with Electricidade de Portugal (EDP), the state-owned electricity utility company. Galp's gas transmission subsidiary, Transgas, will be merged with the country's state-owned transmission company, Rede Electrica Nacional (REN), in which a majority stake will be privatized in 2004. Galp will be left with its oil refining and refined products distribution through its subsidiary Petrogal. GalpEnergia is currently controlled by the Portuguese government (34.81 per cent), followed by Italy's Eni (33.4 per cent), EDP (14.27 per cent), savings bank Caixa Geral de Depositos (13.5 per cent), and Spain's Iberdrola (4 per cent). The Portuguese government also plans to bring forward the full liberalization of the country's energy markets from 2008 to July 2004. Some analysts observe the restructuring as an attempt by the government to improve the competitiveness

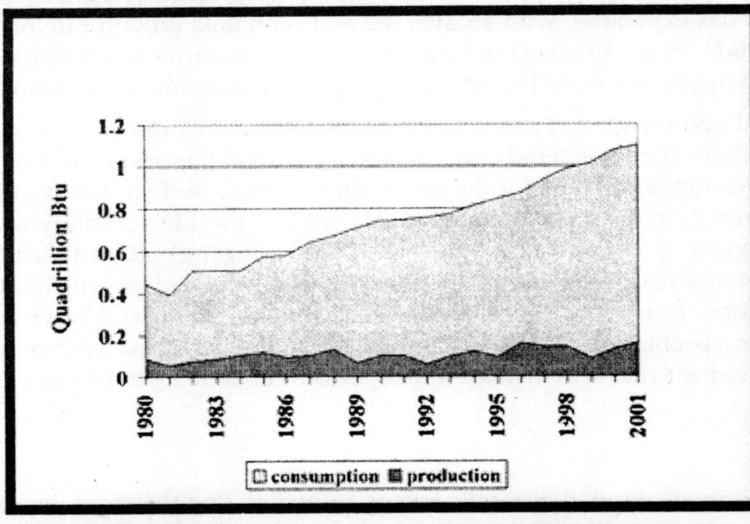

ENERGY PRODUCTION AND CONSUMPTION IN PORTUGAL, 1980-2001

of the Portuguese energy companies ahead of the creation of the Iberian electricity market, Mibel, as well as in the European market.

Oil

Portugal holds no proven commercially viable oil reserves. In 2002, the country consumed 351,000 barrels per day (bbl/d), all of which was imported. Oil made up 64 per cent of the total Portuguese primary energy consumption for that year.

Exploration and Production

Despite decades of exploration activity, Portugal has yet to discover a commercially viable oil deposit. In 2002, the Portuguese government offered 14 deep-water-offshore blocks for exploration and production contracts. After the licensing round expired on December 2, 2002, only Repsol-YPF, in partnership with RWE of Germany, made bids for Blocks 13 and 14. Taurus Petroleum of Sweden is currently the only other company which holds an active offshore license.

Refining/Downstream

Portugal has two refineries, located in the coastal cities of Sines and Porto. Petrogal operates the two refineries, which have a combined capacity of 304,174 bbl/d.

Natural Gas

The Portuguese natural gas sector has grown considerably in the last few years, despite lacking any commercially viable reserves. Annual consumption was nearly non-existent prior to 1997 when the country consumed 4 billion cubic feet (Bcf). In 2001, consumption reached 90 Bcf. The increase in natural gas consumption can be attributed to the completion of the Maghreb-Europe pipeline in 1996, which connected the Iberian Peninsula to Algerian natural gas sources. Transgas spent $35 million to create a link, allowing Portugal to import Algerian gas through Spain. Gas de Portugal

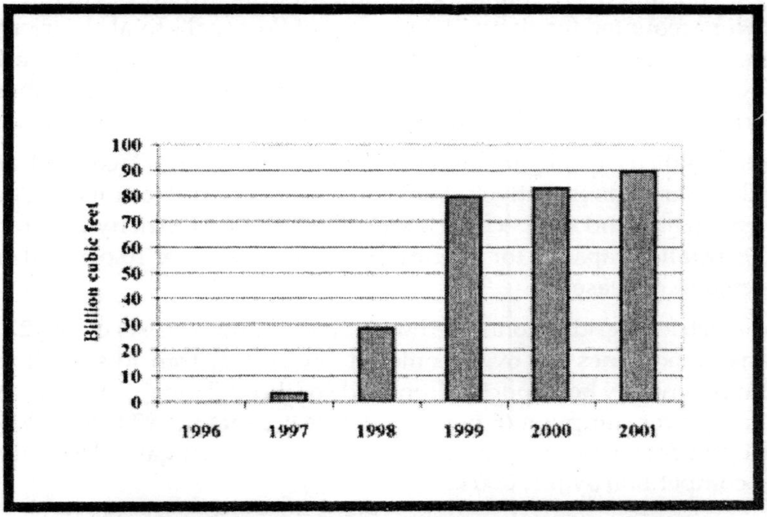

NATURAL GAS CONSUMPTION IN PORTUGAL, 1996-2001

owns the entire Portuguese section, and part of other sections. Portugal (with EU assistance) is making large investments in its natural gas infrastructure, having spent approximately 485 million euros between 1994 and 1999. Since 1999, the government has curtailed this spending somewhat, but there are still plans to expand the natural gas network from 3,761 miles in 1999 to 5,943 miles in 2010.

Exploration and Production

Mohave Oil and Gas Company and Desire Petroleum PLC made Portugal's first significant natural gas discovery at Alijubarrota. However, in June 2001, Desire Petroleum plugged the Alijubarrota 2 sidetrack natural gas well because, according to the company, the zone had "failed to produce commercial quantities of hydrocarbon gases during underbalanced drilling operations."

Liquefied Natural Gas (LNG)

In 1998, Portugal became an LNG purchaser for the first time. In 1999, the country signed a 20-year contract with Nigerian LNG. Portugal is also building its first LNG terminal in Sines, which will go online in late 2003 as a gas-processing facility. Until that time, LNG will continue to be regasified in Spain and piped into Portugal.

Storage

Transgas, Portugal's natural gas transport company, plans to convert two large salt caverns into storage tanks, with construction beginning this year. The first two salt caverns will have a capacity of 2.1 trillion cubic feet (Tcf). Transgas will convert two additional caverns in 2004, doubling the storage capacity.

CoalPortugal has not produced coal since its last mine closed in 1994. It does import relatively small amounts of coal for electricity generation, especially in periods of decreased hydropower. Although hydropower increased in 2001, annual coal consumption fell only slightly in 2001 as electricity demand increased overall.

Electricity

In 2001, Portugal consumed 41.5 billion kilowatthours (Bkwh) and generated 44.3 Bkwh of electricity. Hydropower accounted for about 31.4 per cent of Portugal's total electricity generation in 2001, a substantial increase from 17 per cent in 1999, a year characterized by particularly dry weather. Likewise, Portugal's oil and coal consumption for power generation fell in 2001 to about 64.5 per cent from 79 per cent; thermal power and hydropower tend to exhibit an inverse relationship in Portugal.

During the 1990s, installed capacity for both hydropower and thermal (coal, oil, and natural gas) grew, 36 per cent and 47 per cent respectively. While installed capacity for fuel oil decreased during the period, natural gas, which did not become a major source for electricity generation until 1998, increased. As of 2000, installed capacity for natural gas represented over 10 per cent of total installed capacity and is expected to increase.

The Portuguese electricity grid is connected with Spain's and consists of 44,127 miles of high/ medium voltage transmission lines and 69,640 miles of low voltage transmission lines. Up to 10 per cent of demand could potentially be imported from and/or through Spain (and vice-versa). Despite rapidly increasing energy consumption (6 per cent per year average 1973-1998), Portugal still has the lowest per capita power consumption in the EU, as of 2001. Portugal's electricity market is due to be fully opened to competition by July 2004.

Iberian Integration

In November 2001, Spain and Portugal signed an agreement that calls for complete integration of their electricity markets, creating the *Mercado Ibérico de Electricidade* (MIBEL). MIBEL was scheduled to be opened in early 2003, but both governments agreed to complete incrementally the market in stages by 2006. The integration of the markets includes creating a single market operator to negotiate the sale of electricity, a merger of logistics networks, and increased interconnection of the countries' grids. In addition, all companies will operate under the same competition guidelines, and tariffs will be harmonized.

Sector Organization

Since 1993, there have been independent power producers in Portugal, and by 1998 they accounted for 28 per cent of the country's power production. There are two electricity systems in Portugal, the Public and the Independent. The Public system is heavily regulated to ensure a guaranteed supply. In the Independent system, competition has been introduced (in line with EU directives), but is an option only for large consumers of electricity. In the Public system, the 70 per cent -state-owned/30 per cent - EDP-owned grid operator REN has exclusive and obligatory contracts with suppliers at a price expected to cover costs. In the Independent system, there are no power purchase agreements; consumers can choose their supplier, though the REN transmission system is still used. Erse is the Portuguese independent electricity regulator. Since January 2002, the government no longer has any power over electricity tariffs. Rather, Erse makes the final decision after receiving a recommendation from a council equally composed of consumer groups and the electricity companies.

Electricidade de Portugal (EDP) is Portugal's largest electricity company. EDP has been progressively privatized in recent years. The Portuguese government still retains a 30 per cent stake in the company, but plans to sell its remaining stake in EDP in 2004.

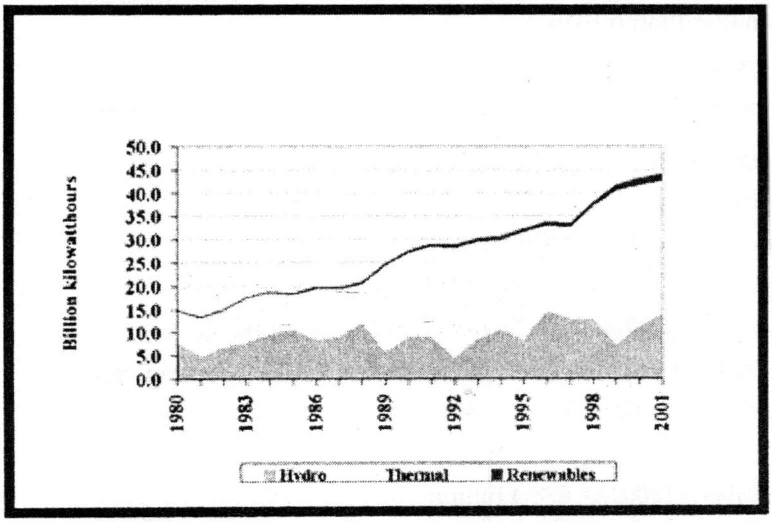

PORTUGUESE ELECTRICITY PRODUCTION, 1980-2001

The signing of the MIBEL in November 2001 spurred a number of deals in the Iberian countries' electricity sectors. In December 2001, EDP acquired a 40 per cent share of the holding company that owns 95 per cent of Spanish utility Hidrocantabrico. EDP, as the senior partner, took control of Hidrocantabrico's operations, giving the Portuguese company a large foothold in the Spanish market. In March 2003, EDP also acquired a 62 per cent stake in Naturcorp, the natural gas distributor in Spain's Basque region. EDP plans to merge Naturcorp with its Gas de Austurias, making EDP one of the largest natural gas companies in Spain.

In October 2002, Spain's largest utility, Endesa and Portugal's industrial holding company Sonae launched a 50:50 joint venture company, *Sonae Comercialización de Energía* (SODESA). The company aims to achieve a 15 per cent share of the large consumer market.

With rising electricity demand and large-scale hydroelectric power already maximized, Portugal will most likely have to build more thermal plants, increase imports, or rely more on renewable sources of electricity such as wind, solar, co-generation, and mini-hydroelectric plants. The Portuguese government aims to increase electricity generation from renewable sources by 40 per cent during the period 2002-2010. Portugal produced 1.8 Bkwh from renewable sources (excluding hydro) in 2001.

Nuclear Power

Portugal has no nuclear power plants. While there has been some interest and discussion regarding the construction of a nuclear plant, there has been no action toward this end.

Country Overview

Independence: 1143; from Spain 1640 (republic proclaimed October 5, 1910)

Capital City: Lisbon

Population (2002E): 10 million

Location/Size: Southwestern Europe, bordering the North Atlantic Ocean, west of Spain/92,391 sq km (slightly smaller than Indiana)

Language: Portuguese

Religion: Roman Catholic 97 per cent, Protestant denominations 1 per cent, other 2 per cent

Economic Overview

Currency: Euro

Exchange Rate (June 2, 2003): 1 US Dollar = 0.854 Euros

Gross Domestic Product (GDP, nominal, 2002E): $121.3 billion

Real GDP Growth Rate (2002E): 0.5 per cent *(2003F):* 0.4 per cent

Inflation Rate (consumer prices, 2002E): 3.6 per cent *(2003F):* 2.9 per cent

Unemployment Rate (2002E): 5.0 per cent

Merchandise Exports (2002E): $25.6 billion

Merchandise Imports (2002E): $38.4 billion

Major Trading Partners: Germany, Spain, France, United Kingdom

Major Export Products: textiles, footwear, leather goods, machinery and transport equipment, wood and cork, agricultural goods and foodstuffs

Major Import Products: Machinery and appliances, agricultural products and byproducts, chemical and plastic products, land transport equipment, mineral products

Energy Overview

Proven Oil Reserves: None

Crude Oil Production (2002E): None

Oil Consumption (2002E): 351,000 (bbl/d)

Crude Oil Refining Capacity (1/1/03E): 304,172 bbl/d

Natural Gas Reserves: None

Natural Gas Consumption (2001E): 90 billion cubic feet

Coal Reserves (2001E): 40 million short tons (Mmst)(no longer mined)

Coal Consumption (2001E): 6.7 Mmst

Coal Production (2001): None since 1994

Electric Generation Capacity (2001E): 10.3 gigawatts

Electricity Generation (2001E): 44.3 billion kilowatthours (thermal 64.5 per cent; hydro 31.4 per cent; biomass, geothermal, solar, and wind 4.1 per cent)

Electricity Consumption (2001E): 41.5 billion kilowatthours

Environmental Overview

Total Energy Consumption (2001E): 1.1 quadrillion Btu* (0.3 per cent of world total energy consumption)

Energy-Related Carbon Emissions (2001E): 16.25 million metric tons of carbon (0.3 per cent of world total carbon emissions)

Per Capita Energy Consumption (2001E): 108.6 million Btu (vs. U.S. value of 341.8 million Btu)

Per Capita Carbon Emissions (2001E): 1.62 metric tons of carbon (vs. U.S. value of 5.5 metric tons of carbon)

Energy Intensity.(2001E): 8,251 Btu/$1995** (vs U.S. value of 10,736 Btu/$1995)

Carbon Intensity (2001E): 0.12 metric tons of carbon/thousand $1995** (vs. U.S. value of 0.17 metric tons/thousand $1995)

Fuel Share of Energy Consumption (2001E): Oil (64 per cent), Coal (11 per cent), Natural Gas (9 per cent) Hydro (13 per cent) Other renewable (3 per cent)

Fuel Share of Carbon Emissions (2001E): Oil (73 per cent), Coal (18 per cent), Natural Gas (9 per cent)

Status in Climate Change Negotiations: Annex I country under the United Nations Framework Convention on Climate Change (ratified December 21st, 1993). Signatory to the Kyoto Protocol (signed April 29th, 1998 and ratified May 31, 2002).

Major Environmental Issues: Soil erosion; air pollution caused by industrial and vehicle emissions and water pollution (especially in coastal areas).

Major International Environmental Agreements: A party to Conventions on Air Pollution, Biodiversity, Climate Change, Desertification, Endangered Species, Hazardous Wastes, Law of the Sea, Marine Dumping, Marine Life Conservation, Ozone Layer Protection, Ship Pollution, Tropical Timber 83 and Wetlands. Has signed, but not ratified: Air Pollution-Persistent Organic Pollutants, Air Pollution-Volatile Organic Compounds, Climate Change-Kyoto Protocol, Environmental Modification, Nuclear Test Ban, Tropical Timber 94.

The total energy consumption statistic includes petroleum, dry natural gas, coal, net hydro, nuclear, geothermal, solar, wind, wood and waste electric power. The renewable energy consumption statistic is based on International Energy Agency (IEA) data and includes hydropower, solar, wind, tide, geothermal, solid biomass and animal products, biomass gas and liquids, industrial and municipal wastes.

67

Puerto Rico Fact Sheet

The U.S. Commonwealth of Puerto Rico is the largest net energy importer in the Caribbean basin. Over the last few years, the government has supported significant new energy projects in order to diversify the country's energy mix.

General Background

In November 2000, Sila Calderón of the Popular Democratic Party (PDP) became the first woman to serve as governor of Puerto Rico. Unlike previous administrations, which promoted tourism as the cornerstone of the country's economy, Governor Calderón has been encouraging the development of high-tech and manufacturing sectors, as well as the decentralization of the economy to outside of San Juan. With the eventual construction of the **Port of the Americas**, Puerto Rican government is expecting a boost in the economy and employment.

Although Puerto Rico's economy contracted by 0.2 per cent in 2002, it is expected to recover in 2003, with a real domestic product growth of 1.9 per cent. Whether Puerto Rico's economy expands relies heavily on the performance of the mainland United States' economy. The United States is Puerto Rico's largest trading partner. Recent high oil prices also have had an adverse affect on the country's economy and inflation, as Puerto Rico is heavily dependent on oil imports to meet its domestic energy needs, particularly electricity generation.

Historically, Puerto Rico's economy has been one of the strongest in the Caribbean basin. Beginning in the 1950s, American firms invested heavily in Puerto Rico's industrial sector, taking advantage of significant tax incentives as well as duty-free access to U.S. markets. As a result, Puerto

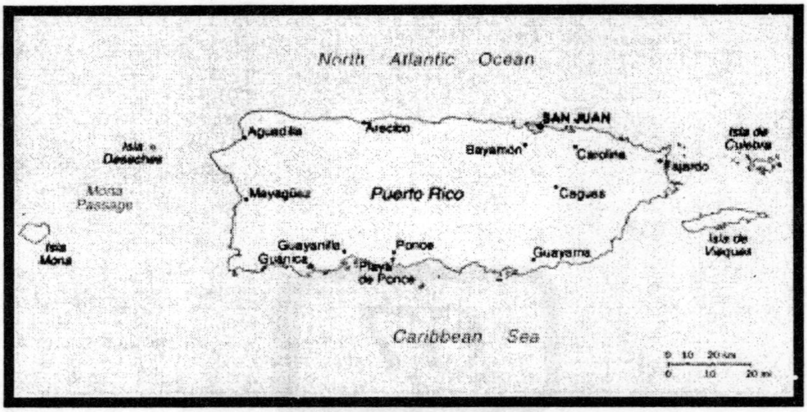

Rico enjoys a highly diversified economy and good trade relations with the United States. In 1996, however, the U.S. Congress repealed the tax incentives that American firms once enjoyed. Since then, Puerto Rico has lost some manufacturing jobs and its economy has suffered.

Energy

Oil is the dominant fuel in Puerto Rico's energy mix, accounting for 93 per cent of total primary energy consumption in 2001. Puerto Rico lacks domestic hydrocarbon reserves and relies on imports for nearly all of its energy requirements. Per capita energy consumption is much lower than in the mainland United States.

Oil

Puerto Rico consumed about 193,000 barrels per day (bbl/d) of oil in 2002, all of which was imported, mostly for transportation and electric power generation. Puerto Rico imports its oil from American and Caribbean suppliers. As of January 2003, Puerto Rico's refining capacity stood at 93,000 bbl/d, coming from two operating refining facilities: the Caribbean Petroleum Refining facility on Bayamon; and Shell Chemical's facility in Yabucoa. Shell Chemical, a subsidiary of Royal Dutch Shell, bought the Yabucoa refinery from Sunoco in December 2001. A third refinery at Guayama is currently used only for storage, with no potential for production in the near future. The Commonwealth also has petroleum storage at its Proterm facility, with a capacity of about 9 million barrels.

Natural Gas

Puerto Rico began importing liquefied natural gas (LNG) in August 2000 to feed the **EcoEléctrica** gasfired power plant in Peñuelas. The EcoEléctrica project consists of a regasification facility and a 540 MW natural gas-fired power plant. Enron and Edison International unit, Edison Mission Energy, own and operate the project. Gas Natural is negotiating to acquire Enron's remaining stake in the project. Enron had previously sold a small portion of its stake to General Electric Capital Corporation.

EcoElectrica has a long-term contract to purchase LNG from Tractebel LNG North America LLC. EcoElectrica consumes approximately 460,000 metric tons of LNG per year, which is approximately equivalent to 23 billion standard cubic feet per year of natural gas.

Coal

Puerto Rico consumes 180,000 short tons of coal annually (based on 2001 data), all of which is imported. A new 454-MW coal-fired plant in Guayama has increased the use of coal since becoming operational in 2002.

Electricity

In 2001, Puerto Rico generated about 19.4 billion kilowatt-hours (Bkwh) of electricity, predominantly from five oil-fired generators, with a fraction coming from small hydroelectric dams. The Puerto Rico Electric Power Authority (PREPA) accounts for a majority of net electricity generation, and is the Commonwealth's sole distributor of electric power. PREPA also purchases excess power generation from cogenerators, primarily in the cement industry. In 2001, installed generation capacity was 4.9 million kilowatts.

With power consumption increasing by over 3 per cent per year since 1990, both PREPA and private power producers have been investing in new capacity not only to meet growing demand but also to diversify energy sources. One of the most important new power plants in Puerto Rico is the aforementioned **EcoEléctrica facility**, a 540-MW, natural gas-fired generator in Peñuelas. EcoEléctrica was world's first independent, private combined LNG terminal/combined cycle natural gas turbine power project. In the 1990s, the commonwealth became interested in the LNG option due to a forecast projecting a power shortfall by 2000, as well as the prospect of meeting the U.S. Environmental Protection Agency tightening clean air standards with a mainly fuel-oil powered electricity sector. The plant came online in March 2000 using initially liquid petroleum gas (LPG), and received its first LNG shipment in August 2000. The project is capable of supplying about 20 per cent of the island's electricity.

Another effort at fuel diversification is the 454-MW coal-fired plant, which started operating on November 29, 2002. The plant has been recognized as one of the cleanest coal-fired plants in the

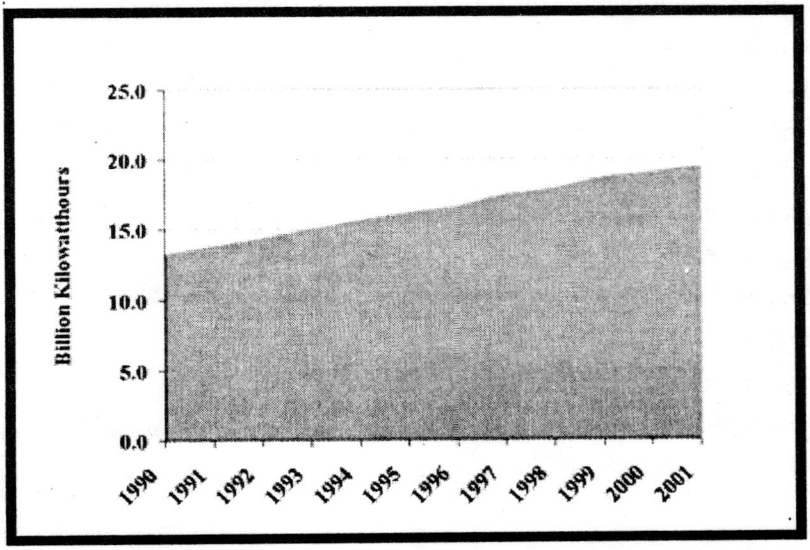

PUERTO RICO ELECTRICITY CONSUMPTION, 1990-2001

world. The plant utilizes **circulating fluid bed (CFB) technology and other emission control systems** to limit the environmental impact of the plant. The Aurora plant in Guayama will provide electricity under a 25-year power purchase agreement with PREPA. In addition, PREPA will sell up to 400,000 pounds per hour of steam to Phillips Petroleum Co. of Puerto Rico.

Renewable Energy

Interstate General Company (IGC), in conjunction with Thermoselect, HDR Engineering, Zachry Construction Company, and Montenay Power, is moving forward with the development of the first **non-incineration waste-to-energy** power plants in the United States in Caguas, Puerto Rico. IGC currently is in negotiations with PREPA over a power purchase agreement. PREPA is also considering whether it will become a partner in the project. The proposed plant will use a gasification process that will break down approximately 3,300 tons of waste per day into basic elements and electricity. The plant was initially proposed in 2000.

Environment

Puerto Rico's 2001 energy-related carbon emissions totaled 7.23 million metric tons. Puerto Rico's long-term energy policy emphasizes diversification of energy sources, energy conservation and transportation initiatives to reduce energy demand. As part of this policy focus, a metrorail system is under construction in metropolitan San Juan. Completion of the $2.3 billion Tren Urbano (Urban Train) project was expected in September 2003, but has been delayed by escalating costs and logistical problems. Another example of long-term environmental sustainability is the planned RENOVA Resource Recovery facility and eco-industrial park. RENOVA would process one-sixth of Puerto Rico's municipal waste, while the energy, metals, and construction materials generated would be used by nearby industries.

Political and Economic Overview

Population (2002E): 3.86 million

Area: 3,500 square miles (almost three times the size of Rhode Island)

Capital: San Juan

Ports: San Juan, Guanica, Punta Guayanilla, Guayama, Playa de Ponce

Languages: Spanish, English

Currency: U.S. Dollar

Gross Domestic Product (GDP, 2002E): $71.1 billion

Per Capita GDP (2002E): $18,409

Real GDP Growth Rate (2002E) -0.2 per cent *(2003 F)*: 1.9 per cent

Inflation rate (2002E): 8.3 per cent

Unemployment rate (2002E): 12.3 per cent

Major Industries: pharmaceuticals, electronics, apparel, food, instruments, cement, tourism

Major Export Markets Outside the United States: Dominican Republic, Japan, Panama

Energy Overview

Oil Consumption (2002E): 193,000 bbl/d

Crude Oil Refining Capacity (1/1/03E): 73,000 bbl/d

Natural Gas Consumption (2001E): 22 billion cubic feet

Coal Consumption (2001E): 180,000 short tons

Electric Generation Capacity (2001E): 4.9 gigawatts

Net Electricity Generation (2001E): 19.4 billion kilowatthours (bkwh)

Net Electricity Consumption (2001E): 19.4 bkwh

68

Qatar

Qatar contains the third largest natural gas reserves and the largest non-associated gas field in the world. Qatar is also emerging as a major exporter of liquefied natural gas. An OPEC member, Qatar currently exports over 745,000 barrels of oil per day.

General Background

Since 1995, Qatar has been ruled by Sheikh Hamad bin Khalifa al-Thani. Sheikh Hamad has implemented several changes in policy, including a limited political liberalization creating an elected council and giving women the right to vote. In the economic sphere, Qatar has suffered from many of the same problems as other oil-dependent Persian Gulf states, especially the need to diversify economic development beyond crude oil exports and scale back the generous state subsidies for consumers, which date from the oil boom of the 1970s and early 1980s.

Qatar's real gross domestic product (GDP) is projected to grow at an annual rate of 3.8 per cent in 2002, after growth of 3.4 per cent in 2001. This follows a phenomenal growth rate of 11.5 per cent in 2000, which was largely the result of a sharp increase in natural gas exports. Inflation in Qatar remains relatively low, projected at 2.8 per cent for 2002.

Qatar's policy of economic diversification has led to a surge in investment in projects for the export of liquefied natural gas (LNG) and petrochemicals. The government expects that it will be able to earn more per barrel of crude oil produced if it can export refined products and petrochemicals, as well as create private sector jobs–in a country which has been heavily dependent on government ministries to provide employment for the population.

Qatar has begun to pay down its large external debt, which peaked at nearly $12 billion in 1999. Qatar accumulated this debt largely for infrastructure investment in oil and gas projects, which sharply increased Qatar's oil production capacity, construction of facilities for the export of LNG, and petrochemical plants. Qatar's external debt currently is around $7 billion, still equal to more than half of the country's GDP.

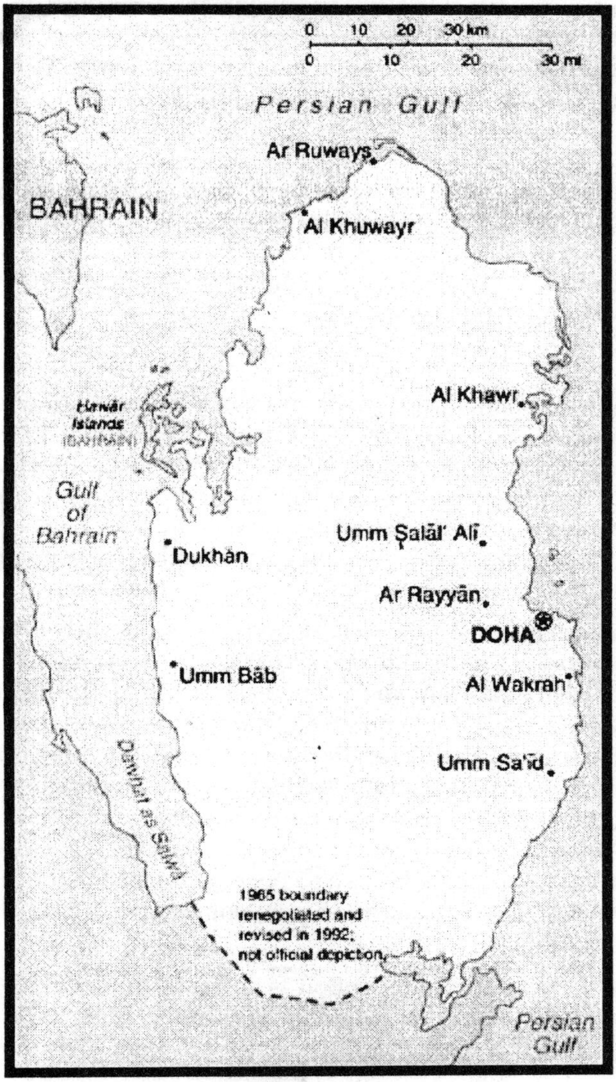

Qatar recorded a large budget surplus for the 2001/2002 fiscal year, which ended in March 2002. The Qatari government for the short term has concentrated on reducing its international debts, rather than increasing public expenditures, and has based its budgets on very conservative projections of oil and gas export revenues.

In a ruling issued in March 2001, the International Court of Justice resolved the dispute between Qatar and Bahrain over the Hawar Islands and neighboring islands. Sovereignty over the Hawar Islands was awarded to Bahrain, while Qatar retained the neighboring islands of Zubarah and Janan.

Oil

Qatar has proven, recoverable oil reserves of 15.2 billion barrels. The onshore Dukhan field, located along the west coast of the peninsula, is Qatar's largest producing oilfield. Qatar also has six offshore fields, Bul Hanine, Maydan Mahzam, Id al-Shargi North Dome, al-Shaheen, al-Rayyan, and al-Khalij. Qatari crude oil has gravities in the 24°-41° API range. The country's two primary export streams are Dukhan (41° API) and Marine (36° API) blend. Despite the country's significant oil production and reserves, oil accounts for less than 15 per cent of domestic energy consumption. Qatar exports almost all of its oil production to Asia, with Japan by far its largest customer. In the first half of 2002, net oil exports totaled 746,777 barrels per day (bbl/d).

In the first half of 2002, Qatar produced 807,777 bbl/d of liquids (including crude oil, natural gas liquids, and condensate), down from 863,918 bbl/d in 2001. The reversal of the previous upward trend was the result of production cuts undertaken in coordination with OPEC, beginning in Januray 2002. As of June 2002, Qatar's monthly crude oil production was averaging 630,000 bbl/d. The country also produces a significant amount of lease condensate and other natural gas liquids (NGLs), both of which fall outside Qatar's OPEC crude production quota, which has been set at 562,000 bbl/d since January 1, 2002.

Following the coup in 1995, Qatar initiated a number of new policies aimed at increasing oil production, locating additional oil reserves before existing reserves become too expensive to recover, and investing in advanced oil recovery systems to extend the life of existing fields. To accomplish this, the government in recent years has improved the terms of exploration and production contracts and production sharing agreements (PSA). The improved terms are designed to encourage foreign oil companies to improve oil recovery in producing fields and to explore for new oil deposits.

Onshore Development

In March 1998, Qatar signed an onshore oil exploration agreement with Chevron (now ChevronTexaco), which is still in effect. It is a five-year PSA and exploration agreement, covering a 4,209 square-mile area known as Block-2. Block-2 covers virtually all of the Qatari peninsula except for the Dukhan field. Seismic surveys were conducted in 1998, and drilling commenced in 1999. ChevronTexaco also holds offshore Block-1 jointly with Hungary's MOL.

Offshore Development

One of Qatar's newer oil fields is al-Rayyan, operated by BP. Wintershall, British Gas, and Gulfstream Petroleum also own significant shares. The field came on stream in November 1996, producing 20,000 bbl/d of heavy oil from four wells. It lies in offshore Block-11 at the southern edge of the North Field near Ras Laffan. The PSA for the field was signed on July 16, 1997. Al-Rayyan has been producing about 25,000 bbl/d in recent months, and BP Amoco and its consortium partners agreed in April 2001 to further development of the field which will bring its production capacity up to 60,000 barrels per day by 2004.

Qatar's latest offshore oil field to come onstream is al-Khalij. Production began in March 1997, after five years of exploration and appraisal work, at an initial rate of 6,000 bbl/d. Al-Khalij is located in Block-6, along Qatar's maritime border with Iran, and to the east of the North Field. Development of the field had been delayed since 1991 as Elf Aquitaine Qatar, the field's operator, sought improved production sharing terms from Qatar Petroleum (formerly QGPC). TotalFinaElf (the result of mergers between the three French oil majors, including Elf Aquitaine) completed a capacity expansion in early 2002 which brought the field's total production capapcity to 60,000 bbl/d. Al-Khalij produces a medium/sweet (28° API) oil with about 1 per cent sulphur. The oil is piped to Halul Island for processing and transportation. TotalFinaElf holds a 55 per cent interest in the 25-year PSA, with an option for a five-year extension. The remaining 45 per cent interest is held by Italy's ENI, which acquired the previous owner Agip.

Maydan Mahzam became operational in 1965, and its production is currently under 50,000 bbl/d, which is down significantly from its peak. However, Qatar Petroleum is currently undertaking a renovation intended to extend the life of the field. This work includes plans for 37 new wells over the next decade.

Bul Hanine came on line in 1973, producing well over 100,000 bbl/d, but production began falling off in the early 1990s. Output 70,000 bbl/d by 2001. Development plans to boost production include drilling 86 new wells. Bul Hanine holds approximately 700 million barrels of recoverable reserves.

Al-Shaheen, operated by Maersk Oil Qatar of Denmark, has become one of Qatar's productive oil fields, with production capacity of around 150,000 bbl/d, though production in recent months has been lower due to Qatar's production cutbacks in coordination with OPEC. Located in Block-5 about 43 miles off Qatar's northeastern coast, al-Shaheen produces a sour (29°-33° API) oil with 1.7 per cent -2 per cent sulphur content. The field is thought to be linked to a section of the North Field.

Maersk agreed in February 2001 to expand the field to a production capacity of 200,000 bbl/d by late 2004. More than 70 new wells are to be drilled.

Id al-Shargi North Dome (ISND), first discovered by Shell in 1960 and now operated by Occidental Petroleum, lies 59 miles east of Qatar's northern tip. In 1994, the field was producing 12,000 bbl/d when Occidental signed a 25-year PSA with Qatar Petroleum, agreeing to invest $700 million in field development, reservoir repairs, gas and water injection systems, and further exploration. Output from ISND stood at about 100,000 bbl/d in 2001.

In December 1997, Occidental signed another PSA with Qatar Petroleum (then known as QGPC) to develop the Id al-Shargi South Dome (ISSD) oil field. ISSD is located 15 miles from ISND, and Occidental will operate ISSD as a satellite of ISND, keeping overall per-unit operating costs lower. The field came onstream in November 1999 at 11,000 bbl/d, and it was producing around 20,000 bbl/d in 2001. Occidental's ownership interest in ISSD is 44 per cent. ISSD is estimated to contain recoverable reserves of 200 million-300 million barrels.

Downstream

Qatar's QP Refinery (formerly known as the National Oil Distribution Company) recently upgarded its refinery at Umm Said. The expansion, completed in early 2002, increased capacity from 57,500 bbl/d to 137,000 bbl/d.

In March 1997, Qatar Petroleum signed a memorandum of understanding (MOU) with the foreign partners of Qatar Liquefied Natural Gas Company (Qatargas) and Ras Laffan Liquefied Natural Gas Company (Rasgas), providing for construction of a condensate refinery at Ras Laffan. While no binding agreement has been concluded, the project is still under active discussion by Qatar Petroleum and TotalFinaElf.

In November 1997, Chevron Phillips Chemical Company signed an $1.1-billion deal with Qatar Petroleum to build a petrochemical plant, Q-Chem, which was completed in 2002. An project to expand the facility, Q-Chem II, is currently under discussion and planned for completion in 2007.

Natural Gas

With proven reserves of 509 trillion cubic feet (Tcf), Qatar's natural gas resources ranks third in size behind Russia's and Iran's. Most of Qatar's natural gas is located in the offshore North Field, which is the largest known non-associated natural gas field in the world. In addition, the onshore Dukhan field contains an estimated 5 Tcf of associated and 0.5 Tcf of non-associated gas. Smaller associated gas reserves also are contained in the Id al-Shargi, Maydan Mahzam, Bul Hanine, and al-Rayyan offhore oil fields. The Qatari government believes that the country's economic future lies in developing this vast natural gas potential. Currently, Qatar has two liquefied natural gas (LNG) exporters: Qatar LNG Company (Qatargas); and Ras Laffan LNG Company (Rasgas).

The Qatargas downstream consortium comprises QGPC (65 per cent), Total (10 per cent), ExxonMobil (10 per cent), Mitsui (7.5 per cent), and Marubeni (7.5 per cent). In December 1996, the Qatargas venture delivered its first shipment of LNG to Japan. The Qatargas LNG plant consists of three, 2-million-ton-per-year (Mmt/y) (97 billion cubic feet–Bcf) trains. The third train was completed in 1999. Qatargas currently is proceeding with a project to debottleneck its existing liquefaction trains, which is to expand total capacity to 9.2 Mmt/y (446 Bcf) when it is completed in 2005. The eventual construction of a fourth train also is planned.

Rasgas is Qatar's second LNG project. The two major shareholders in the project are QGPC and ExxonMobil. Rasgas consists of two 3.3-Mmt/y (163 Bcf) trains. The first train was completed in early 1999, and loaded its first cargo in August 1999 for South Korea's Kogas, which has a long-term supply contract. The second train came onstream in April 2000. The addition of third and fourth trains is planned. Rasgas contracted with Chiyoda, Mitsui, and Snamprogetti in April 2001 for the construction of the third 4.7-Mmt/y (228 Bcf) train, scheduled for completion in 2004, which will be the single largest capacity LNG train in the world.

Qatar's original markets for its LNG exports were Japan and South Korea, the world's two largest LNG importers. India also may eventually become a major market for Qatari LNG. RasGas signed an agreement in July 1999 to supply 7.5 Mmt/y (365 Bcf/y) of LNG to Petronet, an Indian LNG import and gas distribution project. Deliveries under the Petronet contract are now expected to begin in early 2004. Spain's Enagas also has signed a purchase agreement with Italy's Edison, its first term-contract customer in Europe. Deliveries of 3.5 Mmt/y (170 Bcf) are to commence in 2005.

In May 2000, ExxonMobil and QGPC signed a final development and production sharing agreement for the North Field. The Enhanced Gas Utilization (EGU) project will develop upstream infrastructure in a portion of the field for domestic use, export to neighboring Persian Gulf states, and use as a feedstock for petrochemical projects. The initial phase of the project will produce 500 million cubic feet per day (Mmcfd), with eventual capacity slated to rise to 1.75 Bcf/d.

Another significant proposed project will tie Qatar into the United Arab Emirates (UAE) Dolphin Project, an integrated natural gas pipeline grid for Qatar, UAE, and Oman, with a possible subsea connection linking Oman to Pakistan. The United Offsets Group (UOG), a UAE state-owned corporation backing the project, signed preliminary memorandums of understanding with Qatar, Oman, and Pakistan in June 1999. ExxonMobil also signed a preliminary agreement in June 1999 for the natural gas supply from ExxonMobil's production capacity in the North Field. The total project is expected to cost around \$10 billion, including costs associated with the development of more extensive gas distribution networks in the UAE and Oman. Qatar initially will sell around 730 Bcf/y of North Field natural gas, starting in 2005, transported through a subsea pipeline linking the North Field to Abu Dhabi in the UAE. Links between Abu Dhabi, Dubai, and Oman will be added afterwards. UOG announced in March 2000 that TotalFinaElf and Enron had been selected to implement the project, and each would have an equity stake of 24.5 per cent. Enron, however, announced in May 2001 that it was pulling out of the project, and UOG acquired Enron's equity stake, which was resold to Occidental Petroleum in May 2002. The Dolphin Project has been driven in part by the desire of UAE and Oman to use more natural gas for power generation and industrial uses, and the decline in their own production of associated natural gas. Pakistan's participation is highly doubtful, due to its financial condition and the possibility of imports from Iran.

Kuwait also has held discussions with Qatar about the purchase of Qatari gas. A preliminary agreement was signed for gas sales in July 2000, which would source the gas from ExxonMobil's North Field holdings. Details of the project and volumes are still being discussed, and a final agreement has not been reached. Qatar also has held discussions with Bahrain on the possible supply of North Field natural gas, which could be accomplished with a "spur" from the proposed North Field-Kuwait pipeline.

With such vast reserves of natural gas, Qatar has also been interested in potential development of Gas-to-Liquids (GTL) projects. Qatar Petroleum and Sasol signed a preliminary agreement in July 2001 for a 34,000 bbl/d GTL plant at Ras Laffan to be completed by 2005. Foster Wheeler has been engaged to do the initial design work for the plant, but negotiations to secure financing for construction are still underway. ExxonMobil is backing a larger proposed 80,000 bbl/d GTL plant, which could draw on the company's North Field gas production share. A feasibility study is being conducted. Shell and Conoco also have been reported to be studying potential GTL projects in Qatar.

Electricity

Qatar currently has an electric generation capacity of 1,475 megawatts (MW), and produces 9.2 billion kilowatthours of electricity per year. Most of the country's power plants are natural gas-fired. The residential sector accounts for about 80 per cent of Qatar's electricity consumption. In response to financial pressures, the Qatari government announced in 1999 that it would limit the provision of free electricity to Qatari-citizen households, with payment required for consumption above a set threshold.

In May 2000, the Qatari government took a major step towards privatization of its power sector. Assets owned by the Ministry of Electricity and Water (MEW) were transferred to the Qatar General Electricity and Water Corporation (QEWC). QEWC is 57 per cent controlled by local investors and 43 per cent controlled by the government.

The Ras Abu Fontas B-plant is the country's largest and newest power and water desalination plant. It currently has an electric generation capacity of 1,030 MW. France's Alstom Power completed an additional 380-MW generation unit at the facility in June 2002.

QEWC awarded a contract in May 2001 for the Ras Laffan Independent Power and Water Project ("Ras Laffan IWPP"), which will be co-located with the Ras Laffan gas and industrial complex. The plant will have a generating capacity of 750 MW when completed in 2004, with the first 430 MW scheduled to be operational by April 2003. AES will own a 55 per cent equity stake in the project, with QEWC holding 45 per cent.

Country Overview

Independence: September 3, 1971 (from United Kingdom)

Population (2001E): 767,152

Location/Size: Persian Gulf/4,416 square miles, slightly smaller than Connecticut

Major Cities: Doha (capital), Umm Said, Dukhan, al-Khawr

Languages: Arabic (English widely spoken)

Ethnic Groups: Arab (40 per cent), Pakistani (18 per cent), Indian (18 per cent), Iranian (10 per cent), other (14 per cent)

Religion: Muslim (95 per cent)

Defense (8/98): Army (8,500), Navy (1,800), Air Force (1,500)

Economic Overview

Currency: Qatari Riyal

Market Exchange Rate (9/30/02): US$1 = 3.64 Qatari riyals

Nominal Gross Domestic Product (2001E): $16.2 billion *(2002E):* $17.0 billion

Real GDP Growth Rate (2001E): 3.4 per cent *(2002E):* 3.8 per cent

Inflation Rate (consumer prices) (2001E): -0.7 per cent *(2002E)*: 2.8 per cent

Current Account Balance (2001E): $0.5 billion *(2002E)*: $0.0 billion

Major Trading Partners: Japan, United Kingdom, United States, Italy, Germany, France, South Korea

Merchandise Exports (2001E): $7.7 billion

Merchandise Imports (2001E): $6.4 billion

Major Export Products: Crude oil, LNG

Major Import Products: Machinery and transport equipment, manufactured goods, food and live animals

External Debt (2001E): $7.5 billion *(2002E)*: $7.0 billion

Energy Overview

Proven Oil Reserves (1/1/02E): 15.2 billion barrels

Oil Production (1H2002E): 807,878 barrels per day (bbl/d), of which 611,778 bbl/d was crude oil

OPEC Crude Oil Production Quota (effective since 1/1/02): 562,000 bbl/d of crude oil (not including condensate)

Natural Gas Liquids Production (1H2002E): 160,000 bbl/d

Oil Consumption (2002E): 61,000 bbl/d

Net Oil Exports (1H2002E): 746,777 bbl/d

Crude Oil Refining Capacity (1/1/02E): 57,500 bbl/d

Natural Gas Reserves (1/1/02E): 509 trillion cubic feet (Tcf)

Natural Gas Production (2000E): 1,027 billion cubic feet (Bcf)

Natural Gas Consumption (2000E): 532 Bcf

Net Natural Gas Exports (2000E): 496 Bcf

Electric Generation Capacity (1/1/00E): 1.5 gigawatts

Electricity Production (2000E): 9.2 billion kilowatthours

Environmental Overview

Total Energy Consumption (2000E): 0.7 quadrillion Btu* (0.17 per cent of world total energy consumption)

Energy-Related Carbon Emissions (2000E): 9.5 million metric tons of carbon (0.15 per cent of world total carbon emissions)

Per Capita Energy Consumption (2000E): 985.2 million Btu (vs. U.S. value of 351.0 million Btu)

Per Capita Carbon Emissions (2000E): 14.1 metric tons of carbon (vs. U.S. value of 5.6 metric tons of carbon)

Energy Intensity (2000E): 62,814 Btu/$1995 (vs U.S. value of 10,918 Btu/$1995)**

Carbon Intensity (2000E): 0.9 metric tons of carbon/thousand $1995 (vs U.S. value of 0.17 metric tons/thousand $1995)

Sectoral Share of Energy Consumption (1998E): Industrial (81.8 per cent), Transportation (16.5 per cent), Commercial (1.5 per cent), Residential (0.2 per cent)

Sectoral Share of Carbon Emissions (1998E): Industrial (80.4 per cent), Transportation (18.0 per cent), Commercial (1.4 per cent), Residential (0.2 per cent)

Fuel Share of Energy Consumption (2000E): Natural Gas (84.4 per cent), Oil (15.6 per cent), Coal (0.0 per cent)

Fuel Share of Carbon Emissions (2000E): Natural Gas (84.6 per cent), Oil (15.4 per cent), Coal (0.0 per cent)

Renewable Energy Consumption (1998E): 0.04 trillion Btu* (50 per cent decrease from 1997)

Status in Climate Change Negotiations: Non-Annex I country under the United Nations Framework Convention on Climate Change (ratified April 18th, 1996). Not a signatory to the Kyoto Protocol.

Major Environmental Issues: limited natural fresh water resources are increasing dependence on large-scale desalination facilities.

Major International Environmental Agreements: A party to Conventions on Biodiversity, Climate Change, Hazardous Wastes and Ozone Layer Protection. Has signed, but not ratified, Law of the Sea.

The total energy consumption statistic includes petroleum, dry natural gas, coal, net hydro, nuclear, geothermal, solar, wind, wood and waste electric power. The renewable energy consumption statistic is based on International Energy Agency (IEA) data and includes hydropower, solar, wind, tide, geothermal, solid biomass and animal products, biomass gas and liquids, industrial and municipal wastes. Sectoral shares of energy consumption and carbon emissions are also based on IEA data.

Oil and Gas Industries

Organization: Qatar General Petroleum Corporation (QGPC)–exploration and production; National Oil Distribution Company (NODCO)–refining and distribution; Qatar Petrochemical Company (QAPCO)–petrochemical production; Qatar Fertilizer Company (QAFCO)–fertilizer production; Qatar Liquefied Gas Company (Qatargas) and Ras Laffan LNG Company (Rasgas)–production and marketing of liquefied natural gas (LNG)

Major Foreign Oil Company Involvement: BP Amoco Arco, Chevron, Enron, Exxon, Gulfstream, Maersk, Marubeni, Mitsui, Mobil, MOL, Occidental, Phillips Petroleum, TotalFinaElf, Wintershall

Major Ports: Umm Said, Ras Laffan

Producing Oil Fields: Dukhan, Id al-Shargi North Dome, Bul Hanine, Maydan Mahzam, al-Shaheen, al-Rayyan, and al-Khalij

Major Pipelines: Dukhan-Umm Said, an offshore network connecting Halul Island to al-Khalij, Bul Hanine, and Maydan Mahzam, and Das Island (U.A.E.)-al-Bunduq

Major Refineries (capacity–bbl/d): Umm Said (57,500)

69

Regional Indicators: Central America

Although Central America (including Belize, Costa Rica, El Salvador, Guatemala, Honduras, Nicaragua, and Panama) has limited energy resources, it is important to world energy markets as a transit center for oil (via the Panama Canal), and as a potential energy transit center between North and South America.

General Background

Central America is home to some of the world's poorest and most densely populated nations. According to the United States Agency for International Development (USAID), over half of the region's population lives in rural areas, and as many as two-thirds survive on less than $2 per day. Agriculture (*e.g.*, coffee, sugar and bananas) and manufacturing for export and tourism constitute major components of many Central American economies. The last couple of years have been particularly difficult for most Central American countries, as their major trading partner, the United States, has continued to suffer an economic slowdown, and as world commodity prices have remained low. In response, many of these countries are trying to transform their economies by creating new industries and by attracting foreign direct investment. This transition is evident in El Salavador, where, in 1988, more than half of country's export earnings came from coffee, compared to only 3.5 per cent in 2002. In place of traditional industries, countries like Costa Rica, have been successful in attracting large companies, such as Intel Corporation and Proctor and Gamble, to invest.

After spending most of the early 1990s recovering from authoritarian rule or civil wars in Guatemala, El Salvador, Nicaragua and Panama, the Central American countries have begun efforts to improve their economies and cooperation through regional integration. Central American integration first began in 1951, when the Organization of Central American States (ODECA) was

founded. The creations of the Central American Common Market, the Central American Bank for Economic Integration (CABEI) and the Secretariat for Central American Economic Integration followed in 1960. In 1973, ODECA was suspended and the integration process stalled. In 1993, a new integration program, the Central American Integration System (SICA), materialized and reinstated the integration process.

Table 69.1: Economic and Demographic Indicators for Central America

	*Gross Domestic Product (GDP)**				*Population*
	2002E (Billions of U.S. $)	*Real GDP Growth Rate*		*Per Capita GDP, 2002E*	*2002E (Millions)*
		2002 Estimate	*2003 Projection*		
***Belize**	0.8	4.4%	N.A.	$3,078	0.3
Costa Rica	16.9	2.8%	3.2%	$4,299	4.0
El Salvador	14.2	2.1%	2.3%	$2,181	6.5
Guatemala	23.3	2.2%	2.5%	$1,940	11.9
Honduras	6.6	2.5%	2.5%	$979	6.7
Nicaragua	2.5	1.0%	2.1%	$472	5.3
Panama	10.1	0.8%	2.3%	$3,441	2.9
Weighted Ave.	**13.9**	**2.0%**	**2.8%**	**$1,972**	**37.6**

Over the past few years, significant progress has been made in economic integration. In May 2000, after four years of negotiations, the three "northern triangle" countries (El Salvador, Guatemala, and Honduras) signed a free trade agreement with Mexico. Since March 2000, the "northern triangle" countries have been negotiating a trade agreement with the Andean Community (Bolivia, Colombia, Ecuador, Peru, and Venezuela). In March 2002, the Central American Presidents approved an initiative to form a customs union by January 1, 2004. A final agreement to interconnect the electricity networks of six Central American countries was signed in December 2001, allowing for regional power trading among the member states beginning in 2006. This integration of electricity grids is only one of several initiatives by the Inter-American Development Bank's Puebla-Panama Plan, which seeks to promote regional development and integration of Central American countries with Mexico.

Frequent natural disasters, environmental degradation, drug trafficking, and persistent border tensions continue to threaten the region's economic improvement and regional integration. The United States is in the process of negotiating the Central America Free Trade Agreement (CAFTA) with Costa Rica, El Salvador, Guatemala, Honduras, Nicaragua and, as of August 2003, the Dominican Republic. The CAFTA seeks to aid these countries in recovering from the series of natural disasters and economic difficulties, which they have experienced over the past years, as well as in accelerating regional integration.

Energy

All seven Central American countries rely heavily on imported petroleum and indigenous hydropower to meet domestic energy demand. Imported petroleum comes primarily from Venezuela and Mexico under terms of the San Jose Pact and the Caracas Energy Accord. Central America consumes no natural gas and very little coal. Historically, hydroelectric power has dominated Central America's electricity sector. However, since opening up to foreign investors in the middle to late 1990s, the use of thermal generation has grown rapidly. These countries are also attempting to integrate their electricity grids in order to reduce prices and potential disruptions.

Table 69.2: Energy Consumption and Carbon Emissions in Central America, 2001

	Energy Consumption								*Carbon*
	Total Quadrillion Btu)	*Petro-leum*	*Natural Gas*	*Coal*	*Nuclear*	*Hydro-electric*	*Other Electric**	*Net Electric Imports*	*Emissions (Million metric tons)*
Belize	0.011	90%	0%	0%	0%	7.3%	0%	0%	0.2
Costa Rica	0.154	49%	0%	0.01%	0%	39%	16%	-4%	1.4
El Salvador	0.115	69%	0%	0.003%	0%	10%	17.4%	6%	1.5
Guatemala	0.159	78%	0%	3.8%	0%	14.5%	5%	-5%	2.5
Honduras	0.086	70%	0%	3.5%	0%	23%	0%	3%	1.3
Nicaragua	0.058	90%	0%	0%	0%	3%	7%	2%	1.0
Panama	0.138	80%	0%	1.4%	0%	22%	0.5%	0.7%	2.3
Total/Ave.	**0.721**	**75%**	**0%**	**1.2%**	**0%**	**17%**	**6.6%**	**0.4%**	**10.2**

*Other electric includes geothermal, waste, and wood. percentages may not add to 100%due to independent rounding.

Table 69.3: Energy Supply Indicators–Central America

	Fossil Fuel Proved Reserves			Fossil Fuel Production, 2001			Electric Generating Capacity, 1/1/01 (Million Kilowatts)	Crude Oil Refining Capacity, 1/1/03 (Thousand barrels/day)
	Crude Oil, 1/1/03 (Million Barrels)	Natural Gas, 1/1/03 (Trillion Cubic Feet)	Coal (Billion Short Tons)	Petroleum* (Thousand Barrels/ Day)	Natural Gas (Trillion Cubic Feet)	Coal (Million Short Tons)		
Belize	0	0	0	0	0	0	0.07	0
Costa Rica	0	0	0	0	0	0	1.6	15
El Salvador	0	0	0	0	0	0	1.1	22
Guatemala	526	0.11	0	18	0	0	1.7	16
Honduras	0	0	0	0	0	0	0.9	0
Nicaragua	0	0	0	0	0	0	0.7	20
Panama	0	0	0	1	0	0	1.4	60
Total	**526**	**0.11**	**0**	**19**	**0**	**0**	**7.47**	**133**

*Includes crude oil, natural gas plant liquids, other liquids, and refinery processing gain.

Oil Production

Central America produces small volumes of crude oil, most of which either is sent to the United States for refining or consumed domestically. Guatemala, Central America's largest oil producer, produced approximately 24,671 barrels per day (bbl/d) day in 2002. Panama, the region's only other producer, produced 1,000 bbl/d in 2002.

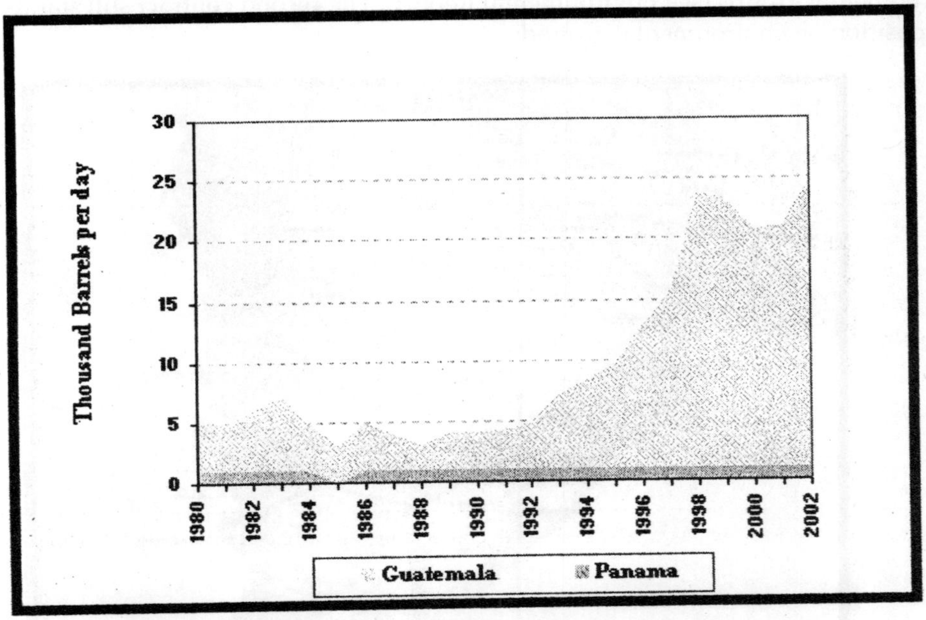

GUATEMALA AND PANAMA OIL PRODUCTION, 1980-2002

Guatemala

Guatemala's 526 million barrels of proven oil reserves are located primarily in the country's northern jungles of the Petén basin, and are most likely associated with those in Mexico's Tabasco formation. For much of the twentieth century, civil war precluded the development of Guatemala's hydrocarbon resources. After the government signed peace accords ending the 36-year internal conflict in December 1996, oil production increased, reaching at 23,767 barrels per day in 1998. Before the peace agreements, petroleum camps suffered from constant attacks from guerrilla groups, which hindered oil exploration and production.

Sector Organization

Since Guatemala's oil industry was first opened to foreign investors, one vertically integrated firm has consistently dominated the industry, the European exploration firm Perenco. In September 2001, Perenco purchased Basic Resources International, a wholly owned subsidiary of Andarko Petroleum Corporation. The sale included all of the country's existing oil fields, a 275-mile crude oil pipeline, a 2,000-bbl/d mini refinery, as well as storage and loading facilities. Since then, Perenco has controlled much of Guatemala's oil sector. There are two other companies conducting oil exploration and activities in Guatemala, Argentina's *Compania General de Combustible,*which is the only other company producing oil in the country and Mepetrol, which is conducting exploration activities.

Exploration

Efforts are also underway to explore and exploit potential reserves near Lake Izabal, Guatemala's largest lake, located in eastern Guatemala near the Gulf of Honduras. In 2001, the government approved two exploration contracts in two adjacent blocks located near the lake. In May 2002, however, Guatemalan President Alfonso Portillo cancelled one of the contracts amid strong pressure from environmental groups (see Environmental Report). The second contract still stands, yet faces similar opposition on environmental grounds.

Nicaragua

In April 2003, a dispute erupted between Colombia and Nicaragua over oil concessions in areas surrounding the San Andres and Providencia islands. The Nicaraguan government was set to issue exploration and production concessions to four U.S. companies (MKJ Exploration International, Hellen Greathouse Year 2000 Trust, Infinity and Arkansas Petroleum). The Colombian government, however, objected to the concessions, claiming that some of the them were located in its territorial waters. Colombia even threatened that it would send in the navy to evict the companies if they commenced operations. As May 2003, Nicaragua had accepted oil exploration bids from four U.S. companies to drill for oil in waters near, but not within, the area being disputed. The Nicaraguan government believes that the region could produce 50,000 bbl/d of crude oil and 2 million cubic feet per day of natural gas. The

Esguerra-Barcena Treaty of 1928 grants Colombia the islands of San Andres, Providencia, and Santa Catalina, in addition to other small islands. In 1979, the Sandinista government denounced the Treaty as void as it had been signed at a time when Nicaragua was under United States military occupation.

In July 2002, Nicaraguan President Enrique Bolanos announced legislation opening the country to foreign oil exploration. The announcement included onshore concessions, as well as offshore blocks in the Atlantic and Pacific Oceans. The announcement marked the first time Nicaraguan hydrocarbon resources have been opened to foreign development since 1979.

Pipeline

In January 2003, U.S.-based Phenix Pipeline and Oleductos Premeri de Nicaragua announced plans to build a pipeline across Nicaragua to transport petroleum products between the coasts. The **Central American Pipeline Project** (282 miles long) would begin at Monkey Point and end at Port of Corinto. The project also includes the construction of two marine terminals, with each having 2 million barrels of storage capacity. Completion of the project is set for 2004.

Oil Consumption

Oil consumption in Central America has doubled since 1980, reaching 259,000 bbl/d in 2002, making the region's total consumption levels roughly comparable to those of Chile (240,000 bbl/d) and Hong Kong (259,000 bbl/d). It is important to note, however, that Central America's population is over twice that of Chile's and six times larger than Hong Kong's. The Central American economies are still among the world's poorest, and significant portions of the population live without access to electricity; hence, per capita oil consumption is low. In 2001, petroleum consumption accounted for approximately 75 per cent of total energy consumption in Central America.

Oil is supplied primarily by Mexico and Venezuela under the auspices of the San Jose Pact and the Caracas Energy Accord. The San Jose Pact, originally implemented in 1980 and renewed annually, commits Venezuela and Mexico to provide all seven Central American countries and four Caribbean countries with a total of 160,000 bbl/d of crude oil and petroleum products under preferential terms. Under the Caracas Energy Accord, agreed to in October of 2000, Venezuela agreed to supply additional oil to Central American and Caribbean countries at preferential prices and terms for the next 15 years. All of the countries except Honduras and Belize have some crude oil refining capacity, with Panama having the largest refinery at 60,000 bbl/d.

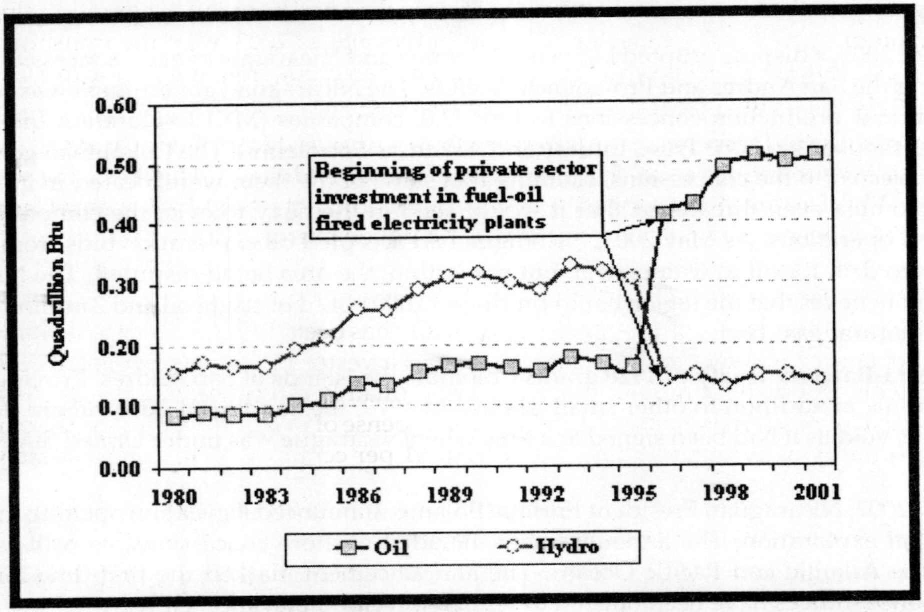

CENTRAL AMERICAN CONSUMPTION OF OIL AND HYDROPOWER, 1980-2001

Natural Gas

Currently, Central America consumes no natural gas. In December 1999, however, Guatemala and Mexico signed a protocol agreeing to construct a natural gas pipeline from Ciudad Pemex, in southern Mexico, to the southern Guatemalan city of Escuintla. The 347-mile, $450 million line would follow the path of an existing oil pipeline in Guatemala's Petén jungle region. The natural gas would be used both for industry and electricity generation. Initial demand is estimated at about 40 million cubic feet per day (MMcf/d). The pipeline, which is expected to be completed by 2004, could eventually be extended to the Honduran and Salvadoran borders, and possibly Nicaragua and Costa Rica as well, as part of a wider Central America natural gas pipeline network.

Panama Canal/Pipeline

At present, Central America's main importance from an energy perspective is as a transit center for oil shipments via the Panama Canal. In 2002, approximately 599,544 bbl/d of crude oil and petroleum products passed through the Panama Canal, with around 63 per cent of total oil shipments moving south from the Atlantic to the Pacific, and oil products dominating southbound traffic. Crude oil accounted for about 44 per cent of northbound oil from the Pacific to the Atlantic. In accordance with the September 1977 Panama Canal Treaty, Panama assumed full responsibility for the Canal at noon on December 31, 1999. At that time, the U.S. Panama Canal Commission was replaced by a new Panamanian entity, the Panama Canal Authority. The Treaty guarantees permanent neutrality of the Canal.

*Trans-Panama Pipeline*In June 2003, Petroterminal de Panama, the owner of the Trans-Panama pipeline, announced that it will reopen the line following the company's conclusion of a deal with

NIC Holding Corporation and Taurus Petroleum. The pipeline had been out of operation since 1996. The resumption of operations is expected to begin in November 2003, with the transportation of 70,000 barrels per day. The pipeline will transport Ecuadorian crude from the Pacific to a Caribbean port for distribution to refineries in the region. The pipeline was

orginally constructed to facilitate the transportation of Alaskan North Slope Crude oil (ANSCO) from Valdez, Alaska to refineries on the Gulf coast of the United States. However, the very large crude carriers (VLCCs) which transported Alaskan crude could not transit the Panama Canal. From 1982 to 1996, the pipeline transported an estimated 2.5 billion barrels of oil.

Electricity

Electricity generation in Central America historically has been dominated by hydropower. The privatization of energy markets, and the entry of foreign investors beginning in the late 1990s have resulted in the development of numerous new thermal (mainly oil) generation plants. As a result, the role of thermal generation is growing rapidly at the expense of hydropower. As a share of regional electricity consumption, hydropower has fallen from 61 per cent in 1980 to approximately 50 per cent in 2001.

Plan Puebla-Panama Initiative for Energy Integration

The goal of this initiative is to unify Central American countries' electricity grids to reduce costs and the frequency of power disruptions, as well as to attract private sector investment for the development of new power plants.

The first stage of the project is the *Sistema de Interconexion Electrica para America Central* (SIEPAC) project. The project entails the construction of transmission lines connecting 37 million consumers in Panama, Costa Rica, Honduras, Nicaragua, El Salvador, and Guatemala. SIEPAC will cost an

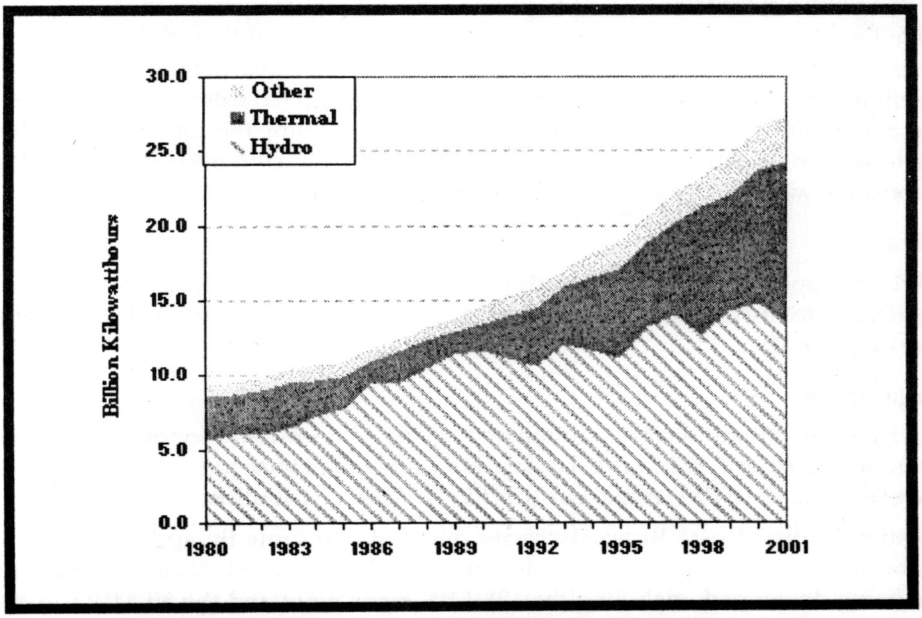

ELECTRICITY GENERATION IN CENTRAL AMERICA, 1980-2001

estimated $320 million and is scheduled for completion in 2006. A second goal of the project is to integrate Mexico with the Central American electricity market by constructing a 62.5-mile, 400kV transmission line between the sub-stations of Tapachula, Mexico and Los Brillantes, Guatemala. On May 20, 2003, Mexico and Guatemala signed an energy integration accord to develop this interconnection. The line is expected to be operational by 2005. The third goal of the project is to link Belize's electricity network with the Central American system. The project entails constructing a 122-mile, 230kV power transmission line between substations in Santa Elena, Guatemala and in Belize City. The initial stages of construction were tentatively set to begin in 2003.

The SIEPAC project will be governed by two new regional institutions, the Regional Electric Interconnection Commission (CRIE) which will regulate the wholesale market, and the Regional Operating Agency, which acts as administrator of regional power transactions.

SIEPAC will allow the member countries to trade electricity regionally. It also will allow countries with severe electricity deficits to purchase power from their neighbors, while enabling countries dependent on thermal power to gain access to Central America's abundant hydropower. Interconnecting the region's national power grids might also make possible the introduction of natural gas into the regional fuel mix, capitalizing upon the proximity of neighboring natural gas producers Mexico, Colombia, and Venezuela.

Belize

Belize Electricity Limited (BEL) is the main generator, distributer, and transmitter of electricity in the country. BEL receives about 85 per cent of its electricty from Comision Federal de Electricidad (CFE), Mexico's national power company, and the Belize Electricity Company Limited (BECOL), a subsidiary of Fortis Incorporated. BECOL owns and operates the 25 MW-Mollejon hydroelectric facility, the only dam currently in operation in Belize. BEL generates the rest of the country's electricity from thermal plants (diesel).

Chalillo Dam For the past couple of years, BACONGO, a coalition of Belizean environmental groups, has been trying to block Fortis from completing the construction of the 5.3-MW Chalillo dam. The environmental group has argued that the dam would have devastating effects on wildlife. The group eventually filed an application for an injunction to stop the construction of the dam with a London-based privy council. On August 12, 2003, the Council denied the injunction, allowing Fortis to continue with the construction of the dam.

Costa Rica

In 2001, hydropower plants generated nearly 90 per cent of Costa Rica's electricity. The country also generates electricity from the Miravalles geothermal plant. Wind energy also accounts for a small percentage of generation.

Sector Organization

Costa Rica's electricity sector is dominated by the state-owned electricity monopoly, *Instituto Costarricense de Electricidad* (ICE), which provides 97 per cent of the country's electricity. A subsidiary of ICE, *Compañía Nacional de Fuerza y Luz* (CNFL) handles distribution.

ICE estimates that Costa Rican electricity demand will grow by approximately 5.7 per cent annually through 2020, requiring an investment of $3 billion by 2011. Numerous new hydropower plants are being developed, including the 128-MW Pirris plant and the 80-MW Cariblanco plant. The Pirris plant is expected to become operation in 2008. The Boruca hydroelectric project is planned to have between 832-942 MW of installed capacity. Currently the government is conducting environmental and feasibility studies on the project, with a potential start to operations in 2012.

El Salvador

El Salvador is Central America's largest producer of geothermal energy. The country has two main geothermal facilities, Ahuachapán, with a generating capacity of 110 MW located in the Ahuachapan province, and Berlín, with a generating capacity of 66 MW, located in the province of Usulutan. Gesal, the state-owned power company, operates the two plants. In 2002, the country produced about 0.935 billion kilowatthours (Bkwh) of geothermal electricity, representing approximately 22 per cent of total electricity consumption, according to government statistics. Thermal generation accounted for 42 per cent of generated electricity while hydropower provided 26 per cent. In January 2002, Gesal began looking for a private partner to develop two new smaller geothermal projects near Ahuachapán and Berlín, which could potentially add approximately 38 MW to the country's capacity.

Sector Organization

In 1995, the Salvadoran National Assembly approved regulation to begin the privatization of the country's state-owned utility company, Comisión Ejecutiva Hidroelétrica del Río Lempa (CEL). In 1998, the government successfully sold majority stakes in CEL's four distribution companies: *Compañía de Luz Eléctrica de Santa Ana* (CLESA) to AES Energy of the United States; *Compañía de Alumbrado Eléctrica de San Salvador* (CAESS) and *Empresa Eléctrica del Orienta* (EEO) to ENERSAL of Venezuela; and *Distribuidora de Electricidad* (DELSUR) to the Chilean power company Emel. In the following year, the government created Gesal to operate the country's geothermal power plants and *Empresa Transmisora de El Salvador* (ETESAL) to maintain the country's transmission lines, as well as the interconnector to Guatemala. The two companies remain state-owned, despite efforts to privatize them. An independent agency to regulate the wholesale market, *La Unidad de Transacciones*, was also created. Finally, the government sold stakes in some of CEL's generation assets, namely 265 MW of thermal power to Duke Energy in 1999.

Guatemala

Electricity generated from thermal sources in Guatemala has increased significantly over the past years. In 1991, hydropower was responsible for around 74 per cent of electricity generation while thermal sources (fuel oils and coal) contributed only 11 per cent. In contrast, 51 per cent of Guatemala's electricity in 2001 was generated from thermal sources and 35 per cent from hydropower. Guatemala has one of Central America's largest coal-fired power plant, the 120-MW San José Power Station, which began operations in January 2000. TECO Power Services (TPS) owns and operates the facility. Numerous other thermal power plants are expected to come online. In March 2003, Duke Energy's 80-MW heavy fuel oil plant (Arizona) began operations in San José. Phase II of the project is almost completed, which will double the plant's capacity to 160 MW. The plant will also be able to run on Orimulsion.

Since 1999, two geothermal powered electricity plants have been in operation. The first plant, Zunil 1, has an installed capacity of 24 MW. The power plant is owned and operated by Orzunil as a joint investment of ORMAT, the International Finance Corporation, Scudder Latin American Power Fund, the Commonwealth Development Corporation, ING Bank and the Netherlands Development Finance Company. The other 5 MW geothermal electric power plant is located in Amatitlan. The

Mexican company, Ingenieros Civiles Asociados, has been running the plant since it began operation. Ormat is planning to build a 20-MW geothermal power plant in the Amatitlan field.

Privatization

In October 1996, the Guatemalan government approved the General Electricity Law, which deregulated the electricity sector, requiring key national enterprises to unbundle generation, transmission, and distribution functions within one year of enactment. The law had an immediate effect on the country's two largest power companies *Empresa Eléctrica de Guatemala Sociedad Anónima* (EEGSA) and *Instituto Nacional de Electrificación* (INDE). In 1997, the government restructured INDE, dividing the company into three different entities responsible for generation, transport and control and distribution. INDE became a holding organization that directed the units. In 1998, INDE's distribution company was split into two regional companies and privatized: *Distribuidora Eléctrica de Occidente* (DEOCSA) and *Distribuidora Eléctrica de Oriente* (DEORSA). Spain's Union Fenosa acquired an 80 per cent share in both the companies. In 1998, a consortium consisting of Spain's Iberdrola (49 per cent share), U.S.-based Teco Energy (30 per cent), and Portugal's Eletricidade de Portugal (21 per cent) acquired 80 per cent of EEGSA distribution assests. The consortium changed the company's name to DECASA, *Distribuidora de Energía de Centroamerica* (Central American Energy Distributor). In May 2003, Mexico and Guatemala signed energy integration accord to develop an electricity interconnection between their national grids. The project entails the construction of a 64-mile, 400kV transmission line. The line is expected to be operational by 2005. This line is an integral part of the larger Pueblo-Panama plan.

Rural Electrification

When the Guatemalan government privatized DEOCSA and DEORSA, it obliged Union Fenosa to help extend the country's rural electrification program, *Programa de Electrificación Rural* (PER). The government offered the two companies an incentive: a $650 subsidy for each verified, eligible residential connection each company makes. As of October 2002, the country had an 85 per cent electrification rate, according to the *Comisión Nacional de Energía Electrica*. When the program is completed in 2004, the government hopes that around 90 per cent of Guatemalan households have electricity.

Honduras

According to *Empresa Nacional de Energía Eléctrica* (ENEE), Honduras' state-owned energy company, the country generated 55 per cent of its electricity from thermal sources, 36 per cent from hydropower, and 0.1 per cent from biomass in 2002. Honduras imported the rest, mainly from Costa Rica. In 2002, over half of Honduras' hydropower was generated from the 300-MW El Cajón hydroelectric dam. The country currently operates 9 hydropower plants, 14 thermal powered plants, and 4 biomass plants.

Electricity demand is expected to increase at about 6 per cent per year until 2015, according to ENEE. In order to meet this demand, ENEE expects that generation capacity will need to more than double by 2014, reaching an estimated 2.9 million kilowatts of installed capacity.

In July 2003, the government created the "Commission for the Development of Hydroelectric Projects" to identify the most efficient mechanisms to develop hydro projects with private sector participation. By the first quarter in 2004, Honduran government plans to call for bids on the 161 MW Patuca III hydroelectric project. The government also plans to call for bids for one of the following three projects in the Sula Valley: El Tablón (20 MW), Los Llanitos (50 MW) and Jicatuyo (170 MW).

The government also expects to increase generation capacity from biomass and wind. ENEE is currently working with the private company, Zond de Honduras, to build a wind park, with 60 MW of installed capacity.

Sector Organization

ENEE controls most the electricity operations in Honduras. In 2002, independent power producers, however, generated more than 50 per cent of country's electricity, while ENEE supplied about 36 per cent. In January 2003, the Honduran government announced plans to begin privatizing ENEE's distribution assets. According to government officials, the country needs $140 million of investment a year in the energy sector in order to meet the country's growing demand through 2015. By further privatizing ENEE, the government hopes to attract private sector investors.

Nicaragua

In 2001, fossil fuels were used to generate 84 per cent of country's electricity generation, with hydropower (7.6 per cent) and other renewables, mainly geothermal (7.6 per cent), accounting for the rest. Nicaragua's electricity sector is heavily dependent on imported petroleum for fuel oil and diesel.

Similar to its neighboring countries, electrification of rural areas remains a major initiative of the Nicaraguan government. In May 2003, the World Bank approved a $12 million loan to finance off-grid rural areas. Around 16,000 households do not have electricity. This is part of the country's National Rural Electrification Program (PLANER).

Geothermal

Geothermal power provides a small but increasing role in Nicaragua's energy mix, particularly since the government would like to decrease its dependence on oil to generate electricity. In February 2003, Ormat International completed a four-year rehabilitation project of the Momotombo geothermal power plant. Through an upgrading process, Ormat increased the plant's installed capacity from 8 MW to 35 MW. In November 2002, San Jacinto Power won a bid to construct a geothermal power plant, which will access steam from the San Jacinto-Tizate field. The initial plans include installing 10 MW which would later be expanded to 66 MW.

Sector Organization

Beginning in March 2001, the Nicaraguan government began privatizing the country's national power company's (*Empresa Nicaragüense de Electricidad*–Enel) generation assets–*Generadora Eléctrica Central* (Gecsa), *Generadora Eléctrica Occidental,* (Geosa), and *Generadora Hidroeléctrica,* (Hidrogesa)– under the auspices of the state privatization agency, URE. In February 2002, Coastal Power, a subsidiary of El Paso, acquired GEOSA and its two thermal power stations, Nicaragua and Chinandega, with a combined installed capacity of 114 MW. Coastal power also placed a bid for Hidrogesa and its two hydropower plants, Santa Bárbara and Centroamérica, but it was not accepted. The Gecsa power stations have yet to attract any bidders. Private generators provide a majority of the electricity in Nicaragua. There are also a large number of small, isolated power producers, which use mainly diesel-fired generators. Spain's Union Fenosa operates Nicaragua's two distribution companies, Disnorte and Dissur.

Panama

Electricity generation accounts for most of Panama's domestic energy production, with hydroelectric generation alone accounting for 75 per cent of the country's total energy production.

There are currently eight electricity generators (2 hydro and 6 thermal) and three distributors (Edemet, Elektra and Edechi) operating in Panama. Transmission remains in the hands of the government through the publicly owned power company, Etesa. Panama also exports electricity to Nicaragua, Honduras and Costa Rica, and imports electricity from Costa Rica. In May 2003, Panama and Colombia signed a memorandum of understanding on energy integration. Possible joint projects include hydroelectric projects in border regions and building an interconnector between the two countries. However, a dense jungle, the Darien Gap, divides the countries and could hinder construction of the interconnector.

Electricity demand is expected to grow significantly in the coming few years, and new projects are planned to help meet this growth. Construction of AES's $200 million, 120-MW Esti hydroelectric project in Panama's Chiriqui province is expected to come onstream in November 2003. Esti is made up of two hydroelectric plants, Guasquitas and Canjilones, on the Chiriqui river. AES also is in the process of expanding and upgrading its 150-MW Bayano hydropower plant. The Bayano III expansion project includes the addition of a third unit with 86 MW of capacity and the rehabilitation of the two existing units to obtain up to 84 MW from each refurbished unit, giving the plant a combined installed capacity of 254 MW.

In June 2002, the electricity distributor Edemet-Edechi, a subsidiary of Union Fenosa, announced plans to invest $80 million over the next 5 years in hydroelectric and wind energy projects. According to Union Fenosa, it has already invested $130 million in transmission lines, infrastructure and distribution systems since entering Panama more than four years ago.

Environment

Central America remains one of the poorest regions on earth; a fact which has led to massive exploitation of the area's natural resource base. Large areas of forest have been cut down and burned for firewood or used in the production of paper, while significant portions of land have been cleared for agricultural use. Oil exploration activities in certain parts of Guatemala, such as the northern Peten rainforest region, have encouraged road construction, accelerating the clearing of land and forested areas. These activities have lead to large-scale erosion and soil loss, leaving many areas vulnerable to flash floods and mudslides as the natural landscape's ability to retain water is jeopardized. Oil is the chief source of energy in most areas of Central America, and pollution from cars, industry, and power generation is a major problem in several areas.

70

Regional Indicators: European Union

The European Union, with increasingly integrated economies and energy sectors, is the world's second-largest energy consumer (behind the United States). EU members include: Austria, Belgium, Denmark, Finland, France, Germany, Greece, Ireland, Italy, Luxembourg, the Netherlands, Portugal, Spain, Sweden, and the United Kingdom.

General Background

The European Union (EU) was founded as the European Economic Community (EEC) by the Treaty of Rome in 1957 to promote economic and political integration in Europe. The founding of the EEC followed the creation of the European Coal and Steel Community, established after World War II as a means of promoting integration among former enemies. The EEC has expanded from its original six members (Belgium, France, the Federal Republic of Germany, Italy, Luxembourg, and the Netherlands) to include the United Kingdom, Ireland, and Denmark in 1973; Greece in 1981; Spain and Portugal in 1986; and Austria, Finland, and Sweden (former members of the European Free Trade Association) in 1995.

All 15 member states delegate a degree of sovereignty to the EU's network of institutions. National governments are represented in the Council of the European Union, while citizens of the member states are elected to the European Parliament. In 1993, the Maastricht Treaty (which renamed the EEC as the European Union), created European citizenship, strengthened the power of the European Parliament, laid out plans for the Economic and Monetary Union (EMU), as well as committed members to negotiate for expansion of the EU to include Central and Eastern European countries. As part of EMU, 12 EU member countries (Belgium, France, Germany, Greece, Italy, Spain, Portugal, Finland, Austria, the Netherlands, Ireland and Luxembourg) adopted a new common European currency, called the "euro". The Euro currency entered into general circulation in January

2002. Monetary policy is overseen by the European Central Bank, which works in conjunction with the national central banks of the 12 euro zone countries.

In 2001, the Treaty of Nice was signed by member governments. This treaty changed the way the institutions of the EU operate in order to make possible the admission of new member states in the future. At its next scheduled meeting in December 2002, the EU Council of Ministers is expected to nominate Poland, the Czech Republic, Slovakia, Hungary, Estonia, Latvia, Lithuania, Slovenia, Malta, and Cyprus for entry in to the EU in 2004. Many other countries also aspire to EU membership, including Romania and Bulgaria, which are expected to join in 2007.

The combined economies of the EU are slightly smaller than the U.S. economy ($9.2 trillion purchasing power parity gross domestic product for the EU in 2001 versus $9.9 trillion for the United States), while the EU population of 376.8 million significantly exceeds the U.S. population of 278 million. The United States has extensive trade relations with the EU. In 2001, 22 per cent of U.S. exports ($159 billion) went to EU members, and 19 per cent of U.S. imports ($220 billion) originated in EU countries.

Table 70.1: Economic and Demographic Indicators for EU Countries

	Gross Domestic Product (GDP) (Purchasing Power Parity)				*Population*
	2001E (Billions of U.S. $)	*Real GDP Growth Rate*		*Per Capita GDP, 2001 (U.S. Dollars)*	*2001E (Millions)*
		2002 Estimate	*2003 Projection*		
Austria	$223.6	1.0%	1.0%	$27,300	8.2
Belgium	$286.7	1.0%	1.0%	$27,800	10.3
Denmark	$151.2	1.0%	1.4%	$28,500	5.3
Finland	$132.8	0.7%	1.2%	$25,500	5.2
France	$1,483	1.8%	1.1%	$25,000	59.0
Germany	$2,113.6	0.7%	0.4%	$25,700	82.4
Greece	$184.9	4.1%	3.8%	$17,400	10.6
Ireland	$123.2	5.9%	3.4%	$32,400	3.8
Italy	$1,414	1.8%	0.4%	$24,500	57.8
Luxembourg	$23.5	3.5%	2.7%	$58,700	0.4
Netherlands	$420.7	1.1%	0.7%	$26,300	16
Portugal	$178.8	1.7%	0.8%	$17,900	10.0
Spain	$804.8	2.7%	2.0%	$20,400	39.5
Sweden	$222.1	1.1%	1.6%	$25,000	8.8
United Kingdom	$1,462.9	1.9%	1.6%	$24,600	59.5
Total	**$9,225.8**	1.6%	1.0%	**$27,100**	**376.8**

Energy

On the whole, the members of the European Union are net energy importers. Import dependency is forecast to grow in the future, as the European Commission estimates that the EU could be forced to import 70 per cent of its total energy, and up to 90 per cent of its oil in the next 20 to 30 years if no new measures are taken. Germany, Italy, and France are the EU's largest net importers of energy; the United Kingdom is the only significant net exporter (see graph). EU oil is imported primarily from the Persian Gulf region, Norway, Russia, and North Africa. Russia also exports significant amounts of natural gas to the EU countries.

EU members possess only about 0.7 per cent of the world's proven reserves of oil and 2.5 per cent of the world's natural gas reserves (see Table 3). However, they have 7.3 per cent of proven coal reserves, 16 per cent of the world's capacity for refining crude oil into petroleum products, and 17 per cent of the world's electric generating capacity. In 2000, they produced 4.5 per cent of the world's crude oil, 9 per cent of the world's natural gas, and 7 per cent of the world's coal.

Energy Consumption

In 2000, the 15 EU countries consumed 63.3 quadrillion British thermal units (Btu) of energy, which represents 16 per cent of the world's total energy consumption and is 35.5 quadrillion Btu's less than the United States energy consumption for the year. EU members consumed about 33 per cent of the world's nuclear power, 19 per cent of the world's oil, 16 per cent of the world's natural gas, and 9 per cent of the world's coal in 2000.

Oil is the dominant fuel, (see Table 71.2), accounting for 43 per cent of total EU energy consumption in 2000, followed by natural gas at 23 per cent. Over the past decade, natural gas has been the fastest growing fuel source in the EU, mainly at the expense of coal, whose share has declined sharply. This is in part due to environmental considerations, but also due to increased availability of natural gas supplies because of pipelines from Algeria, Norway, and Russia. By 2010, natural gas is expected to account for 26 per cent of EU energy consumption.

Nuclear power generation has grown only slightly over the past decade. Eight EU member states are currently operating nuclear power plants, and five of these (Sweden, Spain, the Netherlands, Germany, and Belgium) have announced moratoriums. In May 2002, however, the Finnish Parliament approved plans to build a new nuclear reactor, marking the first expansion of nuclear power in the EU in over ten years. Hydroelectric power consumption has grown by almost 20 per cent over the past decade and accounted for approximately 5 per cent of total EU power consumption in 2000. Other "renewables" (geothermal, biomass, solar, and wind) quadrupled between 1991 and 2000, but still constituted only 1 per cent of total EU energy consumption in 2000. Renewable energy and natural gas are expected to be the two fastest growing fuels in the EU over the next 20 years.

Energy Policy and the Internal Market

Although energy policy was not stipulated in the original treaties establishing the modern European Union, the political and economic policies governed by the EU have found a natural confluence in energy policy. Moreover, many of the EU countries share a similar energy profile, and could face significant import dependency problems in the coming years.

In November of 2000, a European Commission Green Paper on Energy Security outlined the EU's unified energy strategy. The paper identifies four main principles of European energy policy: 1) security of supply, 2) completion of the internal market, 3) environmental responsibility and 4) promoting renewable energy and demand management. To these ends, the European Union has acted

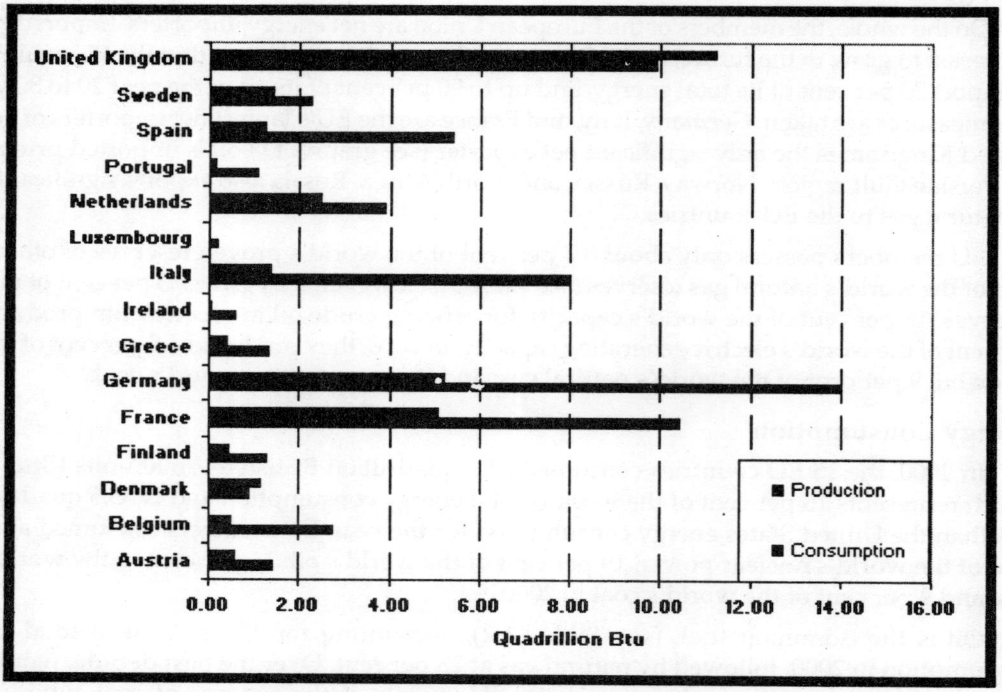

EUROPEAN UNION ENERGY PRODUCTION AND CONSUMPTION, 2000

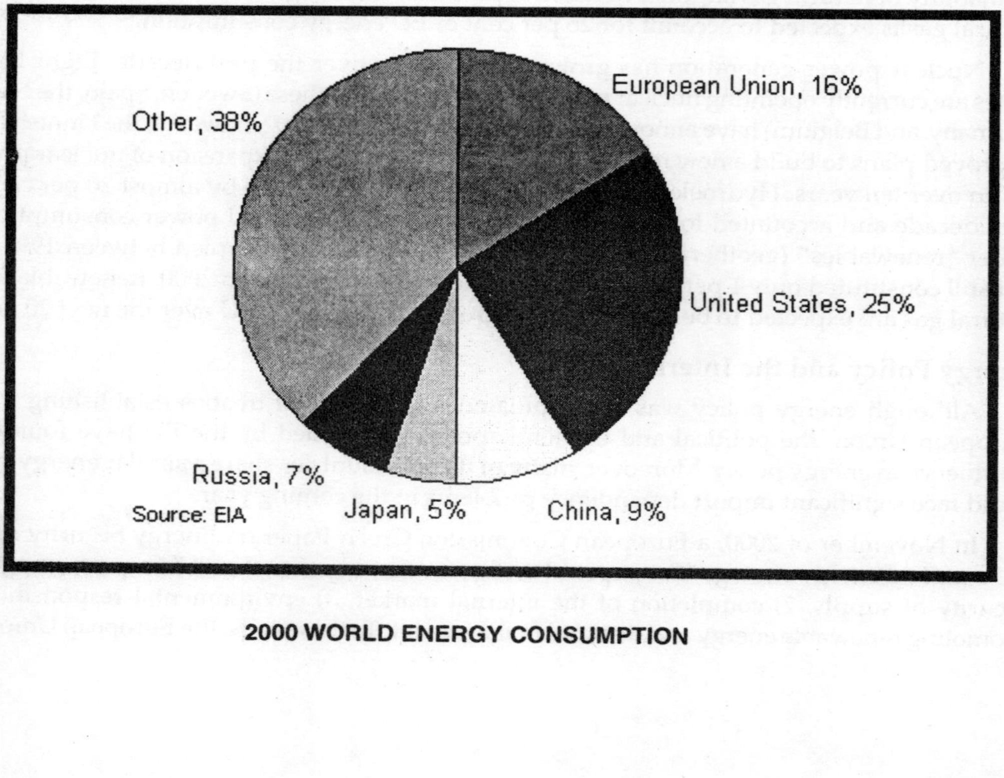

2000 WORLD ENERGY CONSUMPTION

over the years to coordinate the member countries' energy policies and the infrastructure that links them.

Table 70.2: Energy Consumption and Carbon Emissions in EU Countries, 2000

	Energy Consumption								Carbon
	Total Quadrillion Btu)	Petro-leum	Natural Gas	Coal	Nuclear	Hydro-electric	Other Electric	Net Electric Imports	Emissions (Million metric tons)
Austria	1.41	39%	20%	10%	0%	31%	1%	-1%	18
Belgium	2.75	45%	23%	12%	17%	0.2%	0.4%	1.6%	40
Denmark	0.88	51%	23%	19%	0%	0.03%	7%	1%	16
Finland	1.30	32%	12%	11%	17%	12%	7%	9%	13
France	10.41	40%	15%	6%	39%	7%	0.4%	-7%	109
Germany	13.98	41%	22%	23%	12%	1%	1%	0.1%	220
Greece	1.33	63%	6%	28%	0%	3%	0.7%	0.0%	27
Ireland	0.59	60%	26%	13%	0%	1%	0.5%	0.2%	11
Italy	7.96	49%	32%	6%	0%	6%	2%	6%	117
Luxembourg	0.19	50%	15%	3%	0%	1%	0.4%	31%	2
Netherlands	3.91	45%	39%	8%	1%	0.04%	1%	5%	64
Portugal	1.08	64%	8%	14%	0%	11%	2%	1%	17
Spain	5.40	57%	12%	14%	11%	5%	1%	1%	81
Sweden	2.25	30%	1%	4%	24%	36%	2%	2%	13
United Kingdom	9.88	35%	36%	15%	10%	1%	1%	1%	148
Total	**63.32**	**43%**	**23%**	**13%**	**14%**	**5%**	**1%**	**0.7%**	**896**

Community energy policy is developed and implemented by the Energy and Transport Directorate-General, which has its headquarters in Brussels. The Energy and Transport Directorate-General reports to the European Commission, which in turn drafts legislative proposals for the European Parliament. The Parliament then works with the Council of the European Union, (which is made up of government officials from the member states) to amend, and eventually adopt energy legislation for the entire EU.

Integrating the energy sectors of EU members is a work-in-progress. Incremental steps toward enjoining member countries' energy sectors have been taking place since the early 1990s. Because existing treaties already govern the coal and nuclear sectors, and the oil industry was considered to be sufficiently open, most of the EU's recent energy legislation has focused on electricity and natural gas.

In 1996, after years of deliberations, the EU Directive on Electricity was passed, stipulating the deregulation of the production and transport of electricity in EU member states, and setting a schedule for the opening of member countries' markets to free competition. The Electricity Directive originally called for incremental market liberalization, with 32 per cent of the market to be open to free competition by 2003 (Greece, Belgium and Ireland were granted waivers). In 1998, the E.U. Natural Gas Directive was passed with similar provisions, calling upon the member countries to adapt national law to facilitate market opening, and setting a similar schedule of annual market-liberalization benchmarks (with Greece, Belgium, and Ireland again granted waivers).

Table 70.3: Energy Supply Indicators–European Union Countries

	Fossil Fuel Proved Reserves			*Fossil Fuel Production, 2001*			*Electric Generating Capacity, 1/1/01 (Million Kilowatts)*	*Crude Oil Refining Capacity, 1/1/03 (Thousand barrels/day)*
	Crude Oil, 1/1/03 (Million Barrels)	*Natural Gas, 1/1/03 (Trillion Cubic Feet)*	*Coal (Billion Short Tons)*	*Petroleum* (Thousand Barrels/ Day)*	*Natural Gas (Trillion Cubic Feet)*	*Coal (Million Short Tons)*		
Austria	86	0.9	28	22	0.1	1.4	14	209
Belgium	0	0.0	0.0	12	0.0	0.4	14	791
Denmark	1,113	2.7	0.0	367	0.3	0.0	13	176
Finland	0	0.0	0.0	0	0.0	0.0	16	239
France	140	0.4	40	79	0.1	5.2	110	1,896
Germany	364	12	72,753	144	0.8	225.3	109	2,259
Greece	9	0.2	3,168	9	0.0	69.5	10	407
Ireland	0	0.7	15	1	0.0	0.0	4	71
Italy	622	8.1	37	155	0.6	0.0	67	2,283
Luxembourg	0	0.0	0.0	0	0.0	0.0	0	0
Netherlands	107	62.5	548	89	2.6	0.0	21	1,206
Portugal	0	0.0	40	2	0.0	0.0	11	304
Spain	21	18	728	19	0.0	25.8	46	1,294
Sweden	0	0.0	1	0	0.0	0.0	34	424
U.K.	4,930	26	1,653	2,553	3.8	35.3	72	1,784
Total	**7,392**	**131.5**	**79,011**	**3,452**	**8.2**	**362.9**	**541**	**13,343**

Energy market liberalization has gained unprecedented support recently as the European Council in March 2000 called upon the European Commission to expedite the opening of energy markets. In March 2001, the Commission advanced the timetables to achieve full market liberalization by 2005. In March 2002, EU leaders at the Barcelona Summit confirmed their commitment to opening up natural gas and electricity markets, declaring that access to business and commercial customers will be completely open by 2004. The European Parliament has passed a draft resolution on opening electricity and natural gas markets, which will go before EU member energy ministers in November. Approving market opening for residential customers has proven much more contentious, with no deadline set as of yet.

The Energy and Transport Directorate-General also oversees efforts to increase the role of renewable energy sources in the EU fuel mix, as well as demand management programs. The promotion of renewable energy and energy efficiency are handled by the EU's Altener, and SAVE programs respectively. The two programs have been in place since the early 1990s, and in April of 2002 were renewed under the European Commission's proposal, "Intelligent Energy for Europe (2003-2006)." Currently, renewable energy accounts for 5.6 per cent of EU energy consumption. The EU aims to derive 12 per cent of the group's energy consumption from renewable fuels by 2010.

In September 2002, the European Commission proposed that EU member states make arrangements to hold an additional 40 days worth of strategic oil stocks in a joint European reserve pool. These proposed reserves would supplement each state's individual reserves, which are supposed to amount to 90 days worth of consumption. Currently, the EU member states are estimated to hold on average 115 days worth of oil, depending on the country. The Commission aims to raise strategic oil stocks across the EU to 120 days worth of consumption by 2007.

Energy Use and Carbon Emissions

In 2000, EU members generated 896 million metric tons of energy-related carbon emissions, representing 14 per cent of the world total for that year. Of the EU countries, Germany emitted the most carbon (220 Mmt), followed by the United Kingdom (148 Mmt), Italy (117 Mmt) and France (109 Mmt), with each of the countries showing a decline in carbon emissions since last year. Under the December 1997 Kyoto Protocol, the EU is obligated to reduce its greenhouse gas emissions 8 per cent from 1990 levels (in that year, the EU emitted 913 Mmt of carbon) by 2008-2012. All EU member states signed the Kyoto Protocol on April 29, 1998. On June 17, 1998, the EU agreed on how it would meet the 8 per cent reduction. Under this agreement, different EU member states are assigned varying degrees of emission cuts, ranging from a 4 per cent increase in the case of Sweden, to a reduction ot 28 per cent in the case of Luxembourg, with other countries somewhere in between.

71

Russia

Russia is important to world energy markets because it holds the world's largest natural gas reserves, the second largest coal reserves, and the eighth largest oil reserves. Russia is also the world's largest exporter of natural gas, the second largest oil exporter, and the third largest energy consumer.

General Background

In 2002, Russia's gross domestic product (GDP) grew by 4.3 per cent, surpassing average growth rates in all other G8 countries, and marking the country's fourth consecutive year of economic expansion. Russia's economic growth over the last four years has been fueled primarily by energy exports, particularly given the boom in Russian oil production and relatively high world oil prices during this period. But this type of growth has made the Russian economy dangerously dependent on oil and natural gas exports, and especially vulnerable to fluctuations in world oil prices. In 2002, energy accounted for almost 20 per cent of Russia's GDP, 55 per cent of export revenue, and roughly 40 per cent of fiscal revenue.

The Russian government has made decoupling economic growth from commodity exports a priority, and is attempting to restructure and liberalize its energy sector. These reforms have come at the behest of both Russian oil and natural gas producers, who are seeking to grow in a more liberalized marketplace; as well as Russia's external trading partners, who are are pressuring the country to synchronize their policies with those in Western Europe and North America, particularly vis-a-vis Russia's aspiration towards the World Trade Organization (WTO). Key to these efforts will be breaking up the monopolies that control the natural gas and electricity industries.

But in the meantime, the country is also looking to maximize its exports of fossil fuels and sustain economic growth. More than anything, the country's insufficient export infrastructure hinders short term production growth. According to some analysts, Russia could easily produce 10 million barrels per day (bbl/d) by 2007-2010 were it not for the country's insufficient infrastructure.

Accordingly, developing new connections to the outside world, while simultaneously re-organizing the country's domestic industry, will be key to securing sustainable economic expansion into the future. On May 16, 2003, Russian President Vladimir Putin charged the country with doubling GDP within ten years, and on May 22, the Russian government released its energy strategy to 2003-2020, which designates the energy sector as the engine of this growth.

Oil

Russia has proven oil reserves of 60 billion barrels, most of which are located in Western Siberia, between the Ural Mountains and the Central Siberian Plateau. In the 1980s, this prolific region, also known as the "Russian Core," made the Soviet Union a major world oil producer, allowing for peak production of 12.5 million bbl/d in 1988 (most of which came from Russia). However, that level of production ultimately proved to be unsustainable, as the Soviet era's state-mandated production surges are thought to have seriously depleted a number of the country's large fields. The industry was dealt a further blow following the demise of the Soviet Union in 1991, as the state planning system collapsed, throwing the industry into disarray and stunting domestic consumption. By 1996, Russian production had fallen to roughly 6 million bbl/d, less than half of the Soviet-era peak.

A turnaround in Russian oil output began in 1999, which many analysts have attributed to rising world oil prices during this period (oil prices tripled between January 1999 and September 2000), as well as a number of after-effects of the 1998 financial crisis and subsequent devaluation of the ruble in August. Today, Russian oil fields are maintained using modern technologies from around the world, and many of the old command economy institutions have been streamlined.

The rebound in Russian oil production has continued since 1999, resulting in 2002 total liquids production of 7.65 million bbl/d (7.4 million bbl/d of which was crude oil)–a 26 per cent increase over the 1998 level. Accordingly, Russia is now the world's second largest crude oil producer behind only Saudi Arabia. Russia's growing significance in world oil markets was evidenced in September 2003 by an official visit from Saudi Crown Prince Abdullah bin Abdulaziz, during which President Putin and Crown Prince Abdullah pledged to increase communication between the world's two largest oil producers into the future. According to forecasts from the Russian Energy Ministry, oil production will average 8.2 million bbl/d in 2003, and 8.8 million bbl/d in 2004.

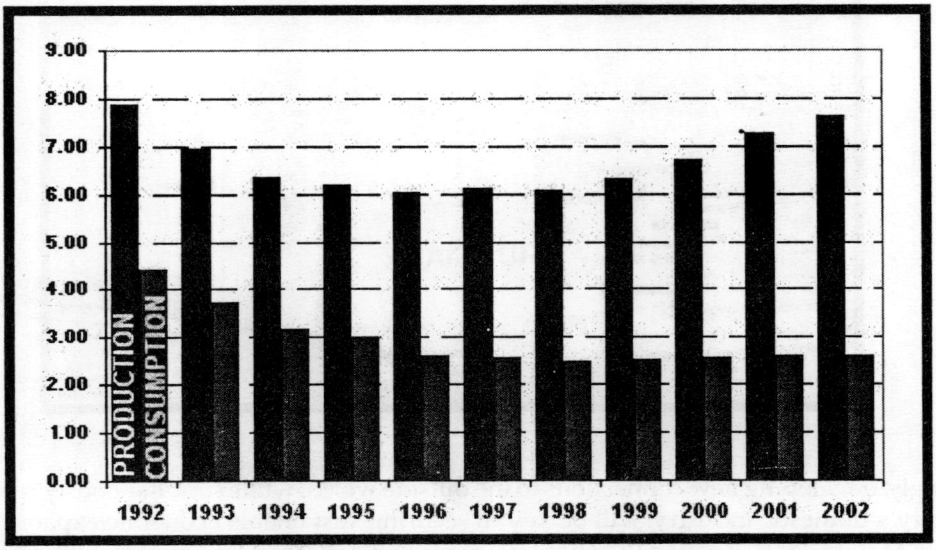

RUSSIAN OIL PRODUCTION AND CONSUMPTION, 1992-2002

Oil Industry Structure

In 2003, two major mergers were announced: Tyumen Oil Company (TNK) with British Petroleum (BP), and Yukos with Sibneft. Both transactions are expected to have a lasting affect on the internal dynamics of the Russian oil market (see chart below), as well as the country's external relations. Prior to these mergers, the Russian oil industry was dominated by five companies: Yukos; LUKoil; Surgutneftegaz Tyumen Oil Company; and Sibneft. Combined, these companies accounted for roughly 70 per cent of the country's oil production. The other 30 per cent belonged to the country's roughly 150 small-to-medium sized oil producers.

TNK-BP

In February 2003, British Petroleum BP announced its intention to purchase a 50 per cent stake in Russia's Tyumen Oil Company (TNK), as well as other assets held by TNK's shareholders, thus creating a new company, TNK-BP. TNK-BP will hold oil reserves of 5.2 billion barrels and have an annual production capacity estimated at 1.2 million bbl/d, making it Russia's third largest oil producer. The deal also includes retail outlets in Russia and Ukraine. The $6.35-billion deal was approved by the European Commission in July 2003 and the Russian Anti-Monopoly Ministry in August 2003. The creation of TNK-BP marks the largest single foreign investment in Russia since the collapse of the Soviet Union, and has been considered by some analysts to mark a dramatic improvement in Russia's investment climate.

YukosSibneft

In April 2003, Yukos and Sibneft announced merger plans to create Russia's largest oil company, YukosSibneft. The $12-$15 billion transaction was approved by Russia's Anti-Monopoly Ministry on August 15, 2003. The new company commands proven oil reserves of 18.4 billion barrels, with

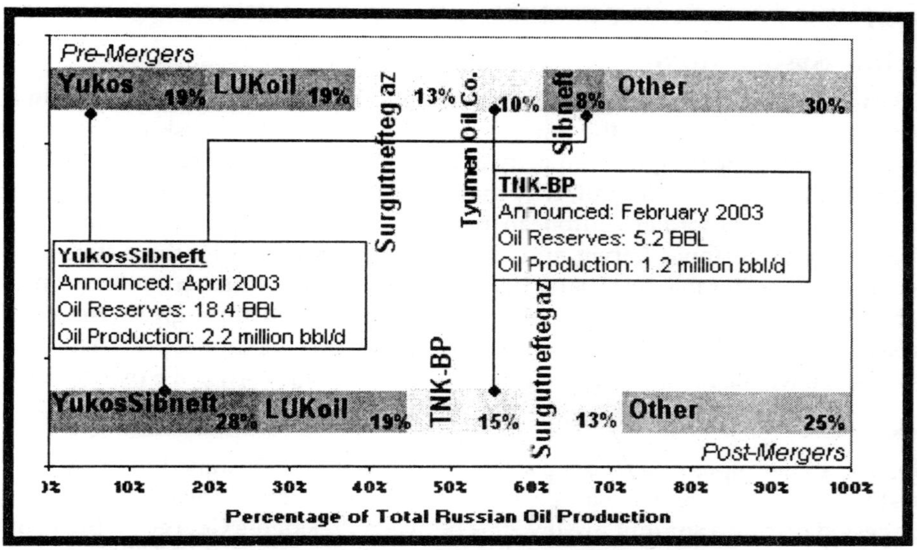

RUSSIAN OIL INDUSTRY MERGERS 2003

production capacity estimated at 2.2 million bbl/d, making it the largest oil producer in Russia, and one of the largest in the world. The company also controls seven oil refineries and 2,500 filling stations across the former Soviet Union. Because of its world-class size, YukosSibneft represents the first Russian "super-major", and could enjoy unique clout with the Russian government.

Oil Exports

More than any other issue, Russian oil production in the near-term is constrained by the country's export capacity. Since the country's oil production boom beginning in 1999, Russia's oil export infrastructure has not expanded to keep pace. As a result, Russia's oil producers are said to be constantly struggling with bottlenecks that restrict their output growth.

Oil exports via pipeline fall under the exclusive jurisdiction of Russia's state-owned pipeline monopoly, Transneft. The Transeft pipeline system currently has a maximum export capacity of roughly 3.5 million bbl/d. However, given Russia's net exports in 2002 of approximately 5 million bbl/d, the Transneft system alone results in a 1.5 million bbl/d deficit. As a consequence, Russian oil producers have recently taken advantage of higher world oil prices and re-directed surplus oil via rail and river barge to external markets. Some of the deficit is also overcome through the exportation of petroleum products. However, these measures come at much greater cost than shipment via pipeline, and this option could become less viable should world oil prices fall. Sales on the domestic market are also less favorable owing to the considerable price discrepancy between the Russian domestic price and prices at world marketplaces. Russian domestic prices are typically just over half of the world market price

The Russian government has acknowledged the problem, and has taken careful steps towards developing new export infrastructure. At issue, however, is not only the direction and scope of enhancements to the country's export infrastructure, but also the potential role that private firms and investors may play in these projects, presumably at the expense of state-owned Transneft.

Baltic Pipeline System (BPS)

The BPS came online in December 2001, and carries oil from Russia's West Siberian and Timan-Pechora oil provinces westward to the newly completed port of Primorsk in the Russian Gulf of Finland. For most of 2003, throughput capacity at Primorsk was 240,000 bbl/d. However, Transneft plans to boost throughput to 360,000 bbl/d by the end of 2003, and to 840,000 bbl/d in 2005. The BPS gives Russia a direct outlet to northern European markets, allowing the country to reduce its dependence on transit routes through Estonia, Latvia, and Lithuania. Unfortunately for these states, the growth of the BPS has come at considerable cost (particularly at Ventspils in Lithuania), as Russian crude which traditionally moved through the Baltic States has been re-routed through Russia's BPS. Russian authorities have stated publicly that when allocating the country's exports, precedence will be given to sea ports in which Russia has a stake over foreign ones; in other words, BPS over other Baltic ports.

Murmansk

The idea for a new pipeline and deepwater tanker terminal, designed to carry crude oil from Russia's West Siberian Basin and Timan Pechora basin westward to Murmansk on the Barents Sea was first suggested during a summit in May 2002 between Presidents George W. Bush and Vladimir Putin. As conceived by a consortium of Russia's leading oil companies, the Murmansk project would allow for between 1.6 and 2.4 million bbl/d of Russian oil exports to reach the United States via tankers. The companies currently are conducting feasibility studies, and expect that initial transit could begin as early as 2007. A liquefied natural gas (LNG) facility at the port has also been suggested, possibly allowing for deliveries to American markets.

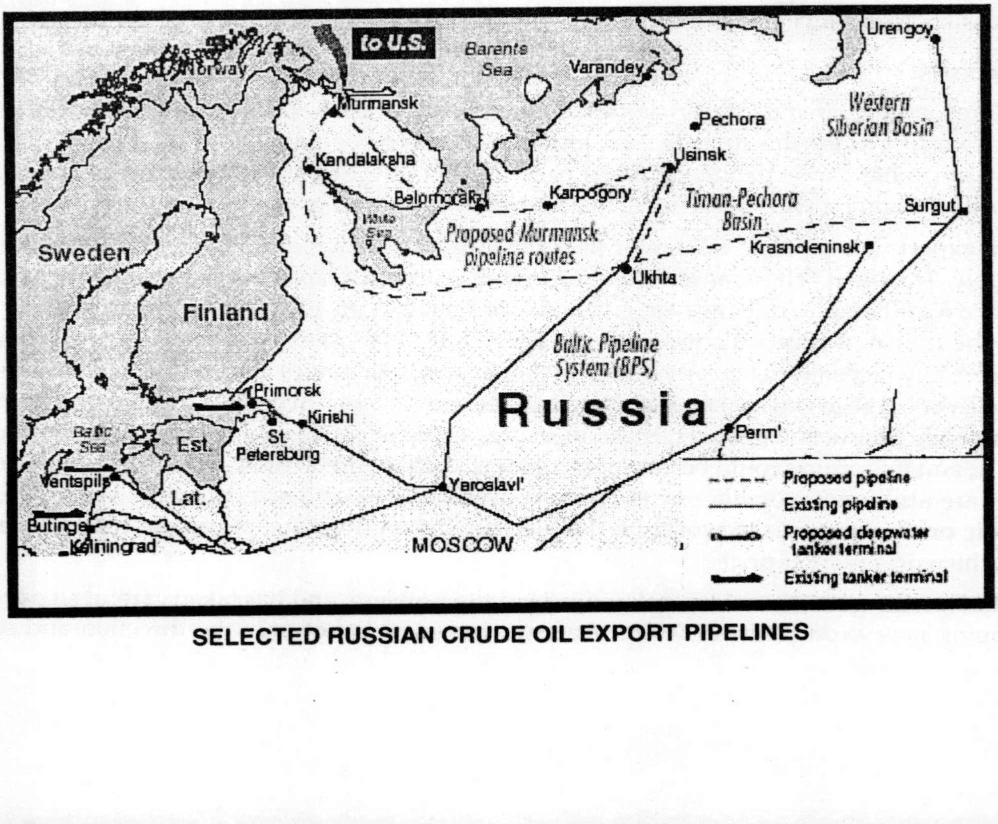

SELECTED RUSSIAN CRUDE OIL EXPORT PIPELINES

But despite vociferous support for the Murmansk proposal from Russian oil companies, American oil companies, and the U.S. government, Transneft (and thereby the Russian government) has approached the project with trepidation. It is unclear whether the Russian government will allow private Russian oil companies to proceed with their plans to construct the port and its associated facilities, or if the project will be handed, either partially or entirely, to Transneft. At stake is not only the Murmansk project, but also the Russian government's ability to control the growing oil industry via Transneft.

Adria Reversal Project

Reversal of the Adria pipeline, which extends between Croatia's port of Omisalj on the Adriatic Sea and Hungary has been under consideration since the 1990s. The pipeline, which was completed in 1974, was originally designed to load Middle Eastern oil at Omisalj, then pipe it northward to Yugoslavia and on to Hungary. However, given both the Adria pipeline's existing interconnection with the Russian system, and Russia's booming production, the pipeline's operators and transit states have since considered reversing the pipeline's flow, thus giving Russia a new export outlet on the Adriatic Sea.

Connecting the pipeline to Russia's Southern Druzhba system would entail the cooperation of six countries (Russia, Belarus, Ukraine, Slovakia, Hungary, and Croatia). In December 2002, these countries signed a preliminary agreement on the project. Since then, however, progress has been slow moving, as the transit states wrangle over the project's details (including tariffs and environmental issues). Given the relative simplicity of reversing the flow should the countries come to an agreement, some analysts expect that the Adria pipeline could begin transiting roughly 100,000 bbl/d of Russian crude in 2004, with an ultimate capacity of approximately 300,000 barrels per day.

Odessa Brody Reversal Project

Since its completion in 2000, the Odessa-Brody pipeline, which extends from Ukraine's Black Sea port of Odessa northward to the city of Brody, has been mostly dormant. The pipeline was initially intended to carry oil from the Caspian Sea region to Europe, but as of yet has transited only small quantities. At Brody, the pipeline is connected to Russia's Druzhba system, and Russia has suggested that the pipeline instead move oil from Russia southwards, to be loaded onto tankers in the Black Sea and shipped to world markets. Dismayed by the possibility of losing access to Caspian Sea region oil, European governments have voiced their opposition to the reversal project. In July 2003, Ukraine and Poland set up a joint venture designed to construct a connecting pipeline from Brody to the Polish city of Plock, thereby allowing for greater access to European markets, and further discouraging the reversal. So far, the Ukrainians have been noncommittal. On August 19, 2003, the Ukrainian government opened a tender for independent investigation of the possibility of reversing the Odessa-Brody pipeline.

Eastern Oil Pipeline Options

Given the public statements of Russian governmental officials and oil company representatives, it is clear that, in the near future, the country plans to construct a pipeline conneccting the Russian city of Angarsk to Asian markets. However, it is as of yet unclear whether the oil will be routed to Japan (the "Angarsk-Nakhodka" option), or to China (the "Angarsk-Daqing" option), or both; and if both, in what order. The Russian government reportedly is weighing the options and has indicated that it will make a decision in September 2003.

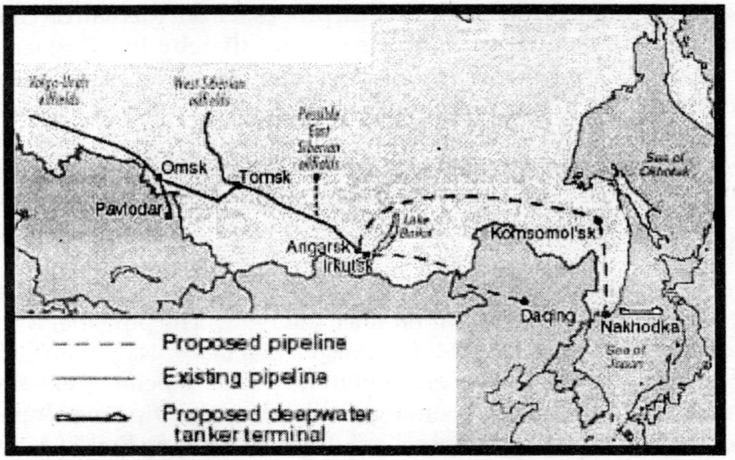

PROPOSED FAR EAST PIPELINES (U)

Angarsk-Daqing

The Angarsk-Daqing option would extend 1,550 miles to the inland Chinese city of Daqing and is estimated to cost $2.5 billion. The project has been backed by Russian oil company Yukos, and could transit 400,000 bbl/d beginning in 2005, with the possibility for 600,000 bbl/d by 2010. In May 2003, President Putin and Chinese President Hu Jintao expressed their mutual support for the pipeline, and Yukos and the China National Petroleum Company (CNPC) signed a 25-year supply deal. However, statements made in June 2003 indicate the Russian government is still seriously considering an alternate route to Japan.

Angarsk-Nakhodka

The Angarsk-Nakhodka option would extend 2,490 miles, from Angarsk, around Russia's Lake Baikal, to the port of Nakhodka where a new export facility would have to be constructed. This option is estimated to cost between $5 and $6 billion, with a capacity of 1 million bbl/d. Although this option is significantly more expensive, as it covers a greater distance and involves more investment, the Angarsk-Nakhodka route would open up a new Pacific port from which Russian oil exports could be shipped by tanker to other Asian markets, and possibly even North America.

Downstream/Refining

Russia has 42 oil refineries–many of which are inefficient, aging, and in need of modernization–with a total processing capacity of 5.4 million bbl/d. With Russian domestic demand of 2.63 million bbl/d in 2002, refining capacity far outstrips demand for refined products. In addition, because a barrel of crude oil on the Russian market typically sells for just over half the world crude oil price, many Russian oil companies prefer to export their crude oil rather than to refine it in Russia. When Russian oil producers do not export their crude oil–often because of the constraints of Russia's pipeline system or the government's limits on each company's exports–many choose to supply their own refineries rather than sell the oil on the open market.

Natural Gas

Russia holds the world's largest natural gas reserves, with 1,680 trillion cubic feet (Tcf)–more than twice the reserves in the next largest country, Iran. Accordingly, in 2001 Russia was the world's largest natural gas produce (20.5 Tcf), as well as the world's largest exporter (6.1 Tcf). But unlike the Russian oil industry, Russia's natural gas industry is not growing. Both production and consumption have remained relatively flat since independence (see graph). Moreover, Russia's energy strategy, released in May 2003, calls for only modest production growth by 2010, even under its most optimistic scenario (see graph). Growth of Russia's natural gas sector has been stunted primarily due to state regulation, Gazprom's monopolistic control over the industry, and insufficient export pipelines.

Gas Industry Structure

Gazprom, Russia's state-run natural gas monopoly, holds nearly one-third of the world's natural gas reserves, produces nearly 90 per cent of Russia's natural gas, and operates the country's natural gas pipeline grid. Gazprom is also Russia's largest earner of hard currency, and the company's tax payments account for around 25 per cent of federal tax revenues.

But despite its terrific size and significance, the company is seriously encumbered by domestic regulation. According to the Russian Gas Law of 1999, Gazprom must supply the Russian natural gas market, regardless of profitability, at government-regulated prices. Thus, the company is forced by the Russian government to sell natural gas to domestic users for less than it costs the company to produce, and only about one-tenth of the export price. The Russian government has recognized this problem and has stipulated that domestic natural gas prices will be freed sometime after 2010. In addition, Gazprom is saddled with chronic non-payments by consumers–although this situation is gradually improving. Accordingly, Gazprom's weak financial position has hindered the company's ability to invest adequately in new projects, and domestic production has remained relatively flat.

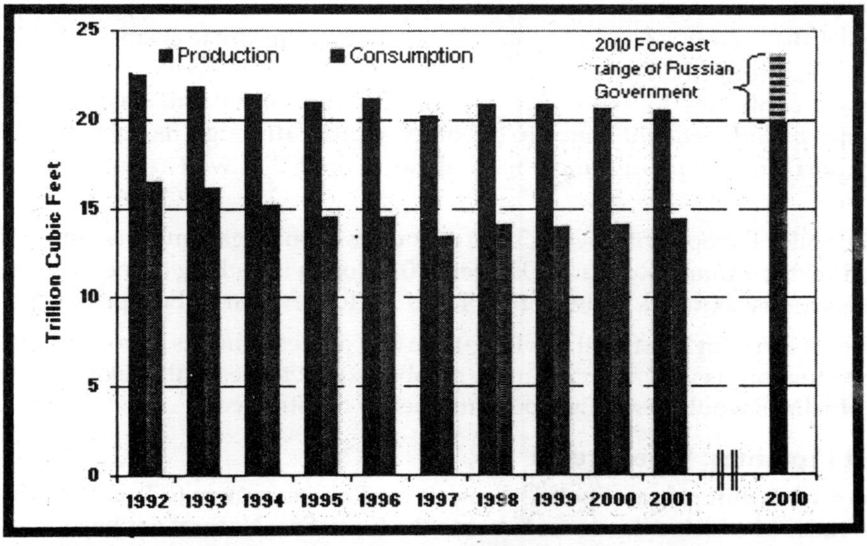

NATURAL GAS PRODUCTION AND CONSUMPTION IN RUSSIA, 1992-2010

Table 71.1: Russia's Natural Gas Exports to Europe 2001 (Top 10 Importers)

	Tcf per Year	*% of Domestic Consumption*
Germany	1.15	35%
Ukarine	1.01	39%
Italy	0.71	28%
Belarus	0.60	94%
France	0.34	27%
Turkey	0.29	70%
Hungary	0.26	61%
Poland	0.26	64%
Czech Republic	0.26	76%
Slovakia	0.26	95%

In June 2003, Gazprom signed its first joint venture for the production of natural gas in Russia with Germany's Wintershall. The partners will develop the Achimovsk formation of Russia's Urengoi field. The partners estimate the formation's recoverable reserves at 55 Tcf of gas, and 2.7 billion barrels of condensate. The first phase of development will begin in 2004, with full production in 2008, to yield 291 Tcf of gas. According to press reports, the consortium, to be known as Achimgaz, plans to sell 75 per cent of their natural gas production on the domestic market, leaving 25 per cent for export.

Many Russian oil companies sit on huge natural gas reserves which they have not developed commercially because they do not have access to lucrative export markets as Gazprom keeps firm control of Russia's pipeline network. When third-party access to Russia's natural gas transportation infrastructure becomes a reality, Russia's natural gas exports could increase dramatically.

Export Markets

Historically, the majority of Russia's natural gas exports were sent to customers in Eastern Europe, but since the collapse of the Soviet Union, Russia increasingly is looking to diversify its export options. Russia continues to export significant amounts of natural gas to customers in the Commonwealth of Independent States (CIS), but Gazprom (through its subsidiary Gazexport) is shifting its export strategy to send more natural gas to the EU as well as Turkey, Japan, and other Asian countries.

Russia supplies Europe with over 25 per cent of its natural gas imports, and with the energy accord signed by the EU and Russia in October 2000, Russia is seeking to increase this percentage. Gazprom already has contracts to deliver 6.2 Tcf to 7.2 Tcf per year to Europe beyond 2007.

However, if Gazprom is to fulfill this long-term aim of increasing its European sales, it will have to boost its production, as well as secure more reliable export routes to the region. Several proposed new export pipelines would serve European markets if constructed.

New Export Pipelines: Blue Stream

The Blue Stream natural gas pipeline connects the Russian system to Turkey through a 750-mile pipeline, 246 miles of which extends underneath the Black Sea. Natural gas began flowing through the pipeline in December 2001, under an initial schedule of 71 Bcf per year, which was to increase by

EXISTING AND PLANNED NATURAL GAS PIPELINES TO EUROPE

71 Bcf annually and plateau at roughly 222 Bcf per year in 2009. However, in March 2003, Turkey halted deliveries through Blue Stream, invoking a clause in the contract allowing either party to stop deliveries for six months. In June 2003, Gazprom announced that it had filed suit against Turkey in Stockholm's International Arbitration court, and the countries are reportedly still in negotiations.

North Trans-Gas Pipeline

The idea of a North Trans-Gas pipeline, extending 746 miles from Russia's Baltic Sea port city of Vyborg (near St. Petersburg), underneath the Baltic Sea to Germany, and then across the Netherlands and under the North Sea to the United Kingdom, was countenanced in June 2003 by Russia and the United Kingdom. The pipeline is expected to cost $5.7 billion and transport approximately 1.05 Tcf per year. A possible spur connection to Sweden has also been considered. Several European oil and gas concerns have reportedly shown interest in the project, however, to-date there is no definite consortium developing the project. Despite great enthusiasm for the project from both sides, some analysts have pointed out that conflicts between the EU's liberalized gas market and Russia's state regulated system could prove to be a hindrance. The pipeline is expected to begin initial operation at a rate of 0.7 Tcf in 2007, and later increase to 1.05 Tcf.

Sakhalin

Sakhalin Island, a former penal colony located off Russia's eastern shore, is home to five oil and gas projects, each operated by a unique international consortium. The five projects are currently in different stages of development, and two of the projects, Sakhalin I and Sakhalin II, aim to bring oil and natural gas production online in the near term. Both projects have targeted Asian markets.

Sakhalin I

The Sakhalin I project is being led by Exxon Neftegaz, in conjunction with consortium members SODECO, ONGC Videsh, Sakhalinmorneftegaz, and RN Astra. The consortium members started drilling in May 2003 and expect preliminary oil production of 250,000 bbl/d in 2005. The project's oil output will be piped westward to the Russian port of De-Kastri and pumped into the Russian system. Sakhalin I's natural gas is expected to be piped southward to Japan via a proposed pipeline. The partners are planning for natural gas exports to Japan to begin in 2008.

Sakhalin II

The Sakhalin II project is being developed by Shell, Mitsubishi, and Mitsui. Unlike Sakhalin I, the Sakhalin II project entails the development of Russia's first liquefied natural gas facility (LNG), to be built on the southern tip of the Sakhalin Island, near the town of Prigorodnoye. The two-train facility will have a capacity of 9.6 million tons per year, with exports to begin in 2007. According to press reports, the Sakhalin II partners have already secured sales contracts with three Japanese utilities worth 2.8 million tons per year, or approximately 30 per cent of the plant's output, for roughly 20 years.

China and South Korea

Gazprom, Rusiya Petroleum, South Korea's state-owned Korea Gas Corp. (Kogas), and the Chinese National Petroleum Company (CNPC) have commenced negotiations on the construction of a pipeline connecting Russia's Kovykta field to China's northeastern provinces and across the Yellow Sea to South Korea. The plan calls for a 706 Bcf per year pipeline which would deliver roughly two-thirds of its gas annually to China, while delivering the rest to South Korea. The partners expect that the pipeline could come online in 2008.

Coal

With 173 billion short tons in estimated recoverable coal reserves, Russia holds the world's second largest coal reserves, behind only the United States which holds roughly 274 billion short tons. However, years of poor management during the Soviet era, and a sharp decline in demand for coal during the early 1990s, significantly undermined the Russian coal sector. In 2001, Russia's total coal production was 299.5 million short tons, less than one-third of U.S. production.

Between 1996 and 2001, Russia worked with the World Bank to restructure the country's coal industry. As a result of the program, the state monopoly, formally known as RosUgol, has been dissolved, and roughly 77 per cent of domestic coal production comes from independent producers. Russian coal production began a three-year upswing in 1999, and preliminary statistics indicate that this trend has continued throughout 2002 and into 2003. According to the government's energy strategy, Russia should produce between 441 and 496 million short tons in 2020.

Electricity

Russia's power sector includes over 440 thermal and hydropower plants (approximately 77 of which are coal-fired) plus 30 nuclear reactors. The system has a total electric generation capacity of 204 million kilowatts, and in 2001 generated approximately 846.5 billion kilowatt hours (bkwh) of electric power. Since the collapse of the Soviet Union, electricity generation has shown both a dramatic decline, (down 18 per cent between 1992 and 1999), and a gradual recovery (up 7 per cent between 1999 and 2001)–see graph. As with similar patterns in oil, natural gas and coal, electricity generation was stunted by the economic slowdown which followed the collapse of the Soviet Union. Also, in the case of the electricity sector, generation was slowed by the shut down of several nuclear reactors during the 1990s. Economic recovery has resulted in an increase in total electricity consumption from 715 Bkwh in 1998, to 773 Bkwh in 2001, resulting in corresponding increases in electric generation.

Thermal power (oil, gas, coal) accounts for roughly 64 per cent of Russia's electricity generation, followed by hydropower (20 per cent) and nuclear (15 per cent)–see graph. The Russian government has stated that it intends to expand the role of nuclear and hydropower in the future in order to allow for greater export of fossil fuels. Currently, Russia has an installed nuclear capacity of 21.2 million kilowatts, distributed across 30 operational nuclear reactors at 10 locations, all west of the Ural Mountains. However, Russia's nuclear power facilities are aging, and the nuclear power industry has been hard hit by Russia's transition to a market economy. Russia already has shut down several reactors that were over 30 years old (the maximum prescribed service life for a reactor), and many more

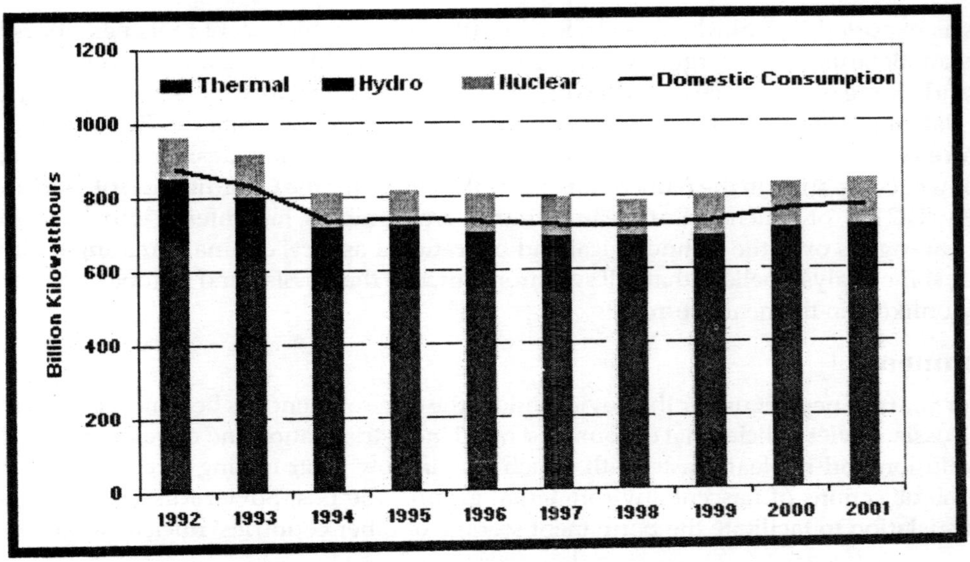

ELECTRICITY GENERATION AND CONSUMPTION IN RUSSIA, 1992-2001

are over 20 years old. By 2010, Russia plans to construct five new units at existing facilities throughout the country. By 2020, the Russian Ministry of Atomic Energy predicts that nuclear generation could reach 300 bkwh per year, more than double the 2001 level.

The Russian government has also made hydroelectric generation a priority, particularly in the country's Far East, where electricity supply can be problematic. In June 2003, a representative from the country's power monopoly, Unified Energy System of Russia (UES), told reporters that the company plans to invest $14 billion in the development of Russia's hydroelectric sector, particularly in Siberia and the Far East. On July 9, 2003, Russia's newest hydroelectric power station, the Bureya dam, located on the Amur river in the Far Eastern Amur Region was officially inaugurated. The dam is expected to reach full capacity in 2007.

Electric Industry Structure

Russia's electricity sector is currently in a transitional period. Since independence, Russia's electricity sector has been dominated by Unified Energy Systems (UES), which is 52 per cent -owned by the Russian government. UES, headed by former privatization minister Anatoly Chubais, controls approximately 70 per cent of the country's distribution system and oversees Russia's 72 regional electricity companies, called energos. In March 2003, Russian President Vladimir Putin signed six bills into law which aim to substantially reform the industry. According to the reform package, tariff rates on the domestic market could be liberalized by July 1, 2005, and UES should be liquidated beginning in 2006. UES's generation and distribution facilities are expected to be privatized, while the country's transmission grid will remain under state control.

According to some analysts, reform of Russia's electricity sector was long overdue. Much of the sector is obsolete by Western standards, and Russia lacks the money to pay for necessary maintenance. In April 2003, UES CEO Anatoly Chubais estimated that $55 billion in investment will be needed over the next 10 years for maintenance and modernization efforts. The March 2003 reform package is expected to boost investment, but only after the reforms are implemented.

Electricity Exports

Russia exported 73.5 bkwh of electricity in2001, most of which went to its neighbors Ukraine, Kazakhstan, Belarus, and Georgia. Russia also exports electricity to Finland, Turkey, China, Latvia, and Poland. Some of this electricity transits Ukraine to Moldova, as well as to Romania, Bulgaria, and the Balkans. There are currently two efforts underway to integrate the Russian and Western European electricity grids. UES is participating in the Baltrel program, designed to create an energy ring of power companies in the Baltic states. Also, the Union for the Coordination of Transmission of Electricity (UCTE), of which 20 European countries are members, has entered into discussions with Russian colleagues over the technological and operational aspects of amalgamating their systems. However, some analysts believe that full synchronization of the Russian and European power systems is highly unlikely in the near future.

Environment

After years of neglect under the Soviet Union, the environment has become a pertinent issue in today's Russia. Soviet policies that encouraged rapid industrialization and development left a legacy of air pollution and nuclear waste with which Russia now is struggling to contend. In addition, despite the objections of nascent environmental groups, the post-Soviet Russian government has passed legislation to facilitate the permanent storage of other countries' nuclear waste on Russian territory. Although environmental awareness in Russia is rising, the cost of remediating the country's environmental hot spots is high, and the Ministry of Natural Resources has a limited budget. As a result, cleanup has been slow, and environmental protection has not been a top priority for the Russian government.

The economic contraction in the aftermath of the Soviet Union's collapse caused a drop in industrial production, resulting in less energy consumption and a drop in Russia's carbon emissions. However, energy and carbon intensities in Russia remain high, and although per capita carbon emissions have fallen over the past 12 years, Russia will need to pursue more sustainable environmental policies in order to maintain this trend, especially with the rebound in industrial production since the August 1998 financial crisis. Russia has abundant fossil fuel resources, but the country will need to pursue more renewable energy options and cleaner environmental technologies in order to preserve its natural wonders and protect its environment for future generations.

Country Overview

Independence: August 24, 1991 (from Soviet Union). National holiday: Russia Day, June 12, 1990

Population (7/02E): 145 million

Location: Eurasia

Size: 6,592,850 square miles, slightly more than 1.8 times the size of the United States

Major Cities: Moscow, St. Petersburg, Yekaterinburg, Irkutsk, Murmansk, Yakutsk, Vladivostok

Languages: Russian, others

Ethnic Groups: Russian 81.5 per cent, Tatar 3.8 per cent, Ukrainian 3 per cent, Chuvash 1.2 per cent, Bashkir 0.9 per cent, Belorussian 0.8 per cent, Moldovan 0.7 per cent, other 8.1 per cent

Religions: Russian Orthodox, Muslim, other

Economic Overview

Currency: Ruble

Market Exchange Rate (9/5/03): $1 = 30.6 rubles

Nominal Gross Domestic Product (GDP) (2002E): $346.5 billion *(2003E):* $415.5 billion

Real GDP Growth Rate (2002E): 4.3 per cent; *(2003E):* 4.3 per cent

Inflation Rate (Change in Consumer Prices, 2002E): 15.8 per cent *(2003E):* 13.7 per cent

Unemployment Rate (2002E): 8.6 per cent *(2003E):* 8.3 per cent

Current Account Surplus (2002E): $32.8 billion *(2003E):* $37.6 billion

Major Trading Partners (2001): Germany, Ukraine, U.S., Belarus, Italy, Netherlands, Kazakhstan

Merchandise Exports (2002E): $107.2 billion; *(2003E):* $119.1 billion

Merchandise Imports (2002E): $61.0 billion; *(2003E):* $67.1 billion

Merchandise Trade Balance (2002E): $46.2 billion; *(2003E):* $52.0 billion

Major Exports: Petroleum and petroleum products, natural gas, wood and wood products, metals, chemicals, and a wide variety of civilian and military manufactures

Major Imports: Machinery and equipment, consumer goods, medicines, meat, grain, sugar, semifinished metal products

External Debt (2002E): $154 billion

Energy Overview

Proven Oil Reserves (1/1/03E): 60 billion barrels

Oil Production (2002E): 7.65 million bb/d (of which 7.4 million bbl/d was crude)

Oil Consumption (2002E): 2.63 million bbl/d

Net Oil Exports (2002E): 5.02 million bbl/d

Major Oil Customers: Europe, Commonwealth of Independent States

Crude Refining Capacity (1/1/03E): 5.4 million bbl/d

Proven Natural Gas Reserves (1/1/02E): 1,680 trillion cubic feet (Tcf)

Natural Gas Production (2001E): 20.5 Tcf

Natural Gas Consumption (2001E): 13.8 Tcf

Net Natural Gas Exports (2001E): 6.7 Tcf

Coal Reserves (1/1/01E): 173 billion short tons

Coal Production (2001E): 299.5 million short tons (Mmst)

Coal Consumption (2001E): 284 Mmst

Electric Installed Capacity (2001E): 204 gigawatts (69 per cent thermal, 21 per cent hydro, 10 per cent nuclear)

Electricity Generation (2001E): 864.5 billion kilowatt-hours (Bkwh)

Electricity Consumption (2001E): 773.0 Bkwh

Net Electricity Exports (2001E): 91.5 Bkwh

Environmental Overview

Total Energy Consumption (2001E): 28.2 quadrillion Btu* (7 per cent of world total energy consumption)

Energy-Related Carbon Emissions (2001E): 440.3 million metric tons of carbon (7 per cent of world carbon emissions)

Per Capita Energy Consumption (2001E): 195.3 million Btu (vs. U.S. value of 341.8 million Btu)

Per Capita Carbon Emissions (2001E): 3.0 metric tons of carbon (vs. U.S. value of 5.5 metric tons of carbon)

Energy Intensity (2001E): 30,239 Btu/$1995 (PPP), vs U.S. value of 10,810 Btu/$1995 (PPP)**

Carbon Intensity (2001E): 0.47 metric tons of carbon/thousand $1995 (PPP) vs U.S. value of 0.17 metric tons/thousand $1995(PPP)

Fuel Share of Energy Consumption (2000E): Natural Gas (52 per cent), Coal (18 per cent), Oil (19 per cent)

Fuel Share of Carbon Emissions (2000E): Natural Gas (48 per cent), Coal (30 per cent), Oil (23 per cent)

Status in Climate Change Negotiations: Annex I country under the United Nations Framework Convention on Climate Change (ratified December 28th, 1994). Under the negotiated Kyoto Protocol (signed on March 11th, 1999, but not yet ratified), Russia has agreed to stabilize greenhouse gases at 1990 levels by the 2008-2012 commitment period.

Major Environmental Issues: air pollution from heavy industry, emissions of coal-fired electric plants, and transportation in major cities; industrial, municipal, and agricultural pollution of inland waterways and sea coasts; deforestation; soil erosion; soil contamination from improper application of agricultural chemicals; scattered areas of sometimes intense radioactive contamination; ground water contamination from toxic waste.

Major International Environmental Agreements: A party to Conventions on Air Pollution, Air Pollution-Nitrogen Oxides, Air Pollution-Sulphur 85, Antarctic-Environmental Protocol, Antarctic-Marine Living Resources, Antarctic Seals, Antarctic Treaty, Biodiversity, Climate Change, Endangered Species, Environmental Modification, Hazardous Wastes, Law of the Sea, Marine Dumping, Nuclear Test Ban, Ozone Layer Protection, Ship Pollution, Tropical Timber 83, Wetlands and Whaling. *Has signed, but not ratified:* Air Pollution-Sulphur 94, Climate Change-Kyoto Protocol.

The total energy consumption statistic includes petroleum, dry natural gas, coal, net hydro, nuclear, geothermal, solar and wind electric power. The renewable energy consumption statistic is based on International Energy Agency (IEA) data and includes hydropower, solar, wind, tide, geothermal, solid biomass and animal products, biomass gas and liquids, industrial and municipal wastes.

72

Saudi Arabia

With one-fourth of the world's proven oil reserves, Saudi Arabia is likely to remain the world's largest oil producer for the foreseeable future. During 2002, Saudi Arabia supplied the United States with 1.5 million barrels per day of crude oil, or 17 per cent, of U.S. crude oil imports during that period.

General Background

With oil export revenues making up around 90-95 per cent of total Saudi export earnings, 70 per cent -80 per cent of state revenues, and around 40 per cent of the country's gross domestic product (GDP), Saudi Arabia's economy remains, despite attempts at diversification, heavily dependent on oil (although investments in petrochemicals have increased the relative importance of the downstream petroleum sector in recent years).

Oil prices have been strong so far in 2003. Combined with increased Saudi oil production to make up for lost Venezuelan, Nigerian, and Iraqi oil supplies, Saudi Arabia is expected to earn around $70 billion in crude oil export revenues, up sharply–$15 billion–from 2002. Saudi Arabia's real GDP is expected to grow by about 3.0 per cent during 2003, up from 2.0 per cent in 2002. The country also is forecast to show a healthy current account surplus. For fiscal year 2003, Saudi Arabia had been expecting a budget deficit of around $10 billion, but this was based on an extremely conservative price assumption of $17.50 per barrel for Saudi oil, and also did not take into account the huge production surge which actually occurred. As a result, the country is more likely to experience a large budget surplus, on the order of $13 billion, in 2003.

In spite of the recent surge in its oil income, Saudi Arabia continues to face serious long-term economic challenges, including high rates of unemployment (around 15 per cent -20 per cent), one of the world's fastest population growth rates, and the consequent need for increased government spending. Saudi Arabia's per capita oil export revenues (in inflation adjusted dollars) remain far below high levels reached during the 1970s and early 1980s (around $2,296 per person in 2002,

versus $23,820 in 1980, for instance). This is in large part due to the fact that Saudi Arabia's young population has more than doubled since 1980, while oil export revenues in real terms have fallen sharply. Meanwhile, Saudi Arabia has faced nearly two decades of heavy budget and trade deficits, the expensive 1990/1991 war with Iraq, and total government debt approaching 100 per cent of Saudi GDP. On the other hand, Saudi Arabia does have extensive–around $110 billion–foreign assets, which provide a substantial fiscal "cushion." Finally, surging oil export revenues removes an incentive for Saudi Arabia to undertake needed economic reforms, such as reduction in the bloated state sector, as urged by the IMF and others.

Movement towards economic reform in Saudi Arabia remains uneven at best. For instance, reducing subsidies and increasing tariffs on electricity has proven problematic, with a rate increase announced in April 2000 subsequently reversed in October in the face of widespread public opposition. Besides economic reform, Saudi Arabia also has made only slow progress on another of its main domestic goals–attracting foreign direct investment, or FDI. Currently, large state corporations, like oil firm Saudi Aramco (which has a monopoly on Saudi upstream oil development and controls 98 per cent of the country's oil reserves) and the Saudi Basic Industries Corporation (SABIC; the world's 11th largest petrochemical producer) dominate the Saudi economy. To date, there has not been a single sale of state assets to private control, and privatization largely has been limited to allowing private firms to take on certain service functions. In May 2002, Saudi Oil Minister Ali Naimi (reappointed in May 2003 for a third, four-year term) stated that the country was considering privatizing some operations of Saudi Aramco. One impetus for Saudi privatization is its desire to join the World Trade Organization (WTO), but progress has been slow towards achieving this goal. A significant recent development on the foreign investment front was the cancellation, in May 2003, of the ambitious, $20 billion "Saudi Gas Initiative".

In general, Saudi Arabia also has moved cautiously and slowly towards government subsidy cuts, tax increases, or financial sector reforms. Saudi leadership (Crown Prince Abdullah, in particular) has indicated that it sees privatization–although controversial–as a "strategic choice," and has created (in August 1999) a "Supreme Economic Council" charged with boosting investment, creating jobs for Saudi nationals, and promoting privatization. Changes to the law governing foreign investment, granting the same basic rights to foreign investors as to Saudi nationals, were approved in April 2000. Among other measures, the new investment law reduced taxes on foreign business profits from a maximum 45 per cent to 30 per cent, and provided greater legal protections against expropriation of investments.

In November 1999, King Fahd stated that "the world is heading for.globalization" and that "it is no longer possible for [Saudi Arabia] to make slow progress." In the context of successfully becoming integrated into the global economy, Fahd also emphasized the importance of regional unity among Gulf states–economically, politically, and militarily. A customs union, for instance, among GCC countries, was agreed upon at the December 1999 GCC summit. The union is to take effect in March 2005. Currently, goods from GCC countries are exempt from all Saudi import duties, as long as 40 per cent of their value has been added within the GCC and the producing company is owned at least 51 per cent by GCC citizens.

In a treaty signed in June 2000, Saudi Arabia and Yemen agreed on the delineation of sections of their common border which had been in dispute since the 1930s. The deal is expected to open up opportunities for increased Saudi trade and investment in Yemen, a possible pipeline across Yemen to the Arabian Sea (see below for more details), and the possible award of oil and gas exploration rights for areas in Yemen adjacent to previously disputed areas of the border. In February 2001, Saudi Arabia and Syria signed a bilateral free-trade agreement. On June 11, 2001, Saudi Arabia announced (in a letter to UN Secretary General Kofi Annan) that it was taking ownership of Iraq's pipeline to the Saudi Red Sea coast (closed since August 1990), citing Iraqi threats and aggressive actions, including (allegedly) a series of cross-border raids in recent months. Saudi Arabia said that Iraq's behavior had "destroyed any rationale for maintaining the [pipeline] facilities."

Oil

Saudi Arabia (including half of the Saudi-Kuwaiti "Neutral Zone") contains 264.2 billion barrels of proven oil reserves (more than one-fourth of the world total) and up to 1 trillion barrels of ultimately recoverable oil. Saudi Arabia is the world's leading oil producer and exporter, and its location in the politically volatile Gulf region adds an element of concern for its major customers, including the United States. During 2002, Saudi Arabia produced around 8.5 million bbl/d of oil (including half of the Saudi-Kuwaiti Neutral Zone's 579,000 bbl/d), compared to production capacity of around 10.0-10.5 million bbl/d. For the first three months of 2003, Saudi total oil production (including crude oil, natural gas liquids, and "other liquids") averaged about 9.9 million bbl/d as the country used its spare production capacity to help compensate for losses from Venezuela, Nigeria, and Iraq. As of June 1, 2003, Saudi Arabia's official OPEC crude oil production quota was raised to 8.256 million bbl/d, up from the country's previous 7.963 million bbl/d quota level.

Although Saudi Arabia has around 80 oil and gas fields (and over 1,000 wells), more than half of its oil reserves are contained in only eight fields, including Ghawar (the world's largest onshore oil field, with estimated remaining reserves of 70 billion barrels) and Safaniya (the world's largest offshore oilfield, with estimated reserves of 19 billion barrels). Ghawar's main producing structures are, from north to south: Ain Dar, Shedgum, Uthmaniyah, Farzan, Ghawar, Al Udayliyah, Hawiyah, and Haradh. Overall, Ghawar alone accounts for about half of Saudi Arabia's total oil production capacity.

In March 2002, Aramco reportedly awarded major turnkey contracts to Italy's Snamprogetti ($630 million) and Technip-Coflexip ($360 million) aimed at increasing total Saudi oil production capacity by 800,000 bbl/d (500,000 bbl/d of Arabian light and 300,000 bbl/d of Arabian medium), by 2005. The $1.2 billion project, known as the Qatif producing facilities development program (QPFDP), is located in the eastern part of the country near Dhahran, and will serve crude oil production from fields in the area. QPFDP involves construction of two gas-oil separation plants (GOSPs), as well as gas treatment and oil stabilization facilities.

Another potential project, at the Khurais field, could increase Saudi production capacity by 800,000 bbl/d by 2005 at a cost of $3 billion. This would involve installation of four gas/oil separation plants (GOSPs), with a capacity of 200,000 bbl/d each, at Khurais, which first came online in the 1960s but was mothballed by Aramco (along with several other fields–Abu Hadriya, Abu Jifan, Harmaliyah, and Khursaniyah) in the 1990s.

In other oil news, in early January 2000, Saudi Arabia announced that it was establishing an 11-member Supreme Petroleum Council (SPC) to oversee oil and gas policies in the country. In mid-October 2000, the government announced that the Council would take over certain powers over Saudi Aramco. The SPC could help push Saudi Arabia's overall goal of accelerating private sector and foreign involvement in the country's oil sector, although there is opposition by conservatives.

Saudi Arabia produces a range of crude oils, from heavy to super light. Of Saudi Arabia's total oil production capacity, about 65 per cent -70 per cent is considered light gravity, with the rest either

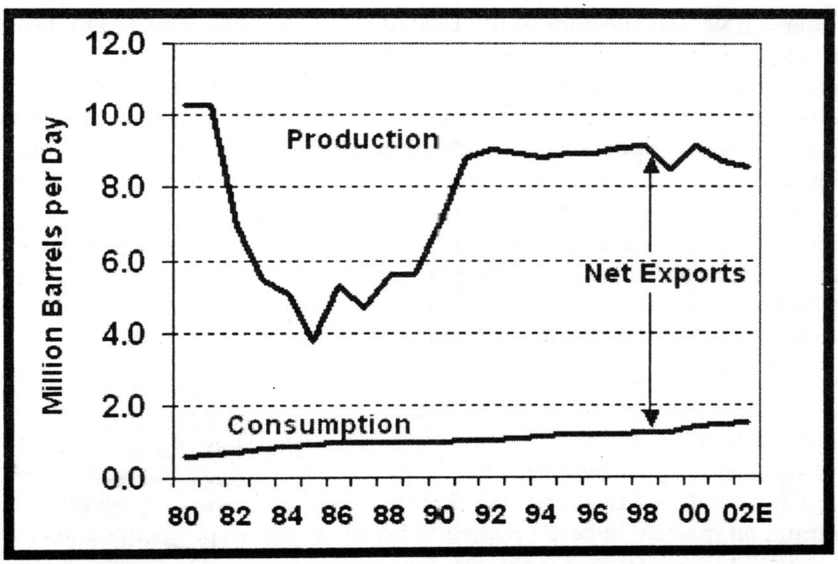

SAUDI ARABIA'S OIL PRODUCTION AND CONSUMPTION, 1980-2002E

Note: Production includes crude oil, natural gas liquids and refinery gain.

medium or heavy. The lightest grades generally are produced onshore, while the medium and heavy grades come mainly from offshore. The Ghawar field is the main producer of 34° API Arabian Light crude, while Abqaiq (a super-giant field with 17 billion barrels of proven reserves) produces 37° API Arab Extra Light crude. Since 1994, the Hawtah Trend (also called the Najd fields), which includes the Hawtah field and smaller satellites (Nuayyim, Hazmiyah) south of Riyadh, has been producing around 200,000 bbl/d of 45°-50° API, 0.06 per cent sulphur, Arab Super Light. Overall, the Najd fields are estimated to contain 30 billion barrels of liquids and major reserves of natural gas. Offshore production includes Arab Medium crude from the Zuluf (over 500,000 bbl/d capacity) and Marjan (270,000 bbl/d capacity) fields and Arab Heavy crude from the Safaniya field.

The Neutral Zone contains about 5 billion barrels of proven oil reserves. Within the Neutral Zone, Japan's Arabian Oil Co. (AOC) traditionally had operated two offshore fields (Khafji and Hout) with 300,000 bbl/d in production, but in February 2000, it lost the concession (in January 2003, AOC reached an agreement with Kuwait on the right to purchase at least 100,000 bbb/d of crude for the next 20 years from Khafji). The offshore Saudi Neutral Zone had represented Japan's most significant upstream oil interest, with 80 per cent of revenues going to AOC and 10 per cent each to Saudi Arabia and Kuwait. Texaco, meanwhile, operates three onshore fields (Wafra, South Fawaris, and South Umm Gudair) in the Neutral Zone. Saudi Arabia had stated that it wanted AOC and Japan to increase their investments in Saudi Arabia (including more than $1 billion in a railway linking remote mining areas to export terminals), as well as their purchases of Saudi oil, as a condition for renewal of AOC's drilling rights in the Neutral Zone. In August 2001, Saudi Arabia rejected a request by AOC to reopen talks on the Saudi Neutral Zone concession. In September 2002, Saudi Oil Minister Naimi proposed that Japan invest in Saudi natural gas fields rather than oil.

Besides the Kuwaiti-Saudi Neutral Zone, Saudi Arabia also produces oil jointly with Bahrain, from the Abu Safa offshore oilfield. Some of the revenues which Saudi Arabia earns from Abu Safa are donated back to Bahrain.

Saudi Arabia is a key oil supplier to the United States, Europe, and Japan; however, in recent years, Western Hemisphere producers (Venezuela, Canada, and Mexico) have challenged Saudi Arabia's dominance in the U.S. market. Asia now takes over 40 per cent of Saudi Arabia's crude oil exports, as well as the majority of its refined petroleum product exports. The United States is Saudi Arabia's second largest oil export market, followed by OECD Europe. During 2002, Saudi Arabia exported 1.55 million bbl/d of oil (mainly crude) to the United States. For this time period, Saudi Arabia ranked second (after Canada, and just ahead of Mexico and Venezuela) as a source of total (crude plus refined products) U.S. oil imports, and first for crude only (ahead of Mexico, Canada, and Venezuela). Saudi Arabia is eager to maintain and even expand its market share in the United States for a variety of economic and strategic reasons. During 2002, Saudi Arabia's share of U.S. crude oil imports was 16.8 per cent (down from 17.3 per cent during 2001).

In October 1999, Oil Minister Naimi stated that Saudi oil policy was based on four facts: 1) the largest oil reserves and among the lowest production costs–around $1.00-$2.00 per barrel–in the world (the country also has extremely low finding costs, estimated at around 10 cents per barrel); 2) maintenance of significant spare oil production capacity; 3) a national economy closely linked to the oil industry; and 4) a stable political and economic system. Naimi also stressed the importance of "a stable international oil market" where "wide and rapid swings in prices are undesirable."

Saudi Arabia's long-term goal is to develop its lighter crude reserves, including the Shaybah field located in the remote Empty Quarter area bordering the United Arab Emirates. Shaybah contains an estimated 7 billion barrels of premium grade 41.6° API sweet crude oil, producing around 500,000 bbl/d of crude oil (and 870 million cubic feet/day of natural gas). Overall, the Shaybah project cost around $2.5 billion, with production starting in July 1998. According to Oil Minister Naimi (October 1999), the development of Shaybah showed that "the cost of adding.capacity–that is, all the infrastructure, producing and transportation facilities–necessary to produce one additional barrel of oil per day in Saudi Arabia is, at most, $5,000 compared to between $10,000 and $20,000 in most areas of the world. So both our current production costs, and the costs for developing more production capacity for the future, are probably the world's lowest." The Shaybah comnplex includes three gas/oil separation plants (GOSPs) and a 395-mile pipeline to connect the field to Abqaiq, Saudi Arabia's closest gathering center, for blending with Arabian Extra Light crude (Berri and Abqaiq streams).

In May 2000, a new law aimed at attracting foreign investment to the Saudi energy sector came into effect. The law permits full foreign ownership of Saudi property and licensed projects, sets up the General Investment Authority (SAGIA) as a "one-stop shop" for foreign investors, and reduces taxes on company profits from 45 per cent to 30 per cent. Previously, foreign companies were limited to a 49 per cent share of joint ventures with Saudi domestic partners. Several important sectors, however, remain closed to 100 per cent foreign ownership, including (as of June 2003): upstream oil, pipelines, media and publishing, insurance, telecommunications, defense and security, health services, wholesale and retail trade, and more. Thus, the new foreign investment law is far less attractive than it appears at first glance. However, in January 2001, SAGIA reported that foreign investment commitments had reached $1.6 billion, including 53 licenses (29 industrial, 24 non-industrial).

Ports and Pipelines

Most of Saudi Arabia's crude oil is exported from the Persian Gulf via the huge Abqaiq processing facility, which handles around two-thirds or so of the country's oil output. Saudi Arabia's primary oil export terminals are located at Ras Tanura (6 million bbl/d capacity; the world's largest offshore oil loading facility) and Ras al-Ju'aymah (3 million bbl/d) on the Persian Gulf, plus Yanbu (as high as 5 million bbl/d) on the Red Sea. Combined, these terminals appear capable of handling around 14 million bbl/d, around 3.5-4.0 million bbl/d higher than Saudi crude oil production capacity (10-10.5 million bbl/d), and about 6 million bbl/d in excess of Saudi crude oil production in 2002. Despite this excess capacity, there have been reports that the Saudis are planning to conduct a feasibility study on construction of an oil pipeline from the Empty Quarter of southeastern Saudi Arabia through the Hadramaut in Yemen to the Arabian Sea.

Saudi Arabia operates two major oil pipelines. The 5-million bbl/d East-West Crude Oil Pipeline (Petroline), operated by Saudi Aramco since 1984 (when it took over from Mobil), is used mainly to transport Arabian Light and Super Light to refineries in the Western Province and to Red Sea terminals for export to European markets. The Petroline was constructed in 1981, with initial capacity of 1.85 million bbl/d on a single, 48-inch line (AY-1). The Petroline was expanded in 1987, during the height of the Iran-Iraq war (and specifically the so-called "tanker war" in the Gulf), to 3.2 million bbl/d, with the addition of a parallel ("looped"), 56-inch line (AY-1L). Finally, in 1993, Petroline capacity was increased to 5.0 million bbl/d by adding significant pumping capability on the line. Reportedly, the Saudis expanded the Petroline in part to maintain Yanbu as a strategic option to Gulf port facilities in the event that exports were blocked at that end. A study in 1997 by the Baker Institute indicated that capacity on the Petroline could be expanded significantly by using so-called "drag reduction agents" (DRAs), and that this could enhance the line's strategic value.

In purely economic terms, Yanbu remains a far less economical option for Saudi oil exports than Ras Tanura. Among other factors, shipments from Yanbu add about 5 days roundtrip travel time for tankers through the Bab al-Mandab strait to Asia compared to Ras Tanura through the Strait of Hormuz. In addition, according to Oil Minister Naimi, the Petroline is only utilized at half capacity. Given this fact, as well as the desire to boost natural gas usage, Saudi Aramco has begun converting the AY-1 (48-inch) line to natural gas pumping capability. The natural gas will supply Yanbu's petrochemical and power facilities.

Running parallel to the Petroline is the 290,000-bbl/d Abqaiq-Yanbu natural gas liquids pipeline, which serves Yanbu's petrochemical plants. The Trans-Arabian Pipeline (Tapline) to Lebanon is mothballed, and the 1.65-million-bbl/d, 48-inch Iraqi Pipeline across Saudi Arabia (IPSA), which runs parallel to the Petroline from pump station #3 (there are 11 pumping stations along the Petroline, all utilizing on-site gas turbine electric generators) to the port of Mu'ajjiz, just south of Yanbu, was closed indefinitely following the August 1990 Iraqi invasion of Kuwait (also, in June 2001, Saudi Arabia seized ownership of IPSA "in light of the Iraqi government's persistence in its stands"). Theoretically, IPSA could be used for Saudi oil transport to the Red Sea, although the Saudis have stated that "there are no plans" to do so. According to Oil Minister Naimi, Saudi Arabia has "surplus oil export and pipelines capacity.[including the] East-West oil pipeline system [which] can carry and deliver 5 million bbl/d" but is being run at "only half capacity."

Refining

Saudi Arabia has eight refineries, with combined crude throughput capacity of around 1.75 million bbl/d. A new, 200,000-bbl/d fractionation unit is scheduled to be completed at Ras Tanura by the end of 2003, raising Saudi refining capacity close to 2 million bbl/d. Also slated for upgrading is the Rabigh refinery on the Red Sea coast. Plans call for boosting capacity at Rabigh, Saudi Arabia's largest domestic refinery, to as high as 400,000 bbl/d, as well as upgrading the refinery's product slate away from low-value heavy products towards gasoline and kerosene at an estimated cost of $1.8 billion. Due to Saudi Arabia's financial difficulties in 1998/1999, the Rabigh project was scaled back by 60 per cent or so, to $800 million. In addition to domestic capacity, Saudi Aramco has around 1.6 million bbl/d of refining capacity overseas.

Saudi Arabia has ambitious plans for expanding petrochemical production using natural gas as a feedstock. State-owned (70 per cent) SABIC, the Middle East's largest non-oil industrial company (and expected to become one of the world's top five ethylene producers by 2005), accounts for around 10 per cent of world petrochemical production. In February 2001, SABIC completed a $1 billion expansion at the Yanbu petrochemical facility, making it the largest polyethylene plant in the world, and in April 2003, Aramco awarded a turnkey contract to Snamprogetti to build to new units at the complex. In early January 2002, SABIC agreed to a $1.15 billion loan to fund a new petrochemicals plant in the eastern Saudi Arabian industrial city of Jubail. The complex is scheduled to come online in 2004, and to produce 1 million tons per year of ethylene, plus olefins, polyethylene, and glycol ethylene. It is uncertain at the present time whether or not the Saudi government will sell off more of its 70 per cent stake in SABIC in the near future.

In August 2002, the Saudi Press Agency reported that Crown Prince Abdullah had formally inaugurated an underground strategic oil storage facility in the city of Jeddah. The facility reportedly has storage capacity of 945,000 barrels of crude oil and refined products at five sites, and is linked by pipeline to refineries.

Natural Gas

Saudi Arabia's proven natural gas reserves are estimated at 224.7 trillion cubic feet (Tcf), ranking fourth in the world (after Russia, Iran, and Qatar), and up about 5 Tcf from 2002. Most (around two-thirds) of Saudi Arabia's currently proven natural gas reserves consist of associated gas, mainly from the onshore Ghawar field and the offshore Safaniya and Zuluf fields. The Ghawar oil field alone accounts for one-third of the country's total natural gas reserves. Most new associated natural gas reserves discovered in the 1990s have been in fields which contain light crude oil, especially in the Najd region south of Riyadh. Most of Saudi Arabia's non-associated gas reserves (Mazalij, Al-Manjoura, Shaden, Niban, Tinat, Al-Waar, etc.) are located in the deep Khuff reservoir, which underlies the Ghawar oil field. Another large gas field, called Dorra, is located near the Khafji oil field in the Saudi-Kuwaiti Neutral Zone and may be developed by Japan's AOC. Natural gas also is located in the countries extreme northwest, at Midyan.

With domestic hydrocarbons demand growing rapidly (for power generation, petrochemicals, and desalination), increasing natural gas production–and replacing oil consumption with natural gas–is a priority for the Saudi government. Gas development is slated to consume a large share of Aramco's budget (in late 1999, Aramco decided to invest $45 billion over 25 years on upstream gas development and processing facilities), and Aramco is aiming to add 3-5 Tcf of new non-associated natural gas reserves per year to meet rapid (5 per cent annual) gas demand growth. Non-associated gas development is desirable in particular because it guarantees a steady flow of gas regardless of oil output, which tends to fluctuate. Currently, non-associated gas accounts for 40 per cent of Saudi Arabia's total gas reserves.

In June 2003, Saudi Oil Minister Naimi officially announced termination of negotiations with foreign energy companies on the $15-$20 billion "Saudi Gas Initiative" (SGI), which had promised to be the first major reopening of Saudi Arabia's upstream hydrocarbons sector to foreign investment since nationalization in the 1970s. [In mid-June 2003, the Dubai-based Gulf News reported that Saudi Arabia was considering inviting international oil companies to bid again on the SGI]. Companies which had been selected (in 2001) for the three "core ventures" under the SGI were: 1) South Ghawar: ExxonMobil (35 per cent), Shell (25 per cent), BP (25 per cent), Phillips (15 per cent); 2) Red Sea: ExxonMobil (60 per cent), plus Marathon (20 per cent) and Occidental (20 per cent); 3) Shaybah: Shell (40 per cent), Total (30 per cent), and Conoco (30 per cent). The SGI had aimed to increase foreign investment and natural gas development in the country, while integrating upstream gas development with downstream petrochemicals, power generation, and water desalination. SGI had been seen as the key to Saudi Arabia's entire foreign investment strategy. However, negotiations broke down over two major stumbling blocks: the extent of gas reserves to be opened to upstream development, and whether or not this should include gas from the Saudi Aramco Reserve Area (SARA); and the rates of return to participating companies (the companies wanted a significantly higher rate than the Saudis were offering). Prior to the cancellation, Saudi Arabia and the foreign oil companies had failed to meet numerous deadlines for agreement on the SGI throughout 2001 and 2002.

Core Venture 1, in South Ghawar, would have been one of the world's largest ($15 billion) integrated natural gas projects, including exploration, pipelines, two gas-fired power plants, two petrochemical plants, two desalination units, and more. Core Venture 2 was to involve exploration in the Red Sea, development of the Barqan and Midyan fields on the Red Sea coast in northwestern Saudi Arabia, as well as construction of a petrochemical plant, a power station, desalination capacity, etc., at a cost of $4 billion. Core Venture 3 would have involved exploration near Shaybah in the Rub al-Khali ("Empty Quarter") of southeastern Saudi Arabia, development of the Kidan gas field, laying of pipelines from Shaybah to the Haradh and Hawiyah natural gas treatment plants east of Riyadh, and construction of a petrochemical plant in Jubail, at a cost of $4 billion.

Additional natural gas production is being encouraged as a feedstock for the country's growing petrochemical industry (at Yanbu and Jubail, for instance), as well as for electricity generation, desalination plants and other industrial establishments, and as a replacement for direct oil burning. Using natural gas instead of oil domestically will help free up additional crude oil for export (OPEC quotas are on production, not exports). Overall, Saudi Arabia aims to triple natural gas output (to 15 Bcf/d) by 2009. To date, Saudi Arabia has not expressed great interest in exporting liquefied natural gas mainly due to doubts regarding economic viability.

Domestic demand is driving a $4.5 billion expansion of Saudi Arabia's Master Gas System (MGS), which was completed in 1984. The MGS feeds gas to the industrial cities of Yanbu on the Red Sea and Jubail, which combined account for 10 per cent of the world's petrochemical production. Prior to the MGS, all of Saudi Arabia's natural gas output was flared.

In October 2002, construction was completed on a $4 billion, 1.4-billion-cubic-feet (Bcf)-per-day, non-associated gas processing plant at Hawiyah, located south of Dhahran and east of Riyadh near the giant Ghawar oil field. Hawiyah represents the largest Saudi natural gas project in more than 10 years, and the first to process only non-associated gas (from the deep Khuff and Jauf reservoirs). Hawiyah was officially inaugurated in October 2002, and reportedly is producing enough natural gas to free up around 260,000 bbl/d of Arabian Light crude oil for export.

Besides Hawiyah, Foster Wheeler is managing a $2 billion project to build a new natural gas processing plant at Haradh, to be completed by late 2003. Haradh is similar in scope to Hawiyah, and is located near Ghawar. When Haradh is finished (reportedly it was nearly complete as of February 2003), total Saudi natural gas processing capability will increase by around 1.6 Bcf/day, to 9.5 Bcf/day. In other news, a key pipeline project was completed in June 2000 to extend the MGS from the Eastern Province (which contains large potential gas and condensate reserves) to the capital, Riyadh, in the Central Province. This is part of a broader expansion of the existing gas transmission system in Saudi Arabia, reportedly to include the construction of around 1,200 miles of additional natural gas pipeline capacity (on top of 10,500 miles of oil, gas, condensate, products, and natural gas liquid pipelines currently in operation) by 2006.

Electricity

Saudi Arabia's rapidly growing population is increasing demand on electric utilities, as power demand grows by 4.5 per cent or more each year. Saudi Arabia's Industry and Electricity Ministry estimates that the country will need up to 66 gigawatts (GW) of power generating capacity by 2020 (nearly triple today's 23.6 GW), at a cost of around $2 billion per year. Most of this will need to come from the private sector, possibly including foreign investors. Also, the vast majority of this capacity will either be natural gas-fired or combined cycle, as part of the government's plans to expand gas utilization in the power sector (and elsewhere) significantly. Meanwhile, new industrial projects have been delayed and brownouts have occurred due to inadequate power supplies, especially during the summer peak cooling demand season.

Privatization of Saudi Arabia's electricity sector is under consideration, and along these lines, an independent Electricity Regulatory Authority (ERA) was established by Saudi Arabia's Supreme Economic Council in November 2001. The new authority is to "regulate electricity services" in the country, as part of a continuing move towards restructuring of the power sector. The Saudi government reportedly is moving towards a significant restructuring of the country's power sector, including moves towards unbundling of power generation, transmission, and distribution, plus an increase in private sector involvement in the sector.

In July 2002, the Supreme Economic Council passed a resolution setting out a framework for private sector involvement in developing independent water and power projects (IWPPs). Saudi Arabia reportedly is hoping to attract private sector investment for up to 60 per cent equity in IWPP projects, and is considering a guarantee that electricity and water produced would be purchased by local Saudi companies. IWPP projects identified for development would serve the capital of Riyadh; plus the cities of Jedda, Mecca, Medina, and Shuaiba on the west coast; and Jubail on the east coast.

On February 16, 2000, Electricity Minister Dr. Hashem Ibn Abdullah Yamani signed a merger agreement between Saudi Arabia's 10 existing power companies (SCECOs), and on April 5, 2000, the long-anticipated Saudi Electric Company (SEC), a joint-stock company owned 50 per cent by the Saudi government, was established. Also in April 2000, Saudi Arabia officially removed subsidies for electricity (although there are reports that, de facto, subsidies continue). SEC was formed from the country's 10 regional power companies (including the four SCECO's–East, West, Central, and South–which controlled 85 per cent of the country's power supplies). The four SCECO companies had operated at a loss because they had been required to sell power below cost to Saudi consumers, as well as due to inefficiencies and difficulties with non-payment of bills. The government for years has subsidized the cost of electricity and has paid a guaranteed dividend to private shareholders. In November 2002, the Saudi cabinet approved a plan to privatize SEC.

Creation of the SEC (and possibly further splitting the company into units dealing with generation, transmission, and distribution companies) could open the door to private sector construction of new power plants on BOO (Build-Own-Operate) and BOT (Build-Own-Transfer) bases. The future of IPP's (Independent Power Producers) in Saudi Arabia remains uncertain, however, with major challenges including: tariffs, legal and operating framework, taxation, and fuel supply.

Several projects now underway employ financing mechanisms that are new to Saudi Arabia's electric power sector. For example, the 1,200-MW, PP9 power station north of Riyadh has been funded with extra revenues generated by a special tariff imposed on heavy users since January 1995. The $1.7 billion, Ghazlan II power project is being financed by an internationally syndicated, $500 million, commercial loan (the first such loan in Saudi history), and being built by a consortium led by Mitsubishi and Bechtel. Ghazlan II consists of four, 600-MW steam turbine units, the first of which came online in the summer of 2001, and the second in July 2002. Combined with the existing 1,600-MW Ghazlan I facility located on the Gulf coast north of Damman, the entire complex–when completed–will have power generating capacity of 4,000 MW and will supply Saudi Arabia's Eastern Province.

Meanwhile, plans for a $1.7 billion, 1,100-MW, gas-fired power plant at Shuaiba on the Red Sea coast apparently are moving ahead, following a groundbreaking ceremony in May 2000, attended by Crown Prince Abdullah. ABB (Asea Brown Boveri) had been awarded the contract on a turnkey basis. Also at Shuaiba, in January 2001, SEC signed a $419 million contract with the Anglo-French engineering company Alstom to expand the Shuaiba oil-fired power plant by 780 MW (units 4 and 5). These two new units should enter service in late 2003 or early 2004. Finally, in May 2001, CMS Energy, along with joint venture partner A.H. al Zamil Group, was chosen to build a 230-MW cogeneration power project for the Saudi Petrochemical Company (Sadaf). This project also represents Saudi Arabia's first privately-owned IPP.

On October 9, 2000, Saudi Arabia approved plans for setting up a new utility company (UCO) in the twin industrial cities of Yanbu and Jubail. The company, named Marafeq, is being founded by the Royal Commission, the Public Investments Fund, Saudi ARAMCO, and SABIC, with local investors also holding a stake in the company. UCO may be privatized when it becomes profitable. In January 2003, Marafeq (UCO) reportedly commenced construction of several water and electricity projects in Jubail and Yanbu, at a cost of around $2 billion.

Besides generation, Saudi Arabia also requires additional investment in power transmission. Currently, for instance, only two of the country's four power regions are connected. Creating a unified national grid could require over 20,000 miles of additional power transmission lines. Currently, Saudi Arabia has around 150,000 miles of transmission lines. For 2003, SEC set a goal of connecting 184,000 new subscribers, including people living in 617 villages and remote habitations.

Environment

Saudi environmental issues are seen related mainly to oil exploration and production. Despite technological advances in exploration and production, offshore oil development continues to have a significant impact on the marine environment, as do oil spills and illegal discharges.

Several air quality initiatives, including the introduction of unleaded gasoline into the country in 2001, should reduce Saudi Arabia's level of air pollution, but rising rates of energy consumption and carbon emissions portend possible future problems for the Kingdom. Oil production and development make the country very energy- and carbon-intensive. Saudi Arabia's plentiful supply of domestic oil and gas reserves has stifled incentives for the country to develop a significant renewable energy sector.

Country Overview

Independence: September 23, 1932 (unification)

Population (2002E): 21.7 million (growing around 3 per cent per year)

Location/Size: Between the Arabian Gulf and the Red Sea/865,000 square miles (about 1/4 the size of the United States)

Major Cities: Riyadh (royal capital), Jeddah (administrative capital), Mecca, Medina, Dammam, Jubayl, Buraydah

Language: Arabic

Ethnic Groups: Arab (90 per cent), Afro-Asian (10 per cent)

Religion: Muslim (100 per cent)–predominantly Sunni

Economic Overview

Currency: Riyal

Market Exchange Rate (6/11/03): US$1 = 3.75 riyals

Gross Domestic Product (GDP–market exchange rate) (2002E): $209.8 billion

Real GDP Growth Rate (1995-2000 average): 1.9 per cent *(2001E):* 1.2 per cent *(2002E):* 2.0 per cent *(2003F):* 3.0 per cent

Inflation Rate (consumer prices) (2001E): -0.4 per cent *(2002E):* -0.5 per cent *(2003F):* 1.2 per cent

Unemployment Rate (Saudi American bank estimate) (2002E): 15 per cent (unofficial estimates are higher)

Current Account Balance (2001E): $14.5 billion *(2002E):* $7.9 billion *(2003F):* $9.9 billion

Major Trading Partners (2002): Japan, United States, European Community

Merchandise Exports (2002E): $69.6 billion (mainly crude oil and petroleum products)

Merchandise Imports (2002E): $30.2 billion (mainly industrial goods, metals, food)

Merchandise Trade Balance (2002E): $39.4 billion

Oil Export Revenues (2002E): $55.0 billion *(2003F):* $70.0 billion

Oil Export Revenues/Total Export Revenues (2002E): 90 per cent -95 per cent

Government Debt (2001E): $168 billion (nearly 100 per cent of GDP; mainly owed to state institutions)

Foreign Exchange Reserves (excluding gold) (2002E): $20.6 billion

Energy Overview

Proven Oil Reserves (1/1/03E): 264.2 billion barrels (includes half of Neutral Zone–NZ)

Total Oil Production (2002E; includes NZ): 8.54 million barrels per day (bbl/d), of which 7.63 million bbl/d was crude oil, 810,000 bbl/d was natural gas liquids (NGLs), and 80,000 bbl/d was "other liquids" (including MTBE)

OPEC Crude Oil Production Quota (effective 6/1/03): 8.256 million bbl/d

Crude Oil Production Capacity (2003E): 10.0-10.5 million bbl/d

Total Oil Consumption (2002E): 1.50 million bbl/d

Net Oil Exports (2002E): 7.04 million bbl/d

Crude Oil Refining Capacity (1/1/03E): 1.75 million bbl/d

Natural Gas Reserves (1/1/03E): 224.7 trillion cubic feet (Tcf) (includes half of NZ)

Natural Gas Production/Consumption (2001E): 1.9 Tcf

Electric Generating Capacity (2002E): 23.6 gigawatts (all thermal)

Net Electricity Generation (2001E): 122.4 billion kilowatthours

Environmental Overview

Total Energy Consumption (2001E): 4.9 quadrillion Btu* (1.2 per cent of world total energy consumption)

Energy-Related Carbon Emissions (2001E): 84.6 million metric tons of carbon (1.3 per cent of world carbon emissions)

Per Capita Energy Consumption (2001E): 233.7 million Btu (vs. U.S. value of 341.8 million Btu)

Per Capita Carbon Emissions (2001E): 4.0 metric tons of carbon (vs. U.S. value of 5.5 metric tons of carbon)

Energy Intensity (2001E): 34,157 Btu/$1995 (vs U.S. value of 10,736 Btu/$1995)**

Carbon Intensity (2001E): 0.59 metric tons of carbon/thousand $1990 (vs U.S. value of 0.17 metric tons/thousand $1995)

Fuel Share of Energy Consumption (2001E): Oil (59.7 per cent), Natural Gas (40.5 per cent)

Fuel Share of Carbon Emissions (2001E): Oil (66.0 per cent), Natural Gas (34.0 per cent)

Status in Climate Change Negotiations: Non-Annex I country under the United Nations Framework Convention on Climate Change (ratified December 28th, 1994). Not a signatory to the Kyoto Protocol.

Major Environmental Issues: Desertification; depletion of underground water resources; the lack of perennial rivers or permanent water bodies has prompted the development of extensive seawater desalination facilities; coastal pollution from oil spills.

Major International Environmental Agreements: A party to Conventions on Climate Change, Desertification, Endangered Species, Hazardous Wastes, Law of the Sea and Ozone Layer Protection.

The total energy consumption statistic includes petroleum, dry natural gas, coal, net hydro, nuclear, geothermal, solar, wind, wood and waste electric power. The renewable energy consumption statistic is based on International Energy Agency (IEA) data and includes hydropower, solar, wind, tide, geothermal, solid biomass and animal products, biomass gas and liquids, industrial and municipal wastes.

Oil and Gas Industries

Organization: The Supreme Petroleum Council governs the nationalized oil industry, including Saudi Arabian Oil Co. (Saudi Aramco) crude production, refining and marketing; Saudi Basic Industries Corp. (SABIC) petrochemicals.

Major Foreign Oil Company Involvement: AOC, BP, Conoco, ExxonMobil, Occidental, Phillips, Shell, Texaco, Total

Major Oil Terminals: Ras Tanura (world's largest offshore oil loading facility, on the Persian Gulf; 6 million bbl/d capacity), Yanbu (on the Red Sea, fed by Petroline; 5 million bbl/d capacity), Jubail, Ras al-Ju'aymah (on the Persian Gulf northwest of Ras Tanura; 3 million bbl/d capacity), Jiddah (on Red Sea south of Yanbu), Jizan (on Persian Gulf, refined products), Ras al-Khafji (on Persian Gulf in the Saudi-Kuwaiti Neutral Zone, crude oil), Rabigh (on Red Sea, south of Jiddah, crude oil and refined products), Zuluf (offshore Persian Gulf, linked to Zuluf oil field)

Major Oil Fields: Ghawar, Safaniya, Najd, Abqaiq, Berri, Manifa, Zuluf, Shaybah, Abu Saafa, Khurusaniya

Major Pipelines (capacity–million bbl/d): Petroline (4.8), IPSA (1.65–closed since August 1990), Tapline (0.5–closed since 1984), Abqaiq-Yanbu NGL line (0.3)

Major Refineries (capacity, 1/1/03E): Aramco–Rabigh 400,000 bbl/d, Ras Tanura 300,000 bbl/d, Yanbu 190,000 bbl/d, Riyadh, 120,000 bbl/d, Jeddah 60,000 bbl/d; Sadui Aramco/Mobil–Yanbu 340,000 bbl/d; Petromin/Shell–al-Jubail 305,000 bbl/d; Arabian Oil Company–Ras al-Khafji 30,000 bbl/d

73

Singapore

Singapore is a major refining center for Southeast Asia, with refining capacity of nearly double its rate of petroleum products consumption. It also is strategically located near the Strait of Malacca, a major route for oil tankers.

General Background

Singapore's strategic location at the entrance to the Strait of Malacca has helped it to become one of the most important shipping centers in Asia. The Port of Singapore, the world's busiest in terms of shipping tonnage, is a key component of Singapore's prosperity and economic health. Singapore also is a leader in new biotechnologies, petroleum refining, and the manufacturing of computer components.

Recognizing that Singapore's future growth depends on overcoming resource limitations and a small domestic market, the Singaporean government has vigorously encouraged local firms to regionalize their operations and to invest abroad. Prime Minister Goh Chok Tong has identified China, India, and the ASEAN countries as priority countries in the regionalization drive.

Singapore's economy has suffered, though, from the global economic slowdown of the last two years. In the short term, its heavy concentration in computer hardware and consumer electronics exports has been a liability. Singapore's real GDP fell by 1.9 per cent in 2001, and growth in 2002 is projected at only 2.2 per cent, well below the phenomenal technology-led growth rates of the 1990's. Real GDP growth is projected at 4.8 per cent in 2003, largely on an expected recovery in demand for computer components and consumer electronics exports.

Oil

Singapore is one of the major petroleum refining centers of Asia, with total crude oil refining capacity of nearly 1.3 million barrels per day (bbl/d). The Asian economic crisis of 1997-98 had a

negative impact on Singapore's refining industry, and Singapore's refining companies lost significant business due to declining demand for oil products in the region. While the region staged a recovery from the financial crisis in 1999 and 2000, the construction of new refineries in Singapore's traditional export markets has had a more enduring negative effect. Recent refinery expansions in several of its traditional markets also are hurting Singapore's exports. New refineries in India, particularly the 540,000-bbl/d Reliance Petroleum refinery at Jamnagar which began production in 2000, have reduced Indian demand for imports of refined products. The Melaka refining complex in Malaysia also has become a competitor. While there has been a short-term recovery in refining margins in 2002, the overall outlook for Singpore's refiners is still uncertain, with so much capacity being built elsewhere in Asia.

In response to these pressures, individual refinery operators in Singapore have been exploring different restructuring measures. For instance, Shell has centralized control of its Asian refining operations in Singapore. Caltex has followed a similar strategy. Other Singaporean refiners are exploring approaches ranging from large run cutbacks to cost cutting in an effort to boost margins.

Some owners of refineries in Singapore reportedly have expressed interest in selling their stakes in 2002. BP has had discussions with Malaysia's Petronas about its interest in the Singapore Refining Company, but no sales agreements have been reached.

Petrochemicals

The rapid growth of Singapore's petrochemical industry has been a direct result of the country's strong base in petroleum refining. A large project to reclaim seven islands to form a 12-square mile petrochemical complex on Jurong Island is in progress. This project will provide more land to support the growth of petrochemicals and chemical industries.

Recent major developments in the petrochemical industry in Singapore include the start up of a second naphtha cracker by the Petrochemical Corporation of Singapore and its downstream partners, Phillips Petroleum, The Polyolefin Company, Hoechst, and Seraya Chemicals. In addition, Germany's Messer Group and U.S.-based Texaco have built a $200-million synthetic gas plant on Jurong Island. The synthetic gas is being used for industrial purposes and as feedstock for petrochemical and refining customers on Jurong Island.

Natural Gas

Singapore imports all of its natural gas, which is mainly used for power generation and as a feedstock for petrochemical production. Natural gas use is rising rapidly, as the Singaporean government promotes policies aimed at reducing carbon dioxide and sulfur emissions, ensuring energy security, and promoting the country as a regional hub for an integrated gas pipeline network. Singapore Power currently imports 155 million cubic feet per day (Mmcf/d) of natural gas through a pipeline from Malaysia, its first natural gas supplier. This pipeline was the first transnational natural gas pipeline built in East Asia.

Singapore has embarked on a diversification strategy so it will not become dependent on a single source for gas imports. In January 1999, the Singaporean gas consortium, SembGas, (which consists of SembCorp Engineering, Tuas Power, EDB International, and Belgium's Tractebel) signed an agreement to purchase West Natuna gas from Indonesian state energy company Pertamina. SembGas has agreed to purchase 325 Mmcfd of natural gas for 22 years, through a pipeline from the West Natuna natural gas fields to Singapore. Deliveries of natural gas through the pipeline began in January 2001.

Another firm contract was signed for supplies to PowerGas from Pertamina in February 2001. The 20-year contract calls for supplies of 150 Mmcf/d to begin in 2003, rising to 350 Mmcf/d by 2009. The natural gas will come from deposits on the Indonesian island of Sumatra. The subsea pipeline linking Sumatra to Singapore is expected to cost $300 million. Construction on the pipeline began in June 2002.

In addition to natural gas imports from Malaysia and the two pipelines from Indonesia, Singapore has plans under discussion to build an LNG import terminal, thereby freeing itself from complete dependence on neighboring states for its gas supply. The Singaporean government announced in September 1999 that it has set aside land at Tuas View for the project. In the last three years, however, the project has made little progress. While it would have obvious energy-security benefits for Singapore, LNG currently would cost more than piped natural gas.

One new use for natural gas in Singapore is as a fuel for motor vehicles. Singapore has launched a program to convert public buses in some areas to compressed natural gas (CNG), and the first SembGas CNG filling station opened in April 2002. The CNG vehicle program may later expand to taxis.

Singapore may eventually become important as a regional natural gas hub for Southeast Asia. The idea of a regional gas grid for members of the Association of Southeast Asian Nations (ASEAN) has been under discussion for several years, and international links already exist or are under construction between Burma and Thailand, between Malaysia and Thailand, and between Indonesia and Singapore. Singapore has an ideal location to function as the hub of such a system if it comes to fruition.

Electricity

Singapore is in the process of restructuring and privatizing its electric power sector, which is to transform what was a monopoly into a competitive market. Two subsidiaries of state-owned Singapore Power, PowerSeraya and PowerSenoko, along with Tuas Power, are currently generating electricity. PowerGrid, another subsidiary of Singapore Power, maintains and operates the country's electricity transmission and distribution system. The Singaporean government currently owns majority stakes in all of these firms through holding companies. The process of privatization has been repeatedly delayed, and current plans call for the Singporean government to divest its stakes in the electric utility sector in 2004.

A regulatory agency for the country's electric utility sector, the Energy Markets Authority (EMA) was created in April 2001. It is working out the details of the privatization process.

Natural gas importer SembCorp already has entered the power generation business as an independent power producer (IPP), completing the construction of a 815-megawatt (MW) gas-fired plant under the name SembCorp Cogen. The facility began operation in September 2001. Malaysia's Tenaga Nasional has expressed interest in entering Singapore's power market after deregulation. It would either sell power from its grid in Malaysia to customers in Singapore, or possibly purchase generation assets in Singapore.

Most of the state-owned utilities' generating capacity has been converted from fuel oil to natural gas as it has become available. Most new planned capacity also will burn natural gas. Tuas Power awarded a contract to Mitsubishi in 2001 for two 367-MW combined cycle generating units, which are to be completed in 2006. The current economic slwodown, however, has largely stalled expansion of electricity generating capacity in Singapore.

Country Overview

Independence: August 9, 1965

Population (2002E): 4.5 million

Location/Size: Singapore lies in Southeast Asia, with Peninsular Malaysia to the north, East Malaysia to the east, and Indonesia to the south. The country consists of one main island and 54 islets located approximately 77 miles north of the equator.

Major Cities: Singapore

Language: Chinese, English, Malay, and Tamil

Ethnic Groups: Chinese (77 per cent); Malay (14 per cent); Indian (8 per cent)

Defense (1998): 53,900 total active armed forces (221,000 reservists). Army: 45,000 personnel (210,000 reservists); Navy: 2,900 personnel (3,600 reservists); Air Force: 6,000 personnel (7,500 reservists)

Economic Overview

Currency: Singapore dollar

Exchange Rate (11/26/02): 1.77 Singapore dollars = 1 U.S Dollar

Real GDP Growth Rate (2001E): -1.9 per cent *(2002E):* 2.2 per cent

Inflation Rate (consumer prices) (2001E): 1.0 per cent *(2002E):* 0.1 per cent

Current Account Balance (2002E): $17.8 billion

Merchandise Trade Balance (2002E): $13.4 billion

Major Exports: Machinery, petroleum and petroleum products, chemicals, telecommunications equipment, computer equipment, food and live animals, crude rubber, beverages, tobacco, clothing.

Major Imports: Machinery and transportation equipment, petroleum and petroleum products, crude materials, foodstuffs, tobacco, textiles, iron and steel, aircraft.

Top Trading Partners: Hong Kong, Japan, Malaysia, Taiwan, Thailand, United States

Total External Debt (2002E): None.

Energy Overview

Oil Consumption (2002E): 722,000 barrels per day (bbl/d) (all imported)

Crude Oil Refining Capacity (1/1/02E): 1.3 million bbl/d

Natural Gas Consumption (2000E): 53 billion cubic feet (Bcf) (all imported)

Electric Generation Capacity (1/1/00E): 6.7 gigawatts (all thermal)

Electricity Generation (2000E): 27.9 billion kilowatt hours

Environmental Overview

Total Energy Consumption (2000E): 1.7 quadrillion Btu* (0.5 per cent of world total energy consumption)

Energy-Related Carbon Emissions (2000E): 31.6 million metric tons of carbon (0.5 per cent of world carbon emissions)

Per Capita Energy Consumption (2000E): 419.1 million Btu (vs. U.S. value of 351.0 million Btu)

Per Capita Carbon Emissions (2000E): 7.9 metric tons of carbon (vs. U.S. value of 5.6 metric tons of carbon)

Energy Intensity (2000E): 14,853 Btu/$1995 (vs U.S. value of 10,918 Btu/$1995)**

Carbon Intensity (2000E): 0.28 metric tons of carbon/thousand $1995 (vs U.S. value of 0.17 metric tons/thousand $1995)

Sectoral Share of Energy Consumption (1998E): Transportation (61.2 per cent), Industrial (30.8 per cent), Commercial (4.8 per cent), Residential (3.2 per cent)

Sectoral Share of Carbon Emissions (1998E): Industrial (58.3 per cent), Transportation (26.5 per cent), Commercial (9.1 per cent), Residential (6.1 per cent)

Fuel Share of Energy Consumption (2000E): Oil (96.7 per cent), Natural Gas (3.3 per cent), Coal (0.0 per cent)

Fuel Share of Carbon Emissions (2000E): Oil (97.5 per cent), Natural Gas (2.5 per cent), Coal (0.0 per cent)

Renewable Energy Consumption (1998E): 2.9 trillion Btu* (60 per cent decrease from 1997)

Number of People per Motor Vehicle (1998): 6 (vs. U.S. value of 1.3)

Status in Climate Change Negotiations: Non-Annex I country under the United Nations Framework Convention on Climate Change (ratified May 29th, 1997). Not a signatory to the Kyoto Protocol.

Major Environmental Issues: Industrial pollution; limited natural fresh water resources; limited land availability presents waste disposal problems; seasonal smoke/haze resulting from forest fires in Indonesia.

Major International Environmental Agreements: A party to Conventions on Biodiversity, Climate Change, Endangered Species, Hazardous Wastes, Law of the Sea, Nuclear Test Ban, Ozone Layer Protection and Ship Pollution.

The total energy consumption statistic includes petroleum, dry natural gas, coal, net hydro, nuclear, geothermal, solar, wind, wood and waste electric power. The renewable energy consumption statistic is based on International Energy Agency (IEA) data and includes hydropower, solar, wind, tide, geothermal, solid biomass and animal products, biomass gas and liquids, industrial and municipal wastes. Sectoral shares of energy consumption and carbon emissions are also based on IEA data.

Energy Industries

State Energy Companies: Singapore National Oil Company; Singapore Petroleum Company; Singapore Power Company; PowerSeraya; PowerSenoko; Tuas Power; PowerGas

Major Refineries (1/1/02 Capacity): Esso Singapore Pty. Ltd. (265,000 bbl/d); Mobil Oil Singapore Pty. Ltd. (300,000 bbl/d); Shell Eastern Petroleum (Pte.) Ltd. (405,000 bbl/d) Singapore Refining Co. Ltd. (285,000 bbl/d). *Major Ports:* Singapore